Simplified Theory of Plastic Zones

Hartwig Hübel

Simplified Theory of Plastic Zones

Based on Zarka's Method

Springer

Hartwig Hübel
Brandenburgische Technische Universität
Cottbus-Senftenberg
Cottbus, Brandenburg
Germany

ISBN 978-3-319-29873-3 ISBN 978-3-319-29875-7 (eBook)
DOI 10.1007/978-3-319-29875-7

Library of Congress Control Number: 2016931606

Translation from the German language edition: *Vereinfachte Fließzonentheorie* by Hartwig Hübel,

Printed on acid-free paper

This Springer imprint is published by SpringerNature
The registered company is Springer International Publishing AG Switzerland

Preface

This book is essentially a translation of the book "Vereinfachte Fließzonentheorie", published by Springer-Vieweg in 2015. However, some changes have been made, for example, to minimize references to literature in German.

It describes the Simplified Theory of Plastic Zones (STPZ), developed in my time as professor at the Fachhochschule Lausitz starting in 1996, based on the simplified elastic-plastic analysis method developed by Prof. Zarka et al. at the École Polytechnique in Palaiseau, often simply referred to as Zarka's method. Its goal is the simplified determination of the plastic behavior of structures, the calculation of which would otherwise be very complicated.

The development of the STPZ was inspired by the fact that I was confronted with a dilemma in my professional practice. On the one hand there was the need to be able to demonstrate a sufficient service life of structures under repeated plastic straining. On the other, the computational effort required for a conventional application of the flow zone theory was so huge that a comprehensive proof of the life of all components of a nuclear power plant subjected to cyclic thermal loading was illusory.

The simplified analysis methods for the estimation of elastic-plastic distortion available in the 1980s had, however, no or at best a very weak theoretical foundation or were applicable only for special configurations of component geometry and loading conditions. Thus the desire arose for better founded simplified elastic-plastic analysis methods.

The need for them has not diminished in recent decades despite massively improved hardware and software. Rather, a reduction of calculation time remains desirable, considering the ever-increasing finite-element models, the development of design codes, which require a better capture of the inevitable nonlinear limit states of a structure, as well as calculating the remaining life in service on the basis of online monitoring systems.

That "simplified analysis methods" are called "simplified" can, in a certain way, be considered a euphemism. In truth, they are in fact more difficult to understand than the "exact" methods because they are derived from those by additional assumptions, which in turn require appropriate justification. Thus the simplification refers not to the fact that complicated theoretical issues could be avoided but to the fact that the amount of computations required may be reduced.

The purpose of this book is to present the STPZ so that a graduate student in civil engineering or mechanical engineering can develop an understanding of it. Other readers may skip some parts of the book, in particular Chaps. 1 and 2, dealing with basic knowledge concerning elastic-plastic material models and the phenomenon of ratcheting. The basics of the STPZ are initially explained in Chap. 3. Here, and after each of the subsequent extension steps of the STPZ (cyclic loading, temperature-dependent material data, multilinear hardening, limit load analysis), examples are presented in detail. Some examples use physical quantities without units, so that they are to be understood in any arbitrary consistent set of units.

Beyond the status of the STPZ described in this book, many extensions appear possible, so that the STPZ cannot be considered as completed.

Cottbus
April 2016

Hartwig Hübel

Contents

1 Introduction to the Behavior of Elastic-Plastic Solids and Structures 1
- 1.1 Plastic Reserve 1
- 1.2 Basic Concepts of Modeling Plastic Material Behavior 3
  - 1.2.1 Elastic Behavior 3
  - 1.2.2 Deviatoric Stress State 4
  - 1.2.3 Yield Surface 5
  - 1.2.4 Flow Law 7
  - 1.2.5 Hardening Law 9
  - 1.2.6 Prager–Ziegler Hardening 11
  - 1.2.7 Besseling Model 13
  - 1.2.8 Chaboche Model 13
- References 16

2 Structural Behavior Under Variable Loading 17
- 2.1 Local and Directional Stress Redistribution 17
- 2.2 Structural Ratcheting Based on a Two-Bar Model 19
- 2.3 Influence of Kinematic Hardening 21
- 2.4 Ratcheting Interaction Diagram 24
- 2.5 Examples of Ratcheting 25
  - 2.5.1 Three-Bar Model 25
  - 2.5.2 Multiaxial Ratcheting 27
  - 2.5.3 Bree Tube 32
  - 2.5.4 Continuous Beam 36
  - 2.5.5 Interaction of Section Forces 39
- 2.6 Myths About Ratcheting 41
- 2.7 Residual Stresses 42
- 2.8 Service Life 44
  - 2.8.1 Fatigue 44
  - 2.8.2 Strain Accumulation 45

2.9 Analysis Methods 47
2.9.1 Fictitious Elastic Analysis and Correction Factors 48
2.9.2 Incremental Elastic-Plastic Analysis 49
2.9.3 Twice-Yield Method 50
2.9.4 Direct Methods 51
References 53

3 STPZ at Monotonic Loading 57
3.1 Transformed Internal Variable 57
3.1.1 Basic Assumptions for the Material Law 57
3.1.2 Reformulation of the Material Law 58
3.1.3 Modified Elastic Analysis 63
3.1.4 Uniaxial Stress State 65
3.1.5 Plane Stress State 66
3.2 Plastic Zone 69
3.3 Transformed Internal Variable at Radial Loading 70
3.4 Examples of Radial Loading 71
3.4.1 Tension Bar with Step-like Change in Cross-Section 71
3.4.2 Bending Beam with Sandwich Cross-Section 74
3.5 Transformed Internal Variable at Non-Radial Loading 79
3.6 Examples of Non-Radial Loading 80
3.6.1 Directional Stress Redistribution Produced by Elastic Incompressibility 81
3.6.2 Directional Stress Redistribution Despite Elastic Incompressibility 85
3.6.3 Disc with a Hole 87
3.6.4 Thick-Walled Cylinder Under Internal Pressure 89
References 94

4 STPZ at Cyclic Loading 95
4.1 Nature of the Shakedown State 95
4.2 Strain Range at Plastic Shakedown 96
4.3 Examples of Strain Ranges at Plastic Shakedown 99
4.3.1 Two-Bar Model 99
4.3.2 Multiaxial Ratcheting 101
4.3.3 Bree Tube 103
4.3.4 Thick-Walled Cylinder Loaded by a Temperature Transient 105
4.4 Accumulated Strains in the State of Elastic Shakedown 109
4.4.1 Identification of the Elastic and the Plastic Subvolumes 109
4.4.2 Estimation of the Transformed Internal Variable 110

4.5 Examples of Accumulated Strains at Elastic Shakedown 114
4.5.1 Two-Bar Model 115
4.5.2 Multiaxial Ratcheting........ 124
4.5.3 Bree Tube 130
4.6 Accumulated Strains at Plastic Shakedown 133
4.6.1 Identification of the Elastic and Plastic Subvolumes..... 134
4.6.2 Estimation of the Transformed Internal Variable 135
4.7 Examples of Accumulated Strains at Plastic Shakedown 138
4.7.1 Two-Bar Model 138
4.7.2 Multiaxial Ratcheting........ 140
4.7.3 Bree Tube 141
4.7.4 Thick-Walled Cylinder Loaded by a Temperature Transient and Internal Pressure 146
4.8 STPZ at Multi-Parameter Loading 150
4.8.1 Identifying the Nature of the Shakedown State 150
4.8.2 Successive Projection at Elastic Shakedown 151
4.8.3 Successive Projection at Plastic Shakedown 152
4.8.4 Example of the Three-Bar Model 152
References 157

5 STPZ with Temperature-Dependent Material Data 159
5.1 Linear Kinematic Hardening 159
5.1.1 Temperature-Dependent Hardening Modulus........ 160
5.1.2 Temperature-Dependent Yield Stress 163
5.1.3 Basic Equations for Tension Bar........ 166
5.2 STPZ at Anisothermal Monotonic Loading 168
5.3 STPZ at Anisothermal Cyclic Loading 169
5.4 Strain Range at Plastic Shakedown 170
5.5 Examples for Strain Ranges at Plastic Shakedown 173
5.5.1 Two-Bar Model 173
5.5.2 Multiaxial Ratcheting........ 175
5.5.3 Bree Tube 175
5.5.4 Thick-Walled Cylinder Loaded by a Temperature Transient 178
5.6 Accumulated Strains at Elastic Shakedown 178
5.6.1 Y^* Is in ω_2 181
5.6.2 Y^* Is in ω_1 182
5.6.3 Y^* Is in ω_4 183
5.7 Examples of Accumulated Strains at Elastic Shakedown 184
5.7.1 Two-Bar Model 184
5.7.2 Multiaxial Ratcheting........ 192
5.7.3 Bree Tube 193
5.8 Accumulated Strain at Plastic Shakedown........ 194

5.9 Examples for Accumulated Strains at Plastic Shakedown 197
5.9.1 Two-Bar Model 197
5.9.2 Multiaxial Ratcheting. 204
5.9.3 Bree Tube 205
5.9.4 Thick-Walled Cylinder Loaded by a Temperature Transient and Internal Pressure 206
References 210

6 Overlay Model 211
6.1 Multilinear Kinematic Hardening 211
6.2 Layer Without Hardening 212
6.3 Layer with Linear Kinematic Hardening 214
6.3.1 Basic Equations for Determining the Material Parameters 214
6.3.2 First Segment of the Total State 219
6.3.3 Second Segment of the Total State 219
6.3.4 Third Segment of the Total State. 221
6.3.5 Material Parameters at Elastic Incompressibility 223
6.3.6 Material Parameters for Trilinear Material Law 224
6.3.7 Material Parameters for Quadrilinear Material Law 226
6.4 Trilinearization of Real Stress–Strain Curves 228
References 232

7 STPZ at Multilinear Hardening 233
7.1 Theory 233
7.1.1 Behavior of the Individual Layers 234
7.1.2 Total State 235
7.1.3 Uniaxial Stress State 243
7.2 Examples 244
7.2.1 Two-Bar Model 244
7.2.2 Multiaxial Ratcheting. 246
7.2.3 Bree Tube 256
7.2.4 Nozzle. 259
References 270

8 Plastic Limit Load 271
8.1 Calculation Steps. 271
8.2 Asymptotic State 272
8.3 Examples of Asymptotic States 274
8.3.1 One Element with Biaxial Stress State 274
8.3.2 Bending Beams and Truss Members 277
8.3.3 Thick-Walled Cylinder Under Internal Pressure. 277
8.3.4 Disc with a Hole. 278
8.4 Completely Plasticized Structure 280

8.5 Examples of Plastic Limit Load for Completely Plasticized Structures 284
8.5.1 One Element with Biaxial Stress State 284
8.5.2 Bending Beam 286
8.5.3 Thick-Walled Cylinder Under Internal Pressure 294
8.5.4 Disc with a Hole 298
8.6 Structures not Completely Plasticized 301
8.7 Examples of Plastic Limit Load for Structures not Completely Plasticized 305
8.7.1 Statically Indeterminate Beam 305
8.7.2 Disc with a Hole and Free Longitudinal Edges 307
8.8 Plastic Load-Bearing Capacity of a Section 308
8.8.1 Fixed Axis of Rotation 308
8.8.2 Axis of Rotation not Fixed 311
References 315

Index 317

About the Author

Hartwig Hübel was born in 1954 in Boppard on the Rhine, grew up in St. Goar, and studied Structural Engineering at the TU Darmstadt from 1972 to 1978. He worked for 16 years in nuclear power plant construction and in the development of rules for the calculation of thermally stressed components. In 1985 he received a Ph.D. at the Institute of Mechanics of the University Kassel, supervised by Prof. Dr.-Ing. O.T. Bruhns, with the dissertation "Plastic strain enhancement factors in design codes and a proposal for the establishment of more appropriate factors." In 1994 he was appointed Professor to the Fachhochschule Lausitz in Cottbus, which merged into the newly founded Brandenburg Technical University Cottbus-Senftenberg in 2013.

Abbreviations

ECM	Elastic Compensation Method
EMAP	Elastic Modulus Adjustment Procedures
ES	Elastic Shakedown
GLOSS	Global Stress–Strain (Method)
HCF	High Cycle Fatigue
LATIN	Large Time Increment (Method)
LCF	Low Cycle Fatigue
LMM	Linear Matching Method
MEA	Modified Elastic Analysis
PS	Plastic Shakedown
RID	Ratcheting Interaction Diagram
RSDM	Residual Stress Decomposition Method
STPZ	Simplified Theory of Plastic Zones
TIV	Transformed Internal Variable

Introduction to the Behavior of Elastic-Plastic Solids and Structures

1

Some basics of plastic material behavior and approaches to its mathematical description by material models are outlined, as well as the impact on the behavior of elastic-plastic structures.

1.1 Plastic Reserve

Because of its high elongation at break of usually more than 20 %, steel material has a large plastic reserve compared to its elastic limit strain of usually far less than 1 %. With respect to the elastic limit stress, the plastic reserve is far lower. Accordingly, the plastic reserve for displacement-controlled loading is much larger than for force-controlled loading.

This is exemplarily illustrated in Fig. 1.1 for a cantilever beam, loaded force-controlled at the free end by a concentrated load or displacement-controlled because of a support displacement causing a concentrated force as a support reaction. The load factor (LF) is shown as a multiple of the elastic limit load over the multiple of the elastic limit strain (lambda). Force-controlled and displacement-controlled loads are indistinguishable from each other in the elastic range (LF $\leq$ 1, lambda $\leq$ 1). In the plastic range, however, a higher load level belongs to a given strain for displacement-controlled than for force-controlled loading.

Especially in the case of displacement-controlled loading, load levels much higher than the elastic limit load may be applied because of the ductility of the material. Displacement-controlled loading is not only represented by the support displacements considered above but can also consist of thermal expansion. Therefore the plastic reserve is particularly exploited in thermal power plants, where the temperature fields resulting from changeable operating conditions may cause considerable strains. Assuming linear elastic material behavior, i.e., calculated fictitious elastic, they can easily exceed 12 times the yield strength of the steel structure.

H. Hübel, *Simplified Theory of Plastic Zones*,
DOI 10.1007/978-3-319-29875-7_1

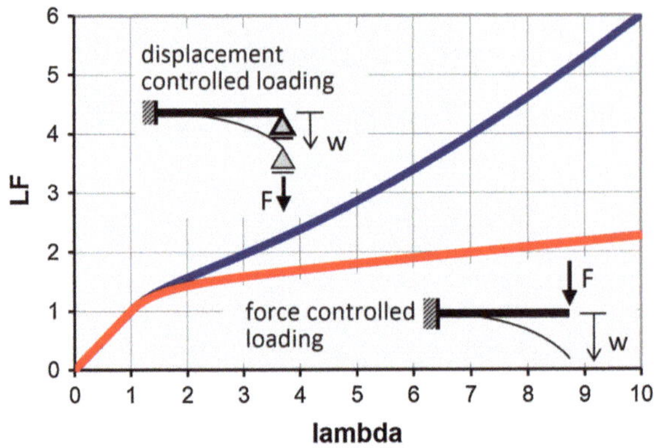

Fig. 1.1 Plastic reserve for a cantilever beam subjected to displacement-controlled and force-controlled loading (applies to a cantilever with rectangular cross-section made of a linear hardening material (hardening parameter $C/E = 0.1$) if limited to the theory of small displacements and small rotations)

Displacement-controlled loading not only results from preventing thermal expansion but may also be attributed to reaction forces and section forces not required for equilibrium but only to maintain continuity of the structure in statically indeterminate structures. Apart from a few special structures such as membrane shells, two- and three-dimensional continua are internally statically indeterminate. There is therefore almost always a plastic reserve, even if these structures are primarily subject to a force-controlled load.

If structures are operated beyond the elastic range, it is not only necessary to ensure that the ductility of the material produced by monotonically increasing loading is not exceeded. It should also be noted that even small variable load portions can set in motion a mechanism which leads to a one-way increase in elongation at any load change, thus limiting the service life of the structure. Such a process is often called ratcheting (frequently also spelled "ratchetting") or referred to as progressive deformation. It is explained in detail in Chap. 2.

In addition, fatigue damage can occur, especially if the strain range is so large that there is alternating plasticity (low cycle fatigue, LCF). However, because portions of the load are always variable, ratcheting should always be taken into account whenever plastic straining occurs, and not only if the variable loading is high enough for alternating plasticity to occur.

The matters covered in this book are not only relevant to thermal power plants, but also to many applications in civil engineering, such as bridge bearings and road surfaces as well as geotechnical engineering. Moreover, ratcheting and LCF can already occur under normal design conditions in mechanical engineering and vehicle construction, in the chemical industry, in gas turbines, and in many other applications. These phenomena may also play a role when guarding against limit states.

To take advantage of the ductility of the material is not restricted to steel but increasingly also to materials with a significantly lower ductility than steel, such as concrete or soil.

1.2 Basic Concepts of Modeling Plastic Material Behavior

In this section, some terms related to plastic material behavior are explained, which are important for the understanding of the STPZ. It is not intended here to explain the thermodynamic or microscopic fundamentals of the theory of plasticity or to represent their extensive phenomenology. However, the following descriptions should suffice to allow interested readers to be able to carry out their own research if necessary. As an introduction to a more detailed treatment of the material-mechanical fundamentals of the theory of plasticity, the reader is referred to the book by Lemaitre and Chaboche [1].

The mechanical behavior of materials is described by material models and captured mathematically by constitutive equations, describing the relationship between the six components of strain and the six components of stress:

$$\varepsilon_i = \begin{pmatrix} \varepsilon_x \\ \varepsilon_y \\ \varepsilon_z \\ \gamma_{xy}/2 \\ \gamma_{yz}/2 \\ \gamma_{xz}/2 \end{pmatrix}; \quad \sigma_i = \begin{pmatrix} \sigma_x \\ \sigma_y \\ \sigma_z \\ \tau_{xy} \\ \tau_{yz} \\ \tau_{xz} \end{pmatrix}. \tag{1.1}$$

1.2.1 Elastic Behavior

The simplest material model is Hooke's law, describing linear elastic behavior. Nonlinear elastic and viscoelastic behavior, which can be used to describe elastic hysteresis effects, is not considered in the following, however. Hooke's law gives a direct relationship between the stress and strain components. In case of a uniaxial stress σ and isotropic material behavior we get for the elastic strain $\varepsilon^{\mathrm{el}}$ acting in the same direction

$$\varepsilon^{\mathrm{el}} = \frac{\sigma}{E}. \tag{1.2}$$

If the isotropic material is subjected to a multiaxial stress we need to make use of a second elastic material parameter beside Young's modulus E, the Poisson's ratio ν:

$$\varepsilon_i^{\mathrm{el}} = E_{ij}^{-1}\sigma_j; \quad i,j = 1\ldots6 \tag{1.3}$$

$$E_{ij} = \frac{E}{(1-2v)(1+v)} \begin{pmatrix} (1-v) & v & v & 0 & 0 & 0 \\ v & (1-v) & v & 0 & 0 & 0 \\ v & v & (1-v) & 0 & 0 & 0 \\ 0 & 0 & 0 & (1-2v) & 0 & 0 \\ 0 & 0 & 0 & 0 & (1-2v) & 0 \\ 0 & 0 & 0 & 0 & 0 & (1-2v) \end{pmatrix} \quad (1.4)$$

$$E_{ij}^{-1} = \frac{1}{E} \begin{pmatrix} 1 & -v & -v & 0 & 0 & 0 \\ -v & 1 & -v & 0 & 0 & 0 \\ -v & -v & 1 & 0 & 0 & 0 \\ 0 & 0 & 0 & (1+v) & 0 & 0 \\ 0 & 0 & 0 & 0 & (1+v) & 0 \\ 0 & 0 & 0 & 0 & 0 & (1+v) \end{pmatrix}. \quad (1.5)$$

In this notation, use is made of the so-called summation convention, which means that summation is to be performed over indices appearing twice in a term, i.e., over j in Eq. (1.3).

Up to 21 different parameters are required to describe its linear elastic behavior if the material is anisotropic.

1.2.2 Deviatoric Stress State

Plastic material behavior usually depends on all six stress components. The arithmetic average of the three normal stresses

$$\sigma_H = \frac{1}{3}\left(\sigma_x + \sigma_y + \sigma_z\right) \quad (1.6)$$

plays a major role for some materials, e.g., soils. At the so-called hydrostatic stress state, all normal stresses consist only of this average stress:

$$\sigma_{i,H} = \begin{pmatrix} 1 \\ 1 \\ 1 \\ 0 \\ 0 \\ 0 \end{pmatrix} \sigma_H. \quad (1.7)$$

In isotropic or anisotropic crystalline materials, however, the plastic behavior is dominated by shear processes, described by so-called deviatoric stresses σ_i', obtained from the stresses σ_i (see Eq. (1.1)) by subtracting the hydrostatic stress state:

$$\sigma_i' = \sigma_i - \sigma_{i,H} = \begin{pmatrix} \sigma_x' \\ \sigma_y' \\ \sigma_z' \\ \tau_{xy} \\ \tau_{yz} \\ \tau_{xz} \end{pmatrix} = \begin{pmatrix} \frac{2}{3}\sigma_x - \frac{1}{3}\sigma_y - \frac{1}{3}\sigma_z \\ -\frac{1}{3}\sigma_x + \frac{2}{3}\sigma_y - \frac{1}{3}\sigma_z \\ -\frac{1}{3}\sigma_x - \frac{1}{3}\sigma_y + \frac{2}{3}\sigma_z \\ \tau_{xy} \\ \tau_{yz} \\ \tau_{xz} \end{pmatrix} \tag{1.8}$$

or

$$\sigma_i' = L_{ij}\sigma_j \quad \text{with} \quad L_{ij} = \begin{pmatrix} \frac{2}{3} & -\frac{1}{3} & -\frac{1}{3} & 0 & 0 & 0 \\ -\frac{1}{3} & \frac{2}{3} & -\frac{1}{3} & 0 & 0 & 0 \\ -\frac{1}{3} & -\frac{1}{3} & \frac{2}{3} & 0 & 0 & 0 \\ 0 & 0 & 0 & 1 & 0 & 0 \\ 0 & 0 & 0 & 0 & 1 & 0 \\ 0 & 0 & 0 & 0 & 0 & 1 \end{pmatrix}. \tag{1.9}$$

It should be noted that

$$\sigma_x' + \sigma_y' + \sigma_z' = 0, \tag{1.10}$$

so that only two of the three deviatoric normal stresses are independent.

1.2.3 Yield Surface

In uniaxial stress states, the material begins to yield when the absolute value reaches the yield strength f_y. In multiaxial stress states, the absolute value is replaced by the equivalent stress. Therefore the yield condition says that yielding is starting to occur if the equivalent stress reaches the yield strength.

There are numerous hypotheses regarding the formation of the equivalent stress. For steel, which behaves very similar under compression and under tension, and which can be regarded as isotropic in good approximation, two hypotheses have become especially popular, the so-called maximum shear stress hypothesis after Tresca and the so-called distortional energy hypothesis according to von Mises. Both only make use of the deviatoric stresses, therefore depend on the five different deviatoric stress components, and are defined in the five-dimensional deviatoric stress space. Considering only the principal stress conditions (σ_1, σ_2, σ_3), the stress space reduces to three dimensions, and, because of Eq. (1.10), the deviatoric stress space reduces to two dimensions, the so-called deviatoric plane. The deviatoric principal stresses (σ_1', σ_2', σ_3') can be conveniently depicted in the deviatoric stress plane by means of a three-legged coordinate system, as shown in Fig. 1.2 for a uniaxial tensile stress.

$$\sigma_i = \begin{pmatrix} \sigma_1 \\ 0 \\ 0 \end{pmatrix} \rightarrow \sigma_i' = \begin{pmatrix} \frac{2}{3}\sigma_1 \\ -\frac{1}{3}\sigma_1 \\ -\frac{1}{3}\sigma_1 \end{pmatrix}$$

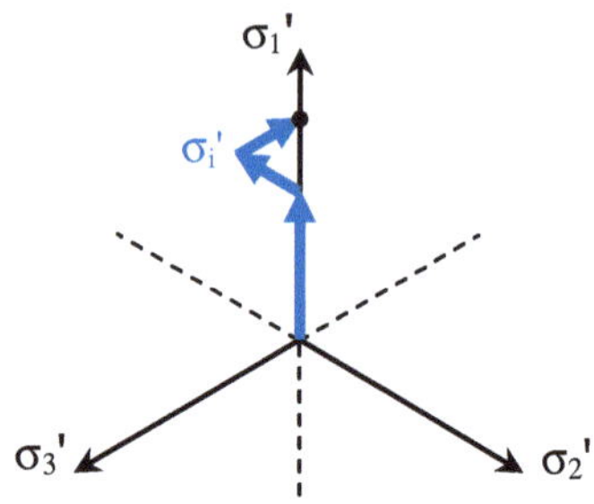

Fig. 1.2 Uniaxial stress state in the deviatoric stress plane

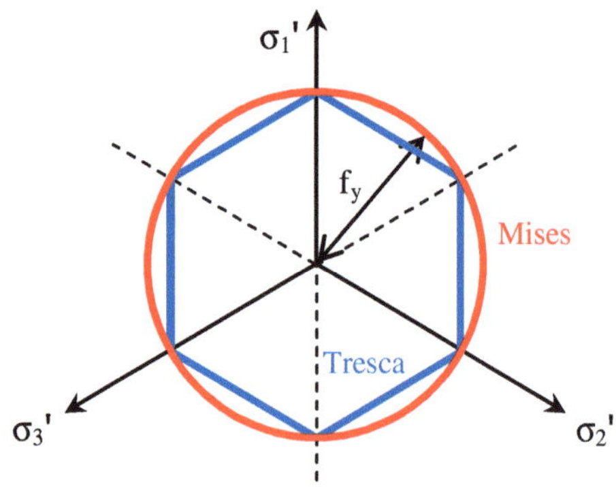

Fig. 1.3 Yield surfaces in the deviatoric stress plane

The line of equal equivalent stresses in the deviatoric stress plane is the so-called yield surface (Fig. 1.3). Because of isotropy, the yield surface must be the same distance from the origin on the three axes, namely the yield strength f_y. This applies in the positive as well as in the negative direction because of the similar behavior in tension and in compression. We therefore have six points on the yield surface where the Tresca and the Mises hypotheses must be identical.

The hydrostatic state of stress is acting perpendicular to the deviatoric plane so that the Mises yield surface can be considered as the surface of a cylinder with the generating line perpendicular to the plane of the drawing as in Fig. 1.3. If the yield surface depends on the hydrostatic stress state, such as in soil, it can be viewed as a cone in the space of non-deviatoric principal stresses in the simplest case (Drucker–Prager yield criterion).

Stress states within the yield surface represent purely elastic behavior. If the stress state reaches the edge with increasing load, the flow initiation occurs. Stress states outside the yield surface are usually not allowed. However, there are also material models which allow stress states lying outside the yield surface. This allows time-dependent (viscous) material behavior such as an increase in strain under constant stress because of creep, stress decrease at constant strain because of relaxation, or strain-rate dependent plasticity. The distance of the stress from the yield surface (overstress) is considered to be a measure of the strain rate. In what follows, only time-independent (inviscid) behavior is considered.

For thermodynamic reasons, which are formulated by Drucker's postulates of stable material behavior, it is essential that a yield surface is convex. Otherwise the plastic region could be left with increasing the load, and the elastic region could be re-entered at a sufficiently high stress again, which is already intuitively to be excluded. Hexagonal symmetrical yield surfaces other than the Tresca and the Mises yield surfaces would also be imaginable by purely geometric considerations. However, because of the convexity condition they cannot differ substantially from the two mentioned.

As follows from simple geometric relationships in Fig. 1.3, the Mises equivalent stress can deviate by up to the factor $2/\sqrt{3}$ with respect to the Tresca equivalent stress, so that both hypotheses differ by a maximum of 15.5 %. Within this difference, both can be considered well established experimentally for the material steel, with slight advantages for the Mises hypothesis. The decision in favor of one or the other hypothesis, therefore, is not necessarily attributed to the greater accuracy, but to the simpler mathematical handling. For some special problems of strength of materials, the Tresca hypothesis has analytical advantages because of piecewise linearity. In addition, it is conservative with respect to the Mises hypothesis. However, singularities present in the vertices cause problems in many other applications, as briefly discussed in the following section. Therefore, the Mises condition is generally preferred.

The Mises yield surface can be generally formulated by the function

$$g\left(\sigma_i'\right) = f_y, \tag{1.11}$$

where $g\left(\sigma_i'\right)$ is the Mises equivalent stress σ_v.

In the deviatoric principal stress space, the Mises yield surface is a circle:

$$\sigma_\text{v}^2 = \frac{3}{2}\left[\left(\sigma_1'\right)^2 + \left(\sigma_2'\right)^2 + \left(\sigma_3'\right)^2\right] = f_y^2 \tag{1.12}$$

or, in non-deviatoric quantities:

$$\sigma_\text{v}^2 = \frac{1}{2}\left[(\sigma_1 - \sigma_2)^2 + (\sigma_1 - \sigma_3)^2 + (\sigma_2 - \sigma_3)^2\right] = f_y^2 \tag{1.13}$$

$$\sigma_\text{v}^2 = \sigma_x^2 + \sigma_y^2 + \sigma_z^2 - \sigma_x\sigma_y - \sigma_y\sigma_z - \sigma_x\sigma_z + 3\left(\tau_{xy}^2 + \tau_{yz}^2 + \tau_{xz}^2\right) = f_y^2. \tag{1.14}$$

1.2.4 Flow Law

As long as the distortions are sufficiently small, say, less than 10 %, which is usually the case for structures in operating conditions, unlike metal forming for example, they can be additively split into elastic and plastic parts (additivity):

$$\varepsilon_i = \varepsilon_i^{\mathrm{el}} + \varepsilon_i^{\mathrm{pl}}. \tag{1.15}$$

The elastic part of the distortions may be obtained from the stress state via Eq. (1.3), but a flow law is required for the plastic part.

The yield surface describes the stress state associated with the onset of plastic yielding. In contrast, the flow law indicates in which direction the plastic strains develop during yielding. Depending on the material, this direction can be quite different at the same state of stress, for example, because of porosity, cohesion, and internal friction. For steel, it is empirically well established that plastic deformations hardly change the volume of the material, so that the incompressibility condition

$$\varepsilon_x^{\mathrm{pl}} + \varepsilon_y^{\mathrm{pl}} + \varepsilon_z^{\mathrm{pl}} = 0 \tag{1.16}$$

can be assumed. It then follows that the plastic distortions (unlike the elastic distortions) are deviatoric in nature so that their directions can be indicated in the deviatoric space. As they have to act in the direction of the stress in a uniaxial stress state and therefore in the direction of the deviatoric stress vector, the plastic strain vector is perpendicular to the Mises yield surface. This is generalized to the so-called normality rule which states that in a stress state $\sigma_i^{'}$ on the yield surface, the plastic strain increment $\mathrm{d}\varepsilon_i^{\mathrm{pl}}$ acts normal to the yield surface and is directed outwards (Fig. 1.4). The flow law is thus associated with the yield criterion:

$$\mathrm{d}\varepsilon_i^{\mathrm{pl}} \propto \sigma_i^{'}. \tag{1.17}$$

The proportionality between plastic strain increment and deviatoric stress state is specified by the Prandtl–Reuss flow law:

$$\mathrm{d}\varepsilon_i^{\mathrm{pl}} = \frac{3}{2}\frac{\mathrm{d}\varepsilon_{\mathrm{v}}^{\mathrm{pl}}}{\sigma_{\mathrm{v}}}\sigma_i^{'}, \tag{1.18}$$

after which the plastic equivalent strain increment $\mathrm{d}\varepsilon_{\mathrm{v}}^{\mathrm{pl}}$ is divided between the different directions proportional to the deviatoric stresses after normalization by σ_{v}. The plastic equivalent strain increment $\mathrm{d}\varepsilon_{\mathrm{v}}^{\mathrm{pl}}$ is unknown for the moment but may be

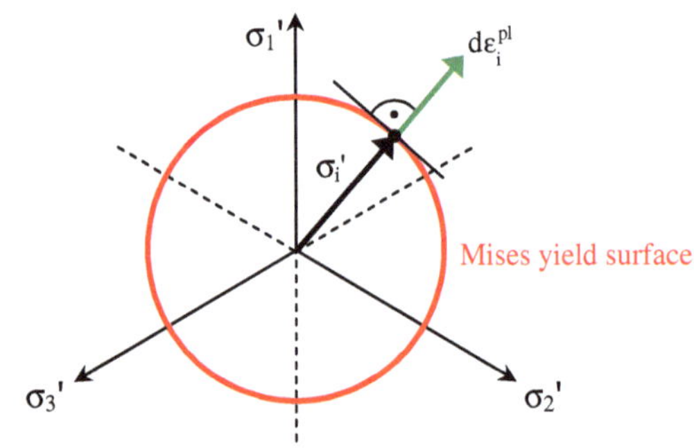

Fig. 1.4 Associated flow law, normality rule

found by the so-called consistency condition stating that the stress vector must remain on the yield surface.

Of great importance for later discussions of the numerical effort required to perform plastic analyses is the fact that Eq. (1.18) cannot usually be integrated in closed form and therefore not exactly. This is because the term σ_i'/σ_v is not usually constant during a plastic process, even at monotonic loading, which is addressed as directional redistribution of stress (cf. Sect. 2.1). This results in a dependency of the plastic state on the load history, i.e., how the actual stress state has been reached, whereas the elastic state only depends on the actual stress state. Only in exceptional cases, such as in uniaxial stress states under monotonic loading, can Eq. (1.18) be integrated exactly to the finite flow law

$$\varepsilon_i^{pl} = \frac{3}{2}\frac{\varepsilon_v^{pl}}{\sigma_v}\sigma_i'. \tag{1.19}$$

If Eq. (1.19) is also used in other cases, as an approximation, we obtain the deformation theory of Hencky.

The normality rule is also the main reason why, as mentioned in the previous section, the use of the Tresca yield surface can be problematic. In its six vertices the normal is not uniquely defined.

It should also be noted that the normality rule is not suitable for some materials or yield surfaces, such as frictional materials, e.g., soils.

1.2.5 Hardening Law

Experiments show that the stress can usually be increased, even after reaching the yield strength. This applies particularly to steels, for which Boller, Seeger, and Vormwald have collected extensive data [2]. A special feature can be observed during the first monotonic loading of structural steel. A further stress increase is then in fact possible only after a plateau-like behavior, in which the stress remains approximately constant, although the strain may increase. In cyclic loading, however, this phenomenon is no longer observed. Instead, the yield stress decreases (Fig. 1.5).

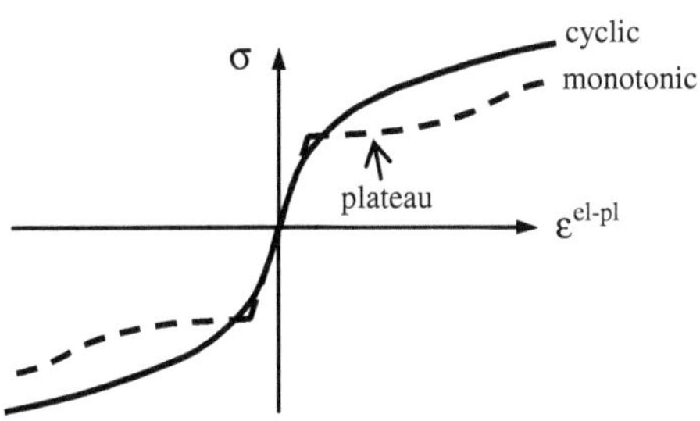

Fig. 1.5 Schematic diagram of the behavior of structural steel, based on [3]

The behavior when exceeding the yield stress is called hardening. In the stress–strain curve, it is always connected to a tangential slope that is less than the yield stress, i.e., in the elastic range. Despite the schematic depiction in Fig. 1.5, you can already guess that hardening is of great importance for cyclic elastic-plastic behavior.

If, on the one hand, the stresses can exceed the yield strength but, on the other, cannot leave the yield surface, then the yield surface has to change during the hardening process. This can be induced by a concentric expansion, by translation, by a change in shape, or by a mixture of these.

A change in shape of the yield surface is indeed confirmed experimentally in principle, but much less pronounced than concentric expansion (isotropic hardening), and this in turn is usually much less pronounced than a translation of the yield surface (kinematic hardening) at monotonic loading. However, in cyclic loading the isotropic hardening often plays a considerable role, but is usually limited to the first few cycles. Some materials may in some circumstances also show cyclic softening so that the yield surface becomes smaller than at first loading [2].

The hardening law describes how the hardening depends on the current plastic strain, the stress history, the change in direction of the stress vector, etc. This dependence is often defined by means of so-called internal variables, which are not directly accessible in experiments, in contrast to external variables such as elongation or temperature. Their development is described as a function of stress or strain history by evolution equations.

In the following sections some common hardening laws are addressed to describe kinematic hardening of a Mises yield surface under isothermal conditions, namely the Prager–Ziegler, the Besseling, and the Chaboche models. For this purpose, a vectorial internal variable ξ_i is introduced, which describes the shift of the center of the yield surface in the deviatoric stress space from the origin of coordinates and thus is referred to as translation vector or backstress—see Fig. 1.6. This creates a plastic anisotropy, because the yield strength is no longer the same in all directions.

The backstress itself is also deviatoric:

$$\xi_x + \xi_y + \xi_z = 0. \tag{1.20}$$

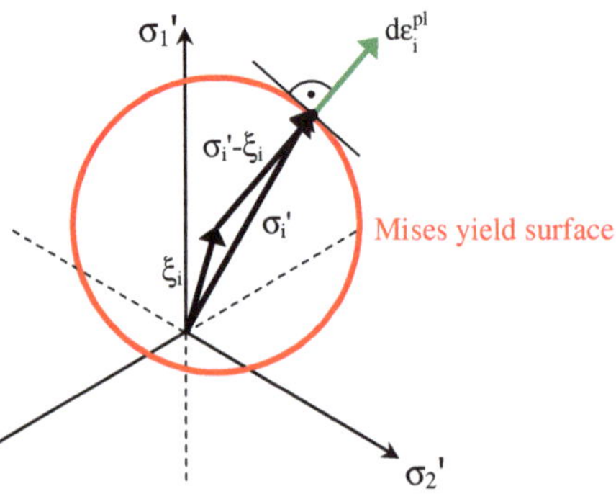

Fig. 1.6 Kinematic hardening of a Mises yield surface

Because of the normality rule, the plastic increment of the distortion is no longer coaxial with the deviatoric stress vector σ_i' (see Eq. (1.19)), but proportional to the reduced stress (σ_i'–ξ_i):

$$\mathrm{d}\varepsilon_i^{\mathrm{pl}} = \frac{3}{2}\frac{\mathrm{d}\varepsilon_{\mathrm{v}}^{\mathrm{pl}}}{f_y}\left(\sigma_i' - \xi_i\right), \tag{1.21}$$

and the yield surface is described in place of Eq. (1.11) by the function

$$g\left(\sigma_i' - \xi_i\right) = f_y \tag{1.22}$$

and for deviatoric principal stresses by

$$\sqrt{\frac{3}{2}}\sqrt{\left(\sigma_1' - \xi_1\right)^2 + \left(\sigma_2' - \xi_2\right)^2 + \left(\sigma_3' - \xi_3\right)^2} = f_y. \tag{1.23}$$

1.2.6 Prager–Ziegler Hardening

According to Ziegler the yield surface is shifted in the direction of the reduced stress, and according to Prager in the direction of the plastic strain increment. In the case of a Mises yield surface, both directions coincide because of Eq. (1.21). For Prager–Ziegler hardening the evolution law for the internal variable can thus be written as

$$\mathrm{d}\xi_i = \frac{2}{3}C\mathrm{d}\varepsilon_i^{\mathrm{pl}}. \tag{1.24}$$

Here the proportionality factor C is the plastic hardening modulus, associated with the elastic-plastic hardening modulus E_t via

$$C = \frac{E\,E_t}{E - E_t} \tag{1.25}$$

(see Fig. 1.7). If C is constant, Eq. (1.24) can be integrated in closed form to

$$\xi_i = \frac{2}{3}C\varepsilon_i^{\mathrm{pl}}, \tag{1.26}$$

which leads to a bilinear relationship between stress and strain in uniaxial stress states for monotonic loading. The relationship between stress and plastic strain is then linear (Fig. 1.7), which is why it is called linear kinematic hardening. For $C = 0$ we get linear elastic-perfectly plastic behavior. Any hardening is missing then, so that f_y cannot be exceeded.

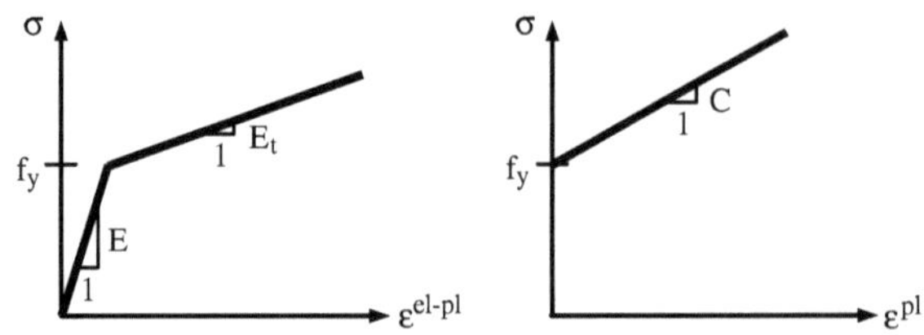

Fig. 1.7 Bilinear stress–strain diagram with linear kinematic hardening for uniaxial stress state under monotonic loading

Fig. 1.8 Linear kinematic hardening (bilinear stress–strain diagram) for uniaxial stress state under cyclic load

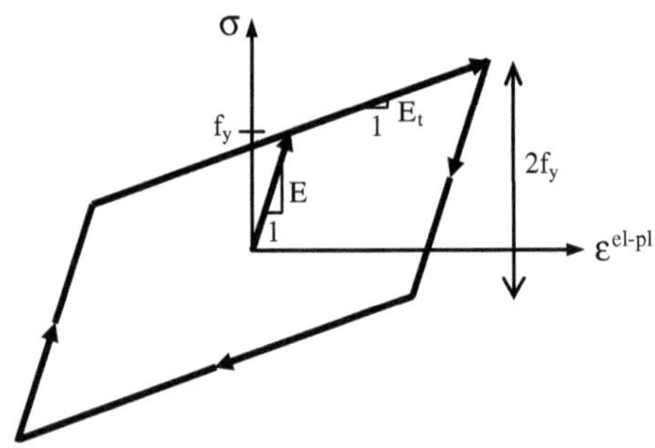

The increment of the plastic equivalent strain in Eq. (1.21), $d\varepsilon_v^{pl}$, initially still unknown, can now be identified because of the unique relationship between stress increment and plastic strain increment as a result of hardening through the so-called consistency condition. This states that the stress must not leave the yield surface even when the yield surface is moved because of the plastic process associated with increasing load. For a Mises yield surface we therefore get according to Eq. (1.22)

$$\mathrm{d}g\left(\sigma_i' - \xi_i\right) = \frac{\partial g(\sigma_i' - \xi_i)}{\partial \sigma_i'} d\sigma_i' + \frac{\partial g(\sigma_i' - \xi_i)}{\partial \xi_i} \mathrm{d}\xi_i = 0. \tag{1.27}$$

After performing the partial derivatives and adopting Eqs. (1.21) and (1.24) we ultimately arrive at

$$\mathrm{d}\varepsilon_v^{pl} = \frac{3}{2} \frac{1}{C f_y} \left(\sigma_i' - \xi_i\right) \mathrm{d}\sigma_i'. \tag{1.28}$$

During unloading, the stress leaves the edge of the yield surface and moves into the interior, which by definition is linked only to an elastic process. Therefore, the yield surface maintains its position unchanged. In the uniaxial stress state it only comes back again plasticizing, in the opposite direction, when the stress has traversed the entire diameter of the yield surface, i.e., at a stress change by $2f_y$ (Fig. 1.8).

An experimentally well verified phenomenon, which is called the Bauschinger effect, is that the reversed flow starts at a lower stress level as a result of the previous plastic straining.

For repeated load reversal the stress–strain curve of the first loading is reached again (Fig. 1.8). During further load cycles, the stress follows the same stress–strain hysteresis loop, which is therefore already stabilized after the first cycle. However, such behavior is not in good agreement with experimental results.

Figure 1.8 is based on stress-controlled alternating loading, i.e., the stresses have the same absolute values at the two extremes in tension and in compression. As a result of the Prager–Ziegler hardening, there is no mean strain associated with zero mean stress.

1.2.7 Besseling Model

The Besseling model is closely related to the Prager–Ziegler model, but offers the possibility of extension to the description of multilinear kinematic hardening by considering the volume of material consisting of several layers and therefore called an overlay model. Each layer is assigned its own nonlinear material law, for example, a linear elastic-perfectly plastic material law with different Young's moduli and Poisson's ratios in different layers. Accordingly, two such layers are needed for a bilinear stress–strain curve. To this end, the Besseling model is identical to the Prager–Ziegler model.

In Chap. 6 the Besseling model is discussed in detail to describe multilinear stress–strain behavior.

1.2.8 Chaboche Model

As mentioned above, material tests show that, under stress-controlled cyclic loading, the same stress–strain curve is not necessarily followed in each load cycle. Then the stress–strain hysteresis is not closed, and there is a net increase in the mean strain from cycle to cycle. This phenomenon is called "ratcheting" or "progressive deformation." It can have different causes as explained in [4]. In particular, a distinction is made between two forms:

- Material ratcheting, which can be observed in a pure material test, i.e., in homogeneous, uniaxial, or multiaxial stress states
- Structural ratcheting, which occurs even in the absence of material ratcheting and as a result is produced by inhomogeneous stress distributions in structures and thus is an effect of field equations in place of constitutive equations

To correct the lack of capacity of the Prager–Ziegler hardening (or, generalized, of the Besseling model) to capture material ratcheting, recovery terms can be included in the hardening law which reduce the dependence of hardening on the degree of hardening achieved previously. As a consequence, hardening becomes nonlinear. As a basic form of a hardening law with a recovery term, Eq. (1.24) can be replaced by the Armstrong–Frederik model for isothermal processes:

$$\mathrm{d}\xi_i = \frac{2}{3} C \mathrm{d}\varepsilon_i^{\mathrm{pl}} - \gamma \xi_i \mathrm{d}p, \tag{1.29}$$

where p is defined by

$$p = \int \mathrm{d}p \tag{1.30}$$

with

$$\mathrm{d}p = \left|\mathrm{d}\varepsilon_i^{\mathrm{pl}}\right| = \frac{\sqrt{2}}{3}\sqrt{\left(\mathrm{d}\varepsilon_1^{\mathrm{pl}} - \mathrm{d}\varepsilon_2^{\mathrm{pl}}\right)^2 + \left(\mathrm{d}\varepsilon_2^{\mathrm{pl}} - \mathrm{d}\varepsilon_3^{\mathrm{pl}}\right)^2 + \left(\mathrm{d}\varepsilon_1^{\mathrm{pl}} - \mathrm{d}\varepsilon_3^{\mathrm{pl}}\right)^2}. \tag{1.31}$$

p is often referred to as "accumulated equivalent strain" and increases monotonically even with alternating plasticity. The term "accumulated equivalent strain" may be perceived as something misleading, as it is not identical with the equivalent strain $\varepsilon_{\mathrm{v}}^{\mathrm{pl}}$ formed by the cyclically accumulated plastic strain components

$$\varepsilon_{\mathrm{v}}^{\mathrm{pl}} = \frac{\sqrt{2}}{3}\sqrt{\left(\varepsilon_1^{\mathrm{pl}} - \varepsilon_2^{\mathrm{pl}}\right)^2 + \left(\varepsilon_2^{\mathrm{pl}} - \varepsilon_3^{\mathrm{pl}}\right)^2 + \left(\varepsilon_1^{\mathrm{pl}} - \varepsilon_3^{\mathrm{pl}}\right)^2} \tag{1.32}$$

with

$$\varepsilon_i^{\mathrm{pl}} = \int \mathrm{d}\varepsilon_i^{\mathrm{pl}}. \tag{1.33}$$

This would in fact not necessarily evolve monotonically.

For a uniaxial stress state and monotonic tension, Eq. (1.29) reduces to

$$\mathrm{d}\xi = (C - \gamma\xi)\mathrm{d}\varepsilon^{\mathrm{pl}}, \tag{1.34}$$

which can be integrated to the exponential hardening law

$$\xi = \frac{C}{\gamma} + c_0 \exp\left(-\gamma \varepsilon^{\mathrm{pl}}\right). \tag{1.35}$$

However, the approach of Armstrong–Frederik still turned out to be insufficient. Therefore Chaboche introduced an expansion by adding multiple Armstrong–Frederik terms (usually k = 2–5) (no summation over j in the last term of Eq. (1.36)):

$$(\mathrm{d}\xi_i)_j = \frac{2}{3} C_j \mathrm{d}\varepsilon_i^{\mathrm{pl}} - \gamma_j (\xi_i)_j \mathrm{d}p \tag{1.36}$$

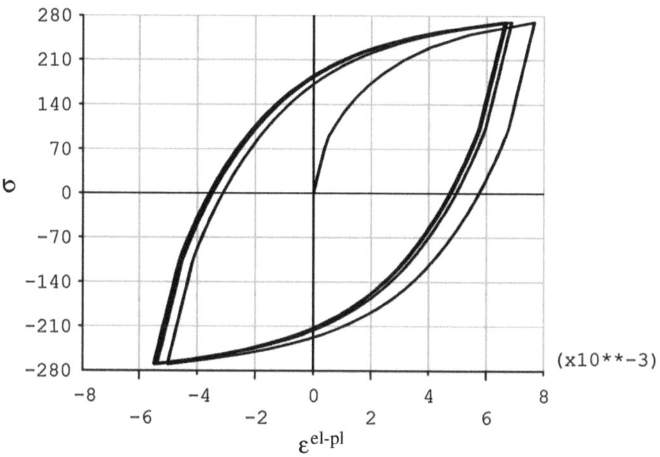

Fig. 1.9 Nonlinear kinematic hardening according to the Chaboche model for a uniaxial stress state under cyclic stress-controlled loading

$$\xi_i = \sum_{j=1}^{k} (\xi_i)_j \quad \text{with} \quad k = 2\ldots5. \tag{1.37}$$

The cyclic stress–strain behavior is exemplarily shown in Fig. 1.9 for the first five cycles of a stress-controlled alternating stress ($\sigma = \pm 270$ N/mm^2) using material parameters given in [5] for $k = 5$ Armstrong–Frederik terms at 200 °C.

A pronounced Bauschinger effect can be seen in Fig. 1.9. After the first loading cycle, the stress–strain curve is not yet closed. After five cycles it is, however, almost stabilized and connected with a positive means strain $\neq 0$.

There exist numerous extensions to Eqs. (1.36) and (1.37) of the Chaboche model (such as bounds for the kinematic hardening, including isotropic hardening, with time dependence) which are not discussed here. Extensive literature is available elsewhere, including the summary work of Dahlberg and Segle [6] and of Abdel-Karim [7] with hundreds of references. Not only is the Chaboche model discussed there, but also the other previously mentioned models (Prager, Besseling) and even the Ohno–Wang model, each from the aspect of cyclic loading. In addition, a comparison with experimental findings is made.

It turns out that material ratcheting is generally overestimated with the Chaboche model and, as mentioned above, underestimated with the Besseling model. A material model that can accurately describe material ratcheting is not yet available, to the knowledge of the author. Perhaps the Ohno–Wang model family has the potential to be used for this purpose, predicting the cyclically accumulated strain usually between the Besseling and Chaboche models. In contrast to the Besseling and Chaboche models, however, the Ohno–Wang model is not generally available in commercial finite element programs such as ANSYS [8]. In view of the considerable scatter observed, even with monotonic elastic-plastic behavior, and especially with ratcheting experiments, we must be aware that there might not exist a material model to describe material ratcheting accurately.

References

1. Lemaitre, J.M., Chaboche, J.-L.: Mechanics of Solid Materials. Cambridge University Press, Cambridge (1990)
2. Boller, C., Seeger, T., Vormwald, M.: Materials Database for Cyclic Loading, Fachgebiet Werkstoffmechanik, TU Darmstadt. http://www.wm.tu-darmstadt.de/mat-db-html/index.html (2008)
3. Klee, S.: Das zyklische Spannungs-Dehnungs- und Bruchverhalten verschiedener Stähle. Technische Hochschule Darmstadt, Fachbereich 14 – Konstruktiver Ingenieurbau, Institut für Statik und Stahlbau, Dissertation D17 (1973)
4. Hübel, H.: Basic conditions for material and structural ratcheting. Nucl. Eng. Des. **162**, 55–65 (1996)
5. Gilbert, R.R.: Investigations aiming at the integration of the simplified yield zones theory (ZARKA's method) in staged elasto-plastic fatigue and shakedown concept for nuclear power plant components subjected to thermo-mechanical loadings. Master-Thesis TU Braunschweig, Institut für Wärme- und Brennstofftechnik (2013)
6. Dahlberg, M., Segle, P.: Evaluation of models for cyclic plastic deformation – A literature study, SSM, Report number: 2010:45, ISSN: 2000-0456. http://www.stralsakerhetsmyndigheten.se/Global/Publikationer/Rapport/Sakerhet-vid-karnkraftverken/2010/SSM-Rapport-2010-45.pdf (2010)
7. Abdel-Karim, M.: Shakedown of complex structures according to various hardening rules. Press. Vessels Pip. **82**, 427–458 (2005)
8. ANSYS Release 14.5, ANSYS Inc. Canonsburg, USA (2012)

2 Structural Behavior Under Variable Loading

Because the Simplified Theory of Plastic Zones to be developed in this book is mainly useful for application to elastic-plastic structural behavior under variable loading, some phenomena of structural behavior relevant to the life of a structure are discussed in this chapter. Particular emphasis is placed on the phenomenon of ratcheting.

2.1 Local and Directional Stress Redistribution

The obvious important points of plastic material behavior at monotonic loading are greater distortions and lower stresses than with linear elastic behavior. Local and directional stress redistributions are often caused in structures by interacting in accordance with field equations, i.e., equilibrium and compatibility conditions (kinematic conditions).

"Local" stress redistribution means that the relationship between the stresses at different locations of the structure is changed as a result of plastic material behavior. Thus, structural parts that appear lowly stressed in an elastic calculation can in fact be highly stressed, or stresses may exhibit different signs at the same location in elastic or plastic calculations. This phenomenon is likely to be familiar to many students from lectures about plastic hinge theory and can, for example, easily be recognized in a *T*-shaped beam cross-section under bending (Fig. 2.1). Cross-section fibers just above the neutral axis are in compression with elastic material behavior, but under tension with plastic material behavior.

In multiaxial stress states it may happen that, in addition to their absolute values, the ratios of the stress components to each other change because of plastic behavior compared to linear elastic behavior. In deviatoric stress space, this manifests itself as a change in angle of the position vector of the stress state and is referred to as directional stress redistribution.

A simple example of this is shown in Fig. 2.2, where a differentially small material volume or finite element is subjected to a plane stress state (stress is zero in the z-direction). In one direction a displacement-controlled loading is applied,

H. Hübel, *Simplified Theory of Plastic Zones*,
DOI 10.1007/978-3-319-29875-7_2

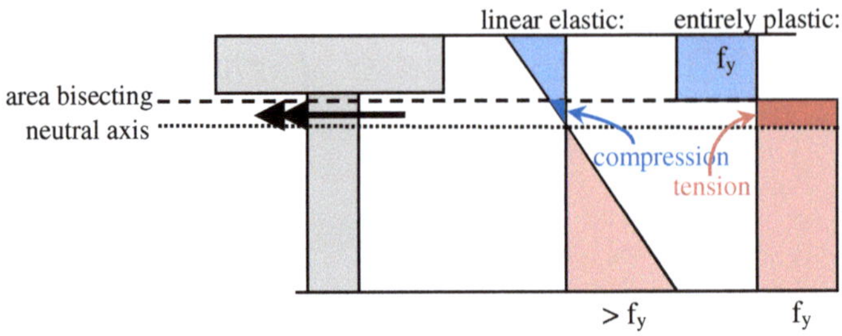

Fig. 2.1 Local redistribution of axial stresses in a *T*-shaped beam cross-section under uniaxial bending (qualitative representation)

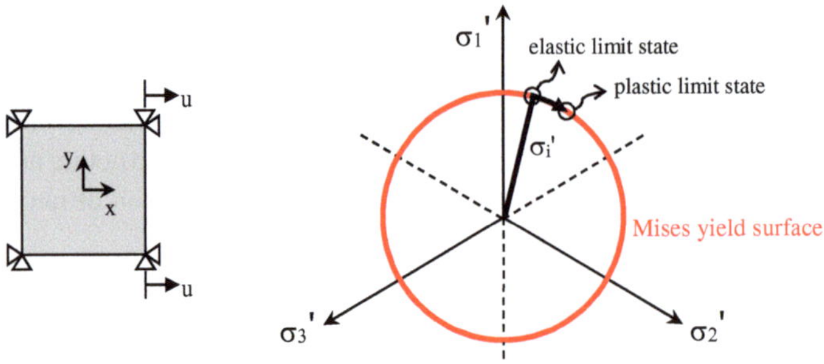

Fig. 2.2 Directional stress redistribution in plane stress state and monotonically increasing displacement-controlled loading in case of non-hardening material

whereas in the other direction any shift of the bearing is prevented, so that a plane strain condition exists (strain is zero in the *y*-direction). Because of the homogeneous stress condition, this is strictly speaking a material problem, not a structural problem.

The stress state is given by

$$\sigma_i = \begin{pmatrix} \sigma_1 \\ \nu\sigma_1 \\ 0 \end{pmatrix} \quad \rightarrow \quad \sigma_i^{'} = \frac{1}{3}\begin{pmatrix} 2-\nu \\ -1+2\nu \\ -1-\nu \end{pmatrix}\sigma_1. \tag{2.1}$$

The Poisson's ratio in steel is approximately $\nu = 0.3$ for elastic behavior. The incompressibility assumption for the plastic portion of the total strain given by Eq. (1.16) corresponds to $\nu^{pl} = 0.5$. After exceeding the yield stress, the effective Poisson's ratio for elastic-plastic behavior increases gradually from 0.3 to 0.5 in the plastic limit state for a non-hardening material at infinitely high load level. Thus, the ratio of the stress components varies from 1 : 0.3 : 0 to 1 : 0.5 : 0, and the ratio of the deviatoric stress components from 1 : –0.235 : –0.765 to 1 : 0 : –1.

It is left to the reader to imagine the impact of hardening for this example of monotonous loading which, because of the directional stress redistribution, is different with kinematic and isotropic hardening.

In the present case, the directional stress redistribution occurs because of the difference in Poisson's ratio of elastic and plastic behavior. However, directional stress redistribution may also occur for a Poisson's ratio of 0.5 (elastic incompressibility). For this purpose, however, an inhomogeneous multiaxial stress state is necessary.

2.2 Structural Ratcheting Based on a Two-Bar Model

If the load is variable rather than increasing monotonically, ratcheting or progressive deformation may occur as already mentioned in Sect. 1.2.8. This phenomenon can be illustrated with reference to a two-bar model (Fig. 2.3) serving as a standard example of structural ratcheting, investigated analytically and experimentally by numerous authors, including dimension variations (different lengths and different cross-sectional areas of the two bars), multiple load variants (pulsating or alternating thermal loading), without and with kinematic and isotropic hardening, for example by Miller [1], Mulcahy [2], Jiang and Leckie [3], and Ponter [4]. The two-bar model is sometimes also referred to as the three-bar model, where the bars are arranged symmetrically, have symmetrical properties, and are loaded symmetrically. The results are then the same as described in the following.

Two parallel bars of the same length, same cross-section, and same material are connected by a rigid plate so that both bars are always equal in length (Fig. 2.3).

Linear elastic-perfectly plastic material behavior with temperature-independent material parameters is considered first. Small strains and small displacements are assumed, so that a change in the cross-sectional area caused by the load does not need to be considered and the equilibrium condition may be formulated at the undeformed system. First of all, a force F is applied, acting constantly in the following and causing a so-called primary stress σ_P of equal magnitude in both bars. In order not to exceed the plastic limit load, this stress must be less than the yield stress. In the left bar, a cyclic temperature change T is then applied whereas the temperature in the right bar remains constant.

In the heating phase, the free thermal expansion strain ε^{th} of the left bar is hindered so that compressive stresses occur whereas the right bar experiences tensile stresses of equal amount for equilibrium reasons. The arithmetic mean of the

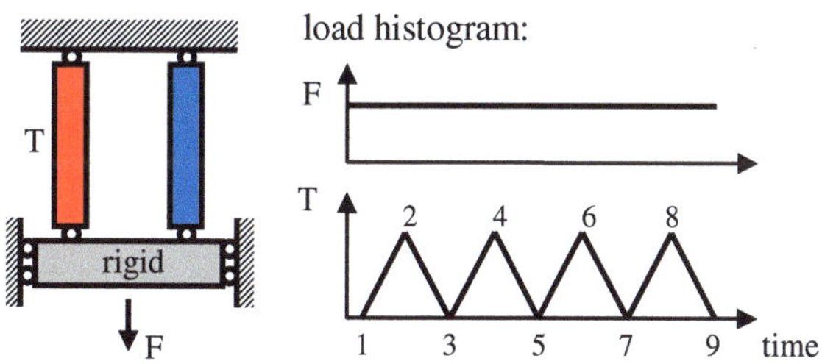

Fig. 2.3 Two-bar model subjected to cyclic loading

stresses in the two bars is the primary stress at any time,[1] so the stresses in the two bars can deviate from this by the same amount with different signs:

$$\Delta\sigma_{\text{left}} + \Delta\sigma_{\text{right}} = 0. \tag{2.2}$$

Because of the lack of hardening, not only is the maximum stress in each of the two bars fixed,

$$\sigma_{\max} = f_y, \tag{2.3}$$

but also the minimum possible stress

$$\sigma_{\min} = 2\sigma_P - f_y. \tag{2.4}$$

Because the rods are always equal in length, the following must apply at all times:

$$\left(\varepsilon^{\text{th}} + \varepsilon^{\text{el}} + \varepsilon^{\text{pl}}\right)_{\text{left}} = \left(\varepsilon^{\text{el}} + \varepsilon^{\text{pl}}\right)_{\text{right}}. \tag{2.5}$$

So we can identify stresses and strains in both bars at each point in the load histogram (Fig. 2.4).

Starting from point 1, where both bars have the same stress σ_P below the yield stress f_y, the free thermal expansion ε^{th} is applied in the left bar (point 2′ in the stress–strain diagram of Fig. 2.4). To maintain equal length in both bars, the mechanical strain must decrease in the left bar and increase in the right. According to the material law, the relief of the left bar can be only purely elastic, causing a reduced stress there, whereas the stress of the right bar increases because of equilibrium, but at most until the yield strength f_y is reached. Because the stress cannot rise any further, the stress in the right bar is known. At time 2, both bars are at the same thermo-elastic strain. The left bar still remains elastic, whereas the right bar is already plasticized.

In the next half cycle, the free thermal expansion ε^{th} in the left bar is withdrawn (point 3′ in the stress–strain diagram). To enforce equal length in both bars, the left bar must become longer, the right bar shorter. The unloading in the right bar occurs purely elastically, whereas the stress in the left bar increases, associated with elastic action, until the stress reaches the yield stress, and then remains constant, whereas its expansion continues to increase. Because at time 3 no free thermal expansion is present, both bars have the same mechanical strain, and there is equilibrium. In this half cycle the left bar is plasticized whereas the right has undergone an elastic relief.

[1]The term "time" is not used here in the sense of a physical quantity, but only as an ordering quantity for successive operations. In the literature it is sometimes referred to as "pseudo-time". The term "histogram" is to be understood in the same sense.

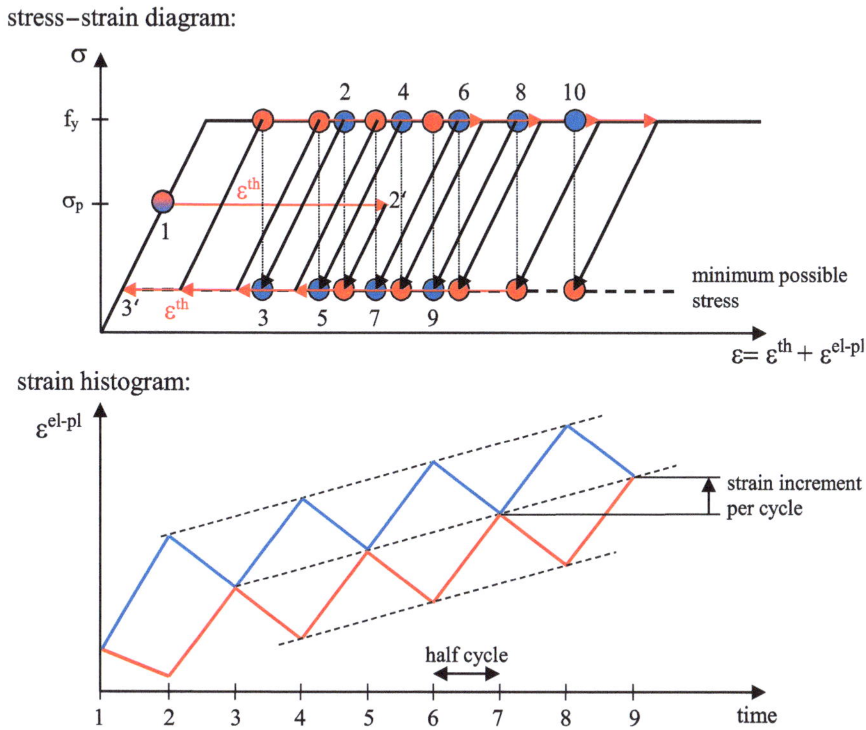

Fig. 2.4 Structural ratcheting with the two-bar model as a result of cyclic loading in case of non-hardening material

In the same way, the development of stresses can be constructed for all further half cycles. After the first load cycle, the strain in the two bars increases with a constant strain increment in every cycle. If the theory of large displacements and large strains were applied, the cross-sectional area would decrease with the number of cycles because of the lateral contraction produced by the axial extension, and the strain would overproportionately increase with the number of cycles.

As a result of the cyclic loading, plastic axial hinges are alternately active in both bars in two successive half cycles, leading to a net increase of plastic strains (a strain accumulation according to Eqs. (1.32) and (1.33)) in each loading cycle.

2.3 Influence of Kinematic Hardening

Taking account of hardening in place of the linear elastic-perfectly plastic material model considered in the previous section, the strain increment per cycle changes in each load cycle. In the case of an unlimited kinematic hardening material, this strain increment decreases monotonically until it completely ceases after a finite or an infinite number of cycles, so that the structural response is eventually periodic.

Strain amplitudes occurring in each half cycle are then of the same absolute values but of opposite sign in two consecutive half cycles. It is said that the system has reached the state of shakedown. If only purely elastic changes of strain take place, we speak of elastic shakedown and of plastic shakedown in the case of elastic-plastic strain amplitudes.

For linear elastic-perfectly plastic material behavior, one needs only to distinguish between elastic shakedown, plastic shakedown, and ratcheting. If ratcheting occurs, theoretically infinitely large strains are achieved because of the constant strain increments after an infinite number of cycles. With hardening, however, we may initially get an accumulation of strain with each load cycle, so that a ratcheting mechanism exists, but coming to a halt after a number of cycles. It is then sometimes called finite ratcheting (Mulcahy [2]).

The two-bar model with linear kinematic hardening treated in the previous section has already been investigated in [1, 2]. For a hardening ratio $E_t/E = 0.05$, a primary stress $\sigma_P/f_y = 0.8$, and two different levels of cyclically applied temperature in the left bar, the histograms of the total elastic-plastic strain and the plastic portion of the strain, respectively, are shown in Figs. 2.5 and 2.6. A pulsating elastically

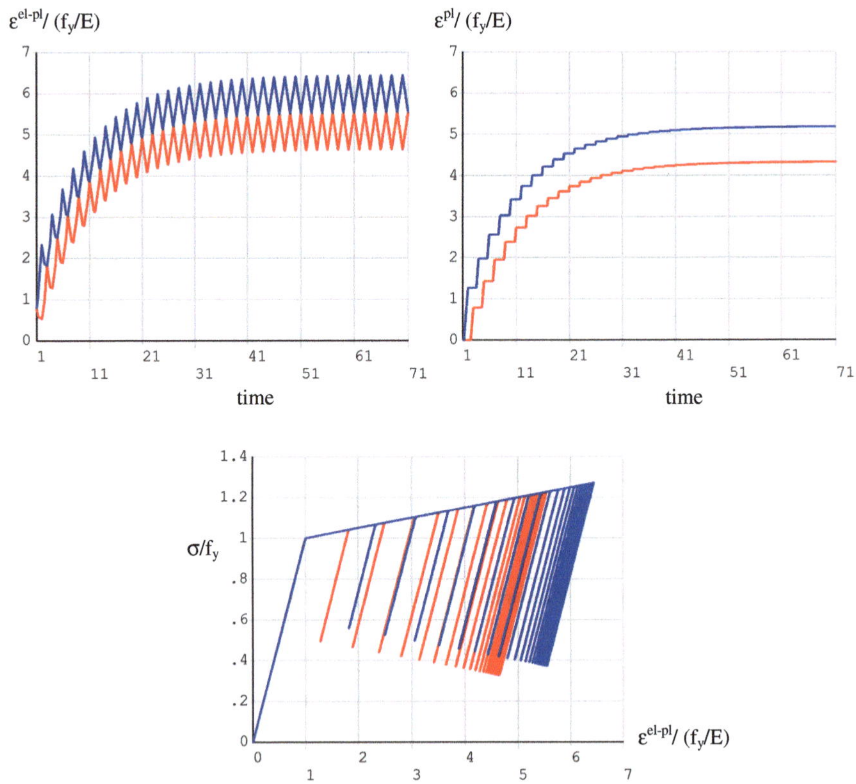

Fig. 2.5 Progressive deformation in the two-bar model with linear kinematic hardening; elastic shakedown ($\sigma_P/f_y = 0.8$; $\sigma_t/f_y = 0.9$; $E_t/E = 0.05$)

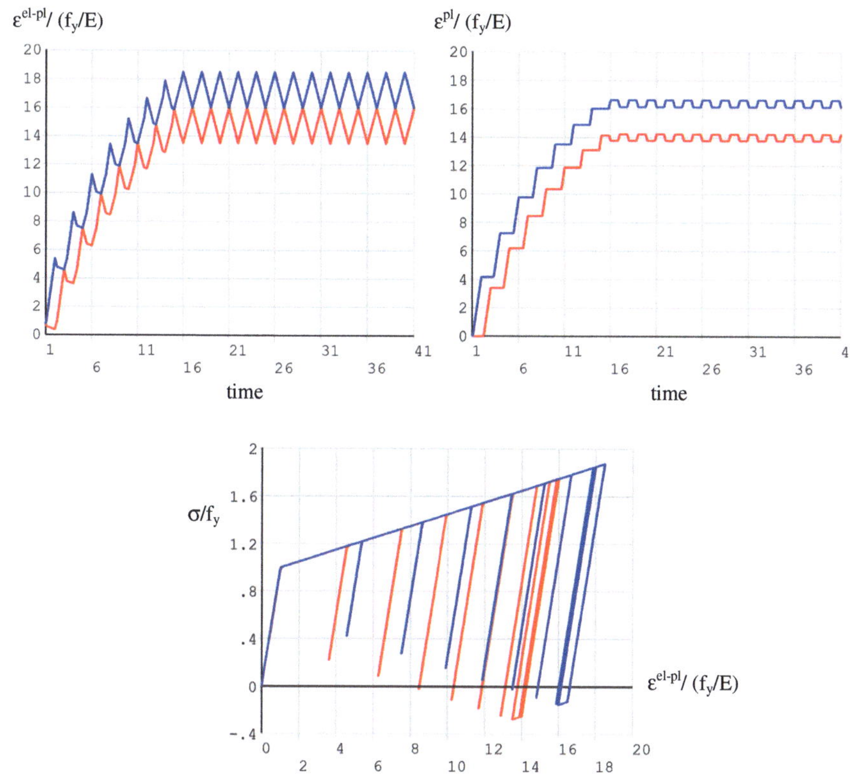

Fig. 2.6 Progressive deformation in the two-bar model with linear kinematic hardening; plastic shakedown ($\sigma_P/f_y = 0.8$; $\sigma_t/f_y = 2.5$; $E_t/E = 0.05$)

calculated thermal stress, also known as "secondary stress", of $\sigma_t/f_y = 0.9$ and 2.5, respectively, was applied there, associated with a negative sign in the left bar and a positive sign in the right bar.

These results were, as were many others in this book, achieved with the finite element program ANSYS [5]. It should be noted that linear and multilinear kinematic hardening is implemented in the KINH material model on the basis of [6], leading to incorrect results in the case of plane stress when adopting the element types PLANE182 or PLANE183. This manifests itself by unequal strains in the two transverse directions with a uniaxial stress state.[2] For all other element types the correct results are obtained with the KINH model, but under plane stress conditions only with the legacy element types PLANE42 or PLANE82, or by using the BKIN material model for linear kinematic hardening. For more details see Chap. 6.

[2]Ansys Inc. says: "`Regarding defect #92325, this is not a defect but a limitation with the sublayer/overlay model according to development`".

The smaller thermal stress (Fig. 2.5) leads to elastic shakedown after theoretically an infinite number of cycles. After about 35 cycles, however, the strain increments per cycle are so small that the shakedown process can be considered virtually complete. The second parameter combination (Fig. 2.6) leads to plastic shakedown after eight cycles. It can be concluded from the stress–strain diagram, by the way, that the same (elastic shakedown) or almost the same (plastic shakedown) level of strain would be reached for isotropic hardening, as reversed plastic straining does not happen at all or, if so, only to a minor degree.

2.4 Ratcheting Interaction Diagram

The stress levels, which result in elastic or plastic shakedown or ratcheting, can be depicted for a given system of component geometry and loading conditions in a so-called ratcheting interaction diagram (RID), sometimes also called a shakedown map. This is shown in Fig. 2.7 for the two-bar model with constant primary and cyclic thermal stress for a linear elastic-perfectly plastic as well as for a linear kinematic hardening material as treated in Sects. 2.2 and 2.3. Because of the pulsating character of the loading (Fig. 2.3), the linear elastic calculated secondary stress σ_t not only marks the maximum value of the stress produced by the thermal load but is also identical to the stress range.

A normalized primary stress $\sigma_P/f_y > 1$ is not possible with non-hardening material, because then the plastic limit load would be exceeded. Although the plastic limit load is defined only for perfectly plastic material, its exceedance is not tolerated in practice even for a hardening material, although the system would then still not be kinematic.

For sufficiently small values of pairs of primary and thermal stress, the structure remains purely elastic (region E). Without hardening (Fig. 2.7, left) the area of plastic shakedown (region PS) degenerates to a line at $\sigma_P/f_y = 0$ and $\sigma_t/f_y > 2$ in this

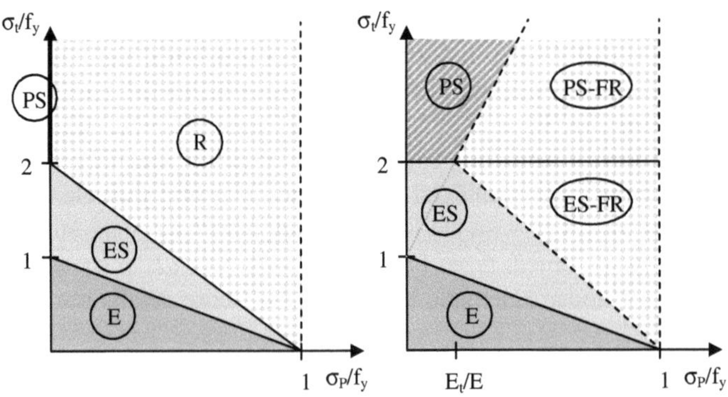

Fig. 2.7 Ratcheting interaction diagram for the two-bar model; *left* with linear elastic-perfectly plastic material; *right* with linear kinematic hardening

example. Later we get to know examples where the region PS can also cover an area despite the lack of hardening. Ratcheting (region R) is only possible in the presence of a primary stress in this example ($\sigma_P/f_y > 0$). An example is shown in Sect. 2.5.1 in which ratcheting occurs even without primary stress. The region ES indicates elastic shakedown.

With unlimited hardening, ratcheting is not possible in the sense of infinitely large strains after infinitely many load cycles. As already mentioned, there is instead guaranteed shakedown, so it comes to limited strains after an infinite number of load cycles. The RID now provides only a qualitative statement about the nature of shakedown.

With kinematic hardening, plastic shakedown is obtained if the elastically calculated range of the thermal stress exceeds twice the yield stress, and elastic shakedown if it falls below twice the yield stress. A quantitative statement about the level of strain, which is connected with the respective state of shakedown, cannot be directly derived from the RID. However, the regions ES and PS can be further subdivided to differentiate certain types of behavior of the structure. In particular, we may be interested in the identification of regions in which shakedown is not achieved within only one loading cycle, but where strains are accumulated over several or many or even an infinite number of cycles, so that larger strains may develop. For such regions, Mulcahy [2] has already used the term "finite ratcheting" (FR). Thus, Fig. 2.7 (right) has regions PS-FR and ES-FR in which finite ratcheting leads to plastic or elastic shakedown. In the case of unlimited isotropic hardening, only the regions E, ES, and ES-FR can exist.

Figure 2.7 shows clearly that ratcheting can occur even at low ranges of loading and therefore in "predominantly" static loading—i.e., without necessarily being linked to an LCF problem.

2.5 Examples of Ratcheting

In the next sections, examples of other configurations of component geometry and load type are treated, allowing insights into the occurrence of ratcheting.

2.5.1 Three-Bar Model

A third bar is added to the two-bar model considered in the previous sections (Fig. 2.8). The thermal stress is no longer applied only in one bar, but one after the other in all three bars, so that unsymmetrical behavior occurs.

This configuration can be viewed as a simple model for a pulsating thermal load on the inner surface of a (pipe) wall and consequently temperature peaks traveling through the wall. The two-bar model treated in the previous sections could, however, at most be regarded as a simple model for a wall with blockwise distributed

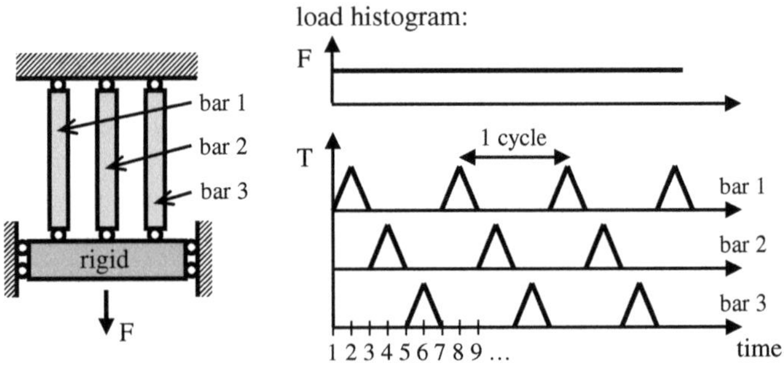

Fig. 2.8 Three-bar model

Fig. 2.9 Ratcheting interaction diagram for the three-bar model

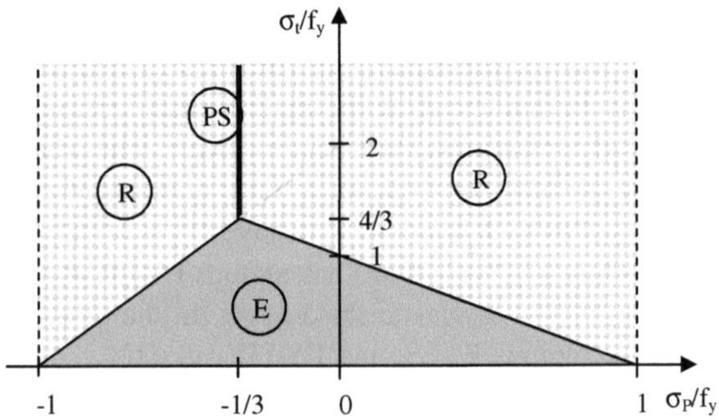

temperatures occurring cyclically. A different configuration of a three-bar model has already been examined by Wolters and Majumdar in [7].

At a time of loading in which one bar is at maximum temperature, causing a linear elastic calculated thermal stress σ_t, the other two bars are stressed by $-\sigma_t/2$. Accordingly, the linear elastic stress range of the thermal stress is $\Delta\sigma_t = 1.5\ \sigma_t$.

Figure 2.9 shows the corresponding ratcheting interaction diagram for the three-bar model for linear elastic-perfectly plastic material.

As with the two-bar model the region of plastic shakedown is degenerated to a line for a non-hardening material. However, there are some other remarkable features:

- There is no region of elastic shakedown
- The RID is not symmetrical with respect to the primary stress
- Even in the absence of a primary stress ($\sigma_P = 0$), a ratcheting mechanism is developing as soon as plastic straining is caused by the thermal load
- Even if the bars are under compression produced by the constant external force F, ratcheting may occur in tension, namely for $\sigma_P/f_y > -1/3$.

2.5.2 Multiaxial Ratcheting

2.5.2.1 Tube Under Internal Pressure and Displacement-Controlled Elongation

So far, structural ratcheting has been represented only in uniaxial stress states. Now a thin-walled tube under constant internal pressure and cyclic heating is considered, where thermal expansion in the longitudinal direction is prevented. The circumferential membrane stresses are then stress-controlled and the axial membrane strains are strain-controlled. At least at a sufficient distance from the ends of the tube there is a biaxial homogeneous state of stress, which can be studied in an isolated material point or at a single finite element: In one direction a constant force F acts, causing the primary stress σ_P, and orthogonal thereto a cyclic strain (Fig. 2.10). Because of the homogeneity of the stress state, this case was categorized under material ratcheting in [8].

With unit lengths chosen as dimensions in Fig. 2.10, the force F causes a biaxial primary stress condition with the Mises equivalent stress

$$\sigma_P = \sqrt{1 - \nu + \nu^2}\, F. \tag{2.6}$$

The elastic limit load is reached at $\sigma_P = f_y$. If the (elastic) Poisson's ratio $\nu = 0$, the force F causes only a uniaxial primary stress σ_P in coordinate direction (1).

The displacement u generates stresses only in coordinate direction (2). For a non-hardening material the yield surface is immutable. At sufficiently high load, a plastic strain increment arises in each extreme value of the cyclic loading that is directed perpendicular to the Mises circle at the position vector of the respective stress state because of the normality rule (see Figs. 1.4 and 1.6). The component of the plastic strain, which acts in the coordinate direction (2) is exactly opposite at the two extreme values of the cyclic loading and is thus cancelled out over an entire load cycle. The component perpendicular to this is, however, equally directed at the two extreme values of the cyclic loading and thus adds up over a whole load cycle, so that each load cycle is creating a net strain increment (Fig. 2.11). This acts in the direction of coordinate (1), as well as with the same intensity but opposite sign in the direction of coordinate (3).

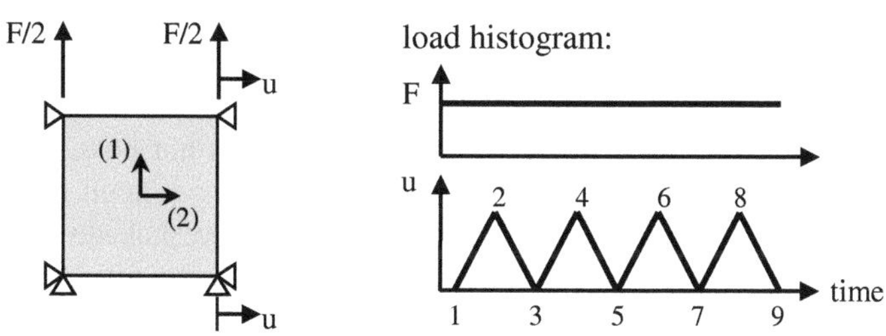

Fig. 2.10 Multiaxial ratcheting: replacement model for a thin-walled tube under internal pressure and displacement-controlled change of length

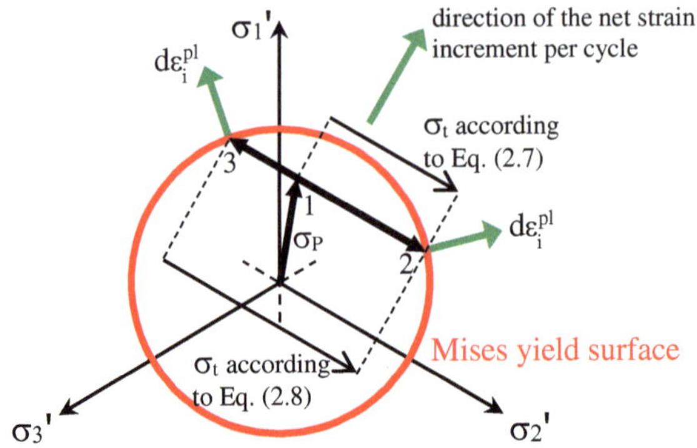

Fig. 2.11 Multiaxial ratcheting with linear elastic-perfectly plastic material: deviatoric stress space

The displacement-controlled load generates the elastic calculated σ_t. The elastic limit load is reached at

$$\frac{\sigma_t}{f_y} = \frac{\left(\frac{1}{2} - \nu\right)\frac{\sigma_P}{f_y} \pm \sqrt{(1 - \nu + \nu^2) - \frac{3}{4}\left(\frac{\sigma_P}{f_y}\right)^2}}{\sqrt{1 - \nu + \nu^2}}. \tag{2.7}$$

If this is exceeded, plastic straining is initiated and the stress vector in Fig. 2.11 does not return from point 2 to point 1 in the subsequent unloading half cycle, but overshoots it. If the displacement-controlled load is not too large, so that the yield point is not reached again at time 3, only purely elastic behavior takes place during the following load cycles. After just one load cycle, elastic shakedown (ES) is achieved (Fig. 2.11).

At sufficiently large displacement-controlled load, the yield surface is again reached at time 3, associated with further plastic action. Thus, the limit state of elastic shakedown is reached. As no directional redistribution can take place in the present system, and no region of plastic shakedown can exist here with linear elastic-perfectly plastic material, the ratcheting limit is reached at

$$\frac{\sigma_t}{f_y} = \sqrt{4 - \frac{3}{(1 - \nu + \nu^2)}\left(\frac{\sigma_P}{f_y}\right)^2}. \tag{2.8}$$

Equations (2.7) and (2.8) are depicted in Figs. 2.12 and 2.13 for $\nu = 0$ and $\nu = 0.5$, respectively.

With linear kinematic hardening, infinite ratcheting is not possible, but either elastic or plastic shakedown is guaranteed. However, it is difficult to identify the boundaries between the regions of ES and ES-FR (elastic shakedown after more than one cycle) or between PS and PS-FR (plastic shakedown after more than one cycle). For this, incremental analysis would be needed, which are discussed later, because the normal direction is constantly changing during the loading process. In addition, the result would be dependent not only on the range of the loading

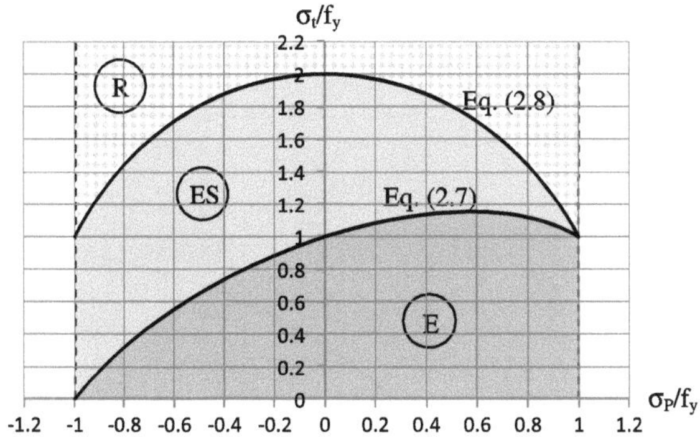

Fig. 2.12 RID for multiaxial ratcheting with linear elastic-perfectly plastic material and $\nu = 0$

Fig. 2.13 RID for multiaxial ratcheting with linear elastic-perfectly plastic material and $\nu = 0.5$

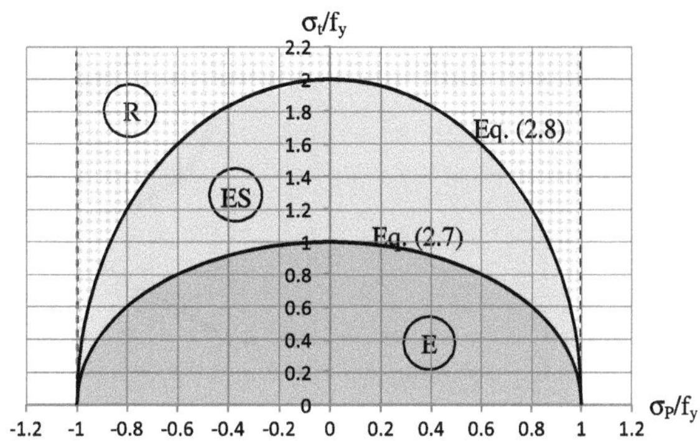

intensity but also on whether the displacement-controlled load is applied swelling as shown in the load histogram of Fig. 2.10 or unequal to zero at minimum load level. Therefore this distinction is omitted in the RID (Fig. 2.14), and is largely omitted in the rest of this book.

The border between ES and PS is given solely by the range of the displacement-controlled load:

$$\frac{\sigma_t}{f_y} = 2. \tag{2.9}$$

If the elastically calculated stress range exceeds twice the yield stress, PS occurs as already noted in the two-bar model (Fig. 2.7). In the present case, this is explained by the fact that the Mises yield surface gradually shifts with the number of cycles in the deviatoric stress space until the cyclically varying stress passes through the centers of the two Mises circles belonging to the extreme loading conditions (marked by crosses in Fig. 2.15). For subsequent cycle numbers n, the system behaves periodically. At loading times $1 + n$ and $2 + n$ the normals to the

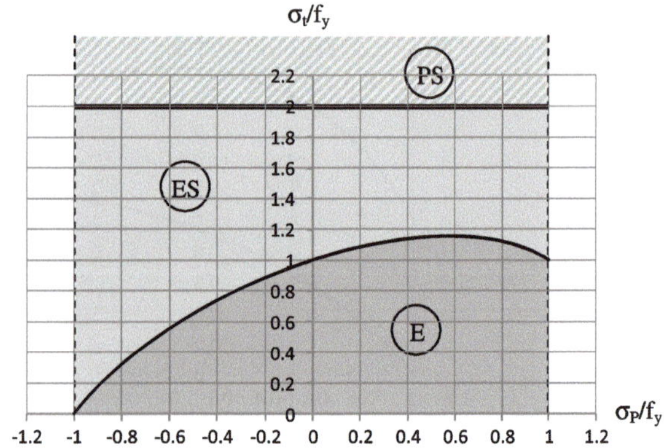

Fig. 2.14 RID for multiaxial ratcheting with linear kinematic hardening material and $\nu = 0$

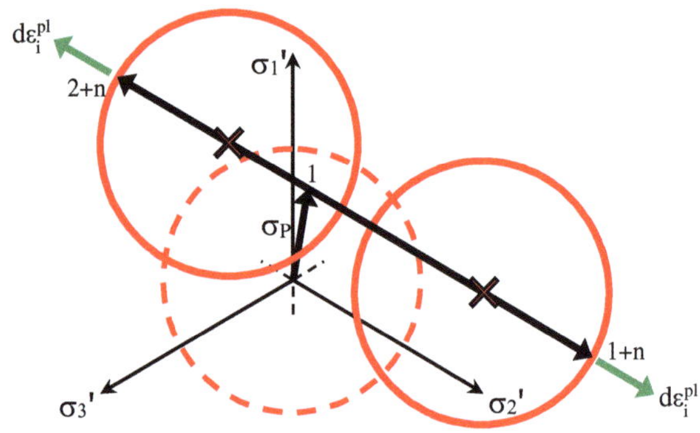

Fig. 2.15 Multiaxial ratcheting with linear kinematic hardening material: PS in the deviatoric stress space

yield surfaces and thus the plastic strain increments are directly opposite, so that they cancel each other out over two consecutive half cycles.

The distance between the two Mises circles can be determined easily in the state of shakedown from the elastic-plastic strain applied as loading and the hardening law (see Eq. (1.26)). However, this does not automatically apply to the position of the Mises circles and thus not to the strain accumulated by finite ratcheting, because the normal direction constantly changes during the loading process with the load level, so that an incremental analysis is necessary.

In the case of elastic shakedown, the Mises yield surface is shifted as shown in Fig. 2.16, until the length of the circle chord corresponds to the elastically calculated stress range. During subsequent cycle numbers, n, their position is fixed. However, this position of the yield surface and thus the strain accumulated by finite ratcheting are only determinable with an incremental analysis.

In Fig. 2.17 the development of plastic strain components over the number of half cycles as derived from an incremental analysis is shown as an example of a loading configuration leading to ratcheting with a non-hardening material, but leading to elastic shakedown with linear kinematic hardening.

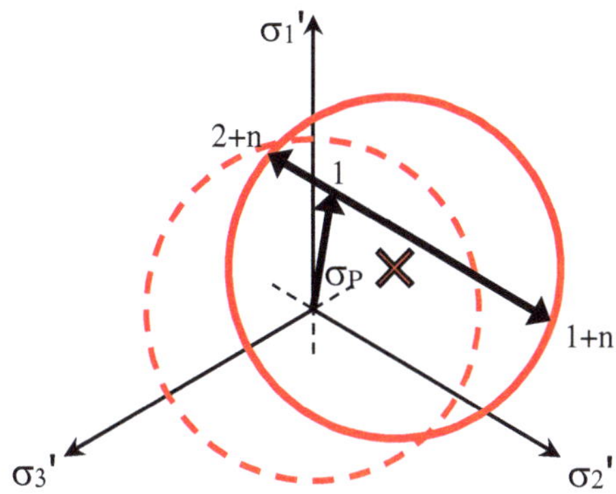

Fig. 2.16 Multiaxial ratcheting with linear kinematic hardening material: ES in the deviatoric stress space

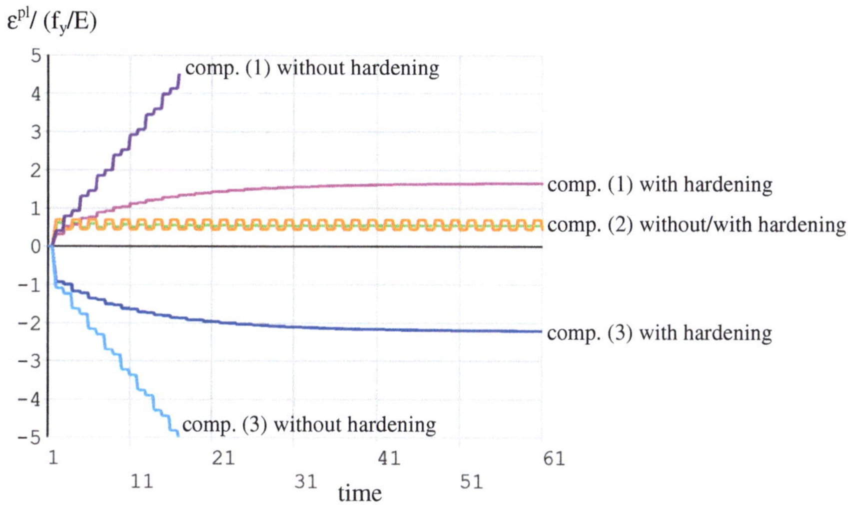

Fig. 2.17 Multiaxial ratcheting: histogram of plastic strains (σ_P/f_y = 0.8; σ_t/f_y = 1.5; E_t/E = 0 and 0.05; ν = 0.3)

2.5.2.2 Tube Under Axial Force and Displacement-Controlled Twist

A situation similar to that for the thin-walled tube with constant internal pressure and cyclic displacement-controlled change in length, and its replacement model shown in Fig. 2.10, is also encountered in a thin-walled tube being pulled in a longitudinal direction with a constant force and being displacement-controlled cyclically twisted (Fig. 2.18)—a situation that has been used several times for ratcheting experiments. This can be simplified to a single material point under constant uniaxial normal stress σ and cyclic distortion γ, causing shear stresses τ.

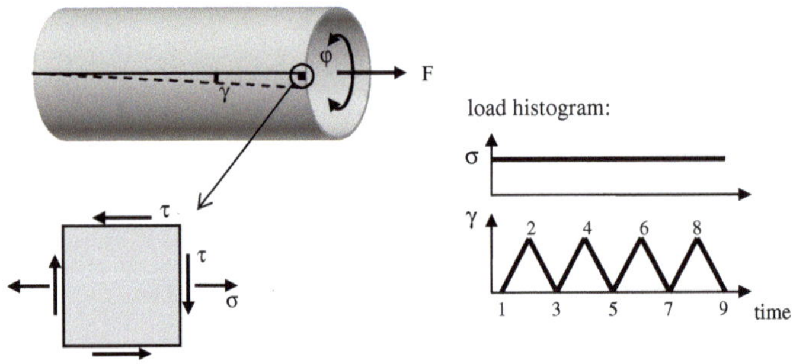

Fig. 2.18 Multiaxial ratcheting: thin-walled tube under axial force and displacement-controlled twist

Fig. 2.19 Multiaxial ratcheting with linear elastic-perfectly plastic material for the system in Fig. 2.18

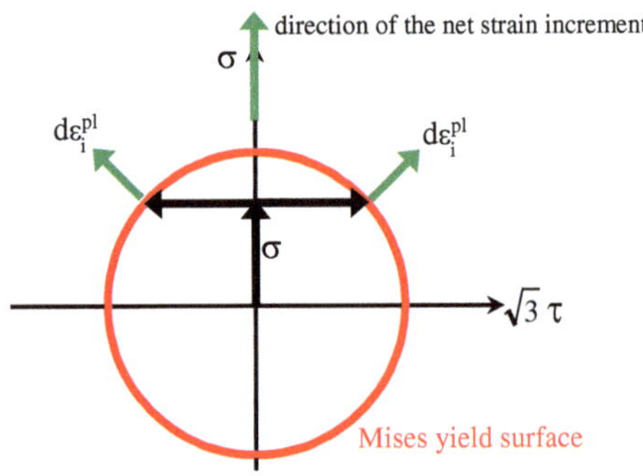

A representation of the stress path in the space of the principal deviatoric stresses is not possible because the principal stress directions do not remain constant during the loading process. Instead, it can be represented in the $\sigma - \sqrt{3}\tau$-plane, where the Mises yield condition

$$f_y = \sqrt{\sigma^2 + 3\tau^2} \tag{2.10}$$

appears as a circle (Fig. 2.19). It can be deduced easily that the RID for non-hardening material is identical with Fig. 2.13 if σ_P is replaced by σ and the $\sqrt{3}$-fold elastically calculated shear stress τ is set in place of σ_t. Accordingly, the boundaries between the regions E, ES, and R are given by Eqs. (2.7) and (2.8) if $\nu = 0.5$ is used.

2.5.3 Bree Tube

The Bree tube is probably the best known example of ratcheting. For some decades it has represented the basis for design rules against ratcheting in many standards, for example, in German nuclear codes [9].

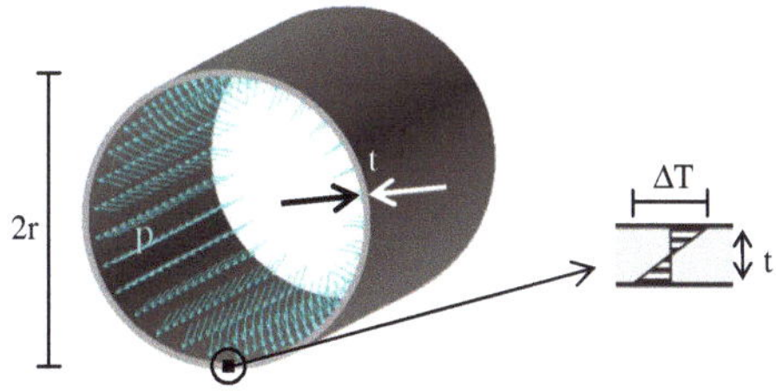

Fig. 2.20 Bree tube with open ends; temperature distribution over the wall thickness

A thin-walled tube (radius r, wall thickness t) is subjected to a constant internal pressure p (Fig. 2.20). This produces the axisymmetric stresses

$$\sigma_p = \begin{cases} \frac{pr}{t} & \text{in circumferential direction} \\ \frac{pr}{2t} & \text{in axial direction} \\ \approx 0 & \text{in radial direction} \end{cases} . \quad (2.11)$$

The axial stress is attributed to the force which acts on the end surfaces of a closed tube. It vanishes, if the tube is to be regarded as open. The stresses only apply at sufficient distances from the ends of the tube (principle of St. Venant). Because of the thin walls, they are membrane stresses, solely attributable to equilibrium requirements and are thus independent of the material behavior.

If a liquid medium moving through the tube changes its temperature, then a temperature distribution is produced through the wall, which is approximately linear in the stationary state because of the thin walls, provided that the outside surface is not insulated but the temperature is kept constant there. The membrane fraction of the thermal expansion strain can develop freely if the tube is supported without axial constraint. However, the temperature difference ΔT between the outer and inner surfaces causes thermal expansion strains that result in equal but oppositely directed mechanical bending strains, because the pipe cannot perform bending deformation across the wall thickness. These bending strains are of equal size in the circumferential and axial directions and are distributed axisymmetrically. For linear elastic material, or fictitiously calculated elastic material, the stresses at the two surfaces become

$$\sigma_t = \begin{cases} \pm\frac{E}{2(1-\nu)}\alpha_t\Delta T & \text{in circumferential direction} \\ \pm\frac{E}{2(1-\nu)}\alpha_t\Delta T & \text{in axial direction} \\ \approx 0 & \text{in radial direction} \end{cases} . \quad (2.12)$$

Miller [1] and, later, Bree [10] have approximated the biaxial stress state caused by internal pressure and radial temperature gradient by a uniaxial stress state as in a flat sheet metal strip, which corresponds to the circumferential direction of the tube. Miller has already pointed out that even Hill [11] examined the steady decrease of the wall thickness of a plate under cyclic bending.

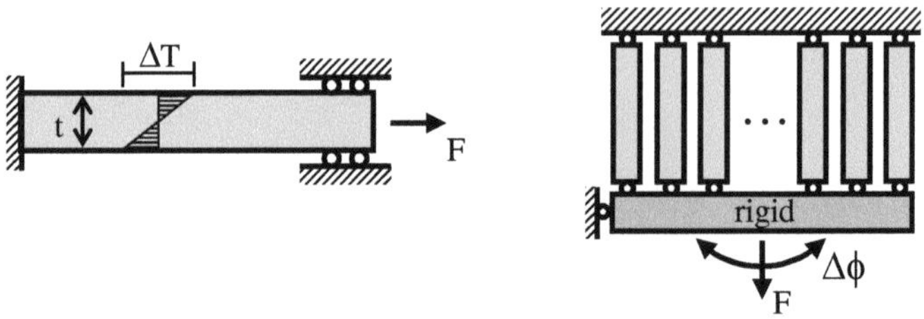

Fig. 2.21 Replacement models with uniaxial state of stress for the Bree tube; *left* beam with linear temperature distribution; *right* coupled rods with displacement-controlled rotation

For this purpose, a beam of height t is supported at both ends against rotation and lateral displacement, but only at one end against axial displacement (Fig. 2.21). At the other end, a constant axial force F is applied. Across the height of the beam a linear temperature distribution with the difference ΔT is applied cyclically. Because this is constant in the axial direction, the length of the beam does not matter. Therefore, it is also sufficient, to replace the beam height t by infinitely many parallel rods, coupled by a rigid plate to satisfy the first Bernoulli hypothesis, stating that plane sections remain plane, and loaded by a cyclic displacement-controlled rotation instead of the thermal loading of the beam (Fig. 2.21).

If restricted to two instead of an infinite number of rods, we get ultimately the two-bar model treated in Sect. 2.2.

Bree first completed the RID for the uniaxial replacement model of the Bree tube subjected to pulsating thermal loads for linear elastic-perfectly plastic material on the basis of the results obtained by Miller, before extending it with regard to linear kinematic hardening material. The corresponding RID (Fig. 2.22) is simply referred to as the Bree diagram. It is often, especially in English language literature, downright eponymous for ratcheting interaction diagrams at all (Bree-like diagrams).

The boundary between plastic shakedown and ratcheting for linear elastic-perfectly plastic material is given by the hyperbola

$$\frac{\sigma_t}{f_y} = \frac{1}{\frac{\sigma_P}{f_y}} \geq 2 \tag{2.13}$$

and for linear kinematic hardening by

$$\frac{\sigma_t}{f_y} = \frac{1 - \frac{E_t}{E}}{\frac{\sigma_P}{f_y} - \frac{E_t}{E}} \geq 2. \tag{2.14}$$

The Bree diagram has been very intensively studied in the literature, for example in terms of its appropriateness to represent the actually biaxial stress state in the original Bree tube, in terms of the difference between an open and a closed tube, in terms of alternating instead of swelling thermal load, with regard to cyclic primary stress either in or out of phase with the cyclic temperature, and in terms of

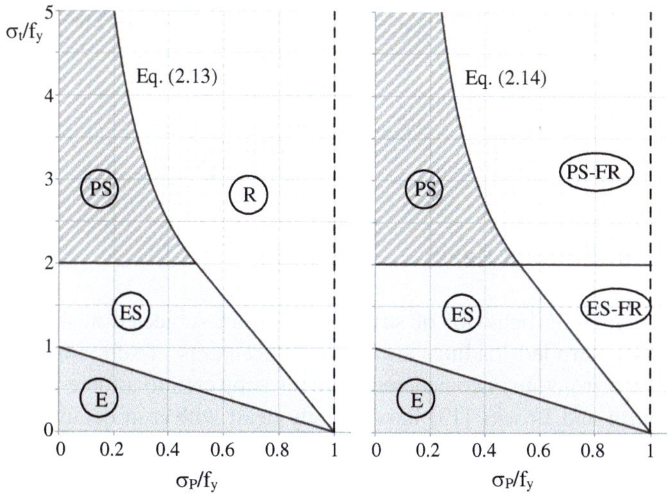

Fig. 2.22 Bree diagram; *left* linear elastic-perfectly plastic material; *right* linear kinematic hardening material ($E_t/E = 0.05$)

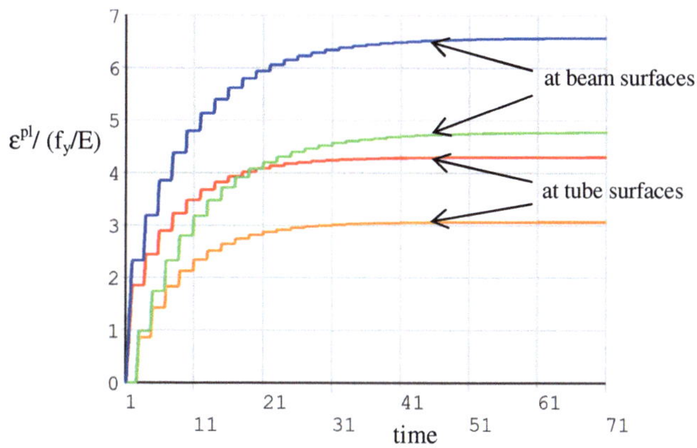

Fig. 2.23 Plastic strain histogram at each of the two surfaces in ES ($\sigma_P/f_y = 0.8$; $\sigma_t/f_y = 1.9$; $E_t/E = 0.05$): Bree tube (multiaxial stress state, $\nu = 0.3$, circumferential component) and replacement beam with uniaxial stress state

additional stresses from piping forces. Solutions for parabolic [1] or bilinear [12] instead of linear temperature distribution over the wall thickness have also found their way into the regulatory framework (such as [9]).

For a configuration of loading parameters in region ES-FR, Fig. 2.23 indicates exemplarily that the replacement beam investigated by Bree (Fig. 2.21) is a conservative simplification of the original tube (Fig. 2.20) with respect to the plastic strains. The same maximum secondary stress was chosen for both models ($\sigma_t = 1.9\, f_y$), for which a lower temperature gradient had to be applied at the tube than at the beam:

$$\Delta T_{\text{tube}} = 2\sigma_t \frac{1-\nu}{E\alpha_t} \tag{2.15}$$

$$\Delta T_{\text{beam}} = 2\sigma_t \frac{1}{E\alpha_t}. \tag{2.16}$$

2.5.4 Continuous Beam

So far, the ratcheting behavior of structures that are subjected to thermal or other displacement-controlled loadings such as displacements of supports has been discussed. Now, a structure is considered which is subjected to a force-controlled load only and Burth and Brocks [13] have already dealt with it in detail: A continuous beam of two equally long fields is loaded in the middle of the first field by a constant force F_1 and in the middle of the second field by a cyclically pulsating force F_2 ($F_2 = 0$ to F_1) (Fig. 2.24).

Only bending is considered, so that shear deformation produced by the lateral forces is disregarded. To understand the ratcheting process it is sufficient to make use of the plastic hinge theory, according to which plastic straining is confined to discrete sections. The cross-section behaves linearly elastic below the plastic moment M_{pl}, and after the plastic hinge is developed, the section cannot carry a moment higher than M_{pl}. This behavior corresponds to a linear elastic-perfectly plastic material, but not in terms of stress and strain, but in terms of generalized stress and strain, i.e., for a moment–curvature relationship.

At time 1, only the force F_1 is active. Its value is chosen so that the plastic moment is exceeded by 12 % at location 1 if the theory of elasticity is applied (which is the case with $F = 2.5\ M_{pl}/L$), so that a plastic hinge is produced there, in which the full plastic moment is acting (Fig. 2.25).

Accordingly, a plastic kink $\varphi_{1(t1)}$ is created at location 1. At time 2 an additional second force F_2 (same intensity as F_1) is applied in the second field. This causes the plastic hinge at the location 1 to be relieved, so that although the kink $\varphi_{1(t1)}$ is maintained and acting as a constraining load, the moment is reduced there. Simultaneously, a new plastic hinge is formed at the inner support so that the plastic moment M_{pl} is acting there with a negative sign, associated with a plastic kink $\varphi_{s(t2)}$.

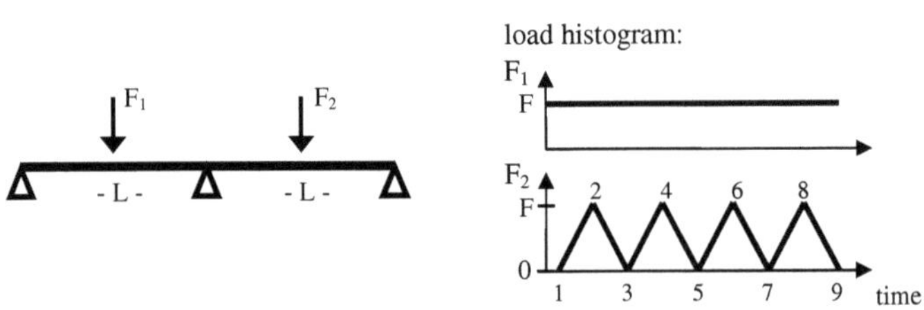

Fig. 2.24 Continuous beam

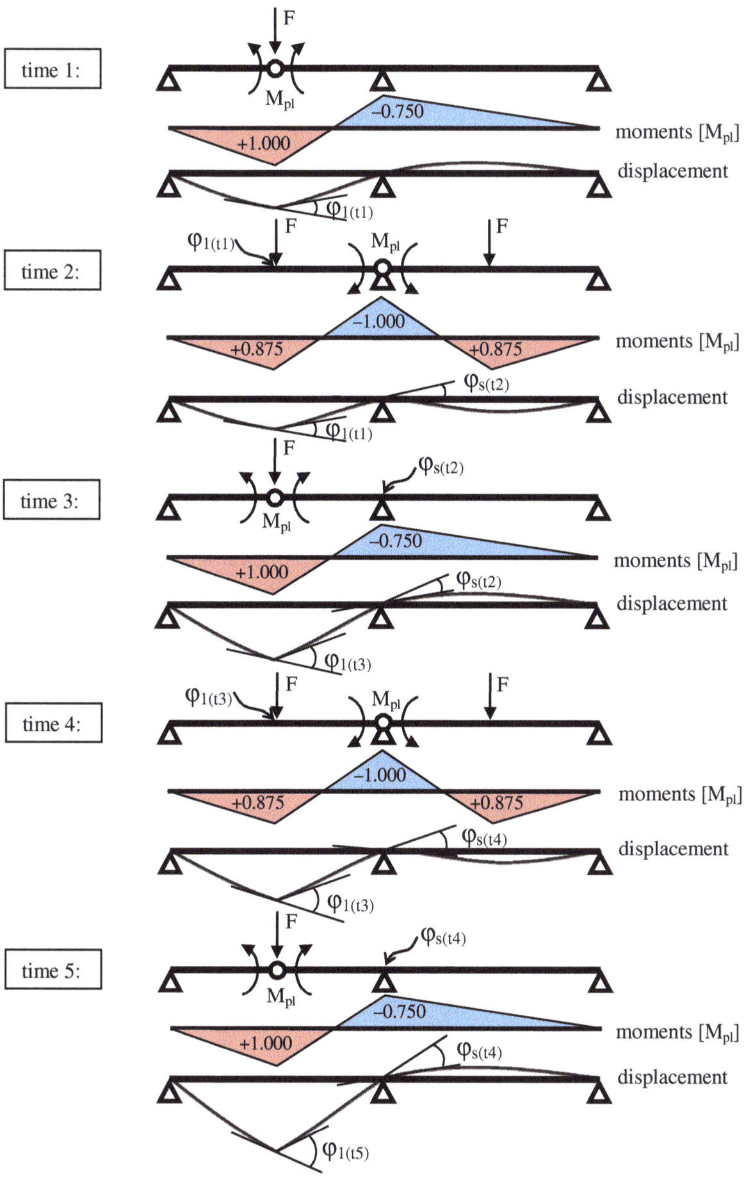

Fig. 2.25 Ratcheting of the continuous beam

If at time 3 the force F_2 is now removed, the plastic hinge is relieved at the inner support, but the plastic kink $\varphi_{s(t2)}$ is retained there, so only the magnitude of the moment decreases. Instead, the moment grows at the position of force F_1 until the plastic moment is again reached, so again the plastic hinge is activated with a corresponding plastic kink $\varphi_{1(t3)}$. The structural system is now the same as at time 1, but an additional constraining load is introduced in terms of kink $\varphi_{s(t2)}$. Because it is a statically determinate system because of the plastic hinge, this constraining load causes no internal forces, but merely deformation. The moments are equally distributed at times 1 and 3, although the displacements are not!

Upon renewed application of force F_2 at time 4, and maintaining the kink $\varphi_{1(t3)}$ at location 1, the moment at the inner support increases in magnitude again until again the plastic moment is reached and the plastic kink $\varphi_{s(t4)}$ is formed, whereas at location 1 the moment decreases. The structural system and the moments are now therefore the same as at time 2, but the constraining load at location 1 has changed and, consequently, the displacements as well. Accordingly, the kink in the plastic hinge is now different from the time 2 ($\varphi_{s(t4)} \neq \varphi_{s(t2)}$).

Thus another starting point exists for the following removal of the load F_2 (time 5) as for the removal of the load at time 3. Accordingly, the displacements at time 5 and also the plastic kink at location 1 differ from those at time 3.

In turn, the displacements at time 6 are different than at time 4, at time 7 different than at time 5, etc. After each load cycle only the displacements are different in the present case, although the magnitudes of the moments are always the same. The change of displacements is equal in each cycle, so that the deflections in the left field increase linearly with the number of cycles (Fig. 2.26).

In Fig. 2.27 the histogram of the bending moments at the two positions of the plastic hinge is shown for the first five cycles. So it is easy to see that there is no alternating plasticity and thus there is no pronounced fatigue problem.

Ratcheting arises here because two plastic hinges are alternately active. It is worth noting that the applied load is only force-controlled. However, because of the static indeterminacy of the system, constraints are present which have displacement-controlled character.

Figure 2.28 shows the RID for the loading configuration of constant force F_1 and pulsating force F_2. The axes are scaled to the plastic limit loads F_{1T} and F_{2T}, obtained for the sole effect of $F_1 = F$ or $F_2 = F$, respectively. Three different regions of elastic shakedown are recognizable, depending on whether a plastic hinge is formed at one of the two locations of the applied forces or at the inner

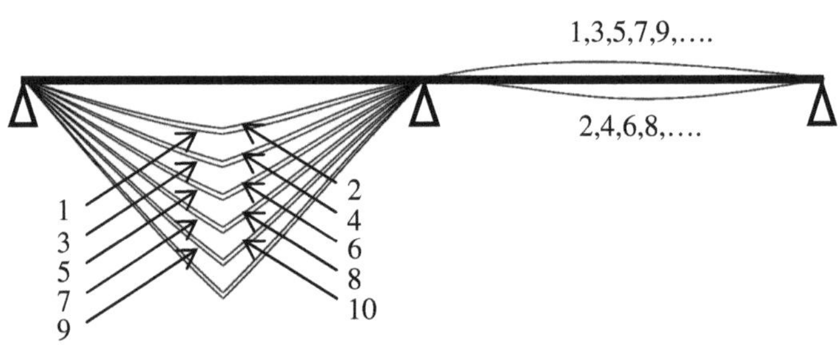

Fig. 2.26 Displacements of the continuous beam at different times because of ratcheting

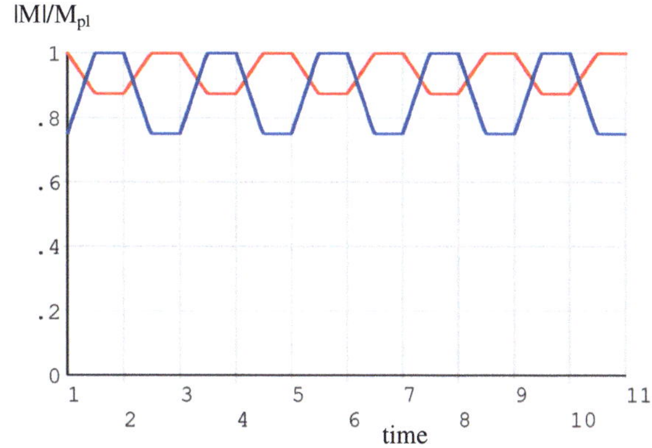

Fig. 2.27 Histogram of the bending moments at the plastic hinges of the continuous beam (*red* in the middle of the left field, *blue* at the support)

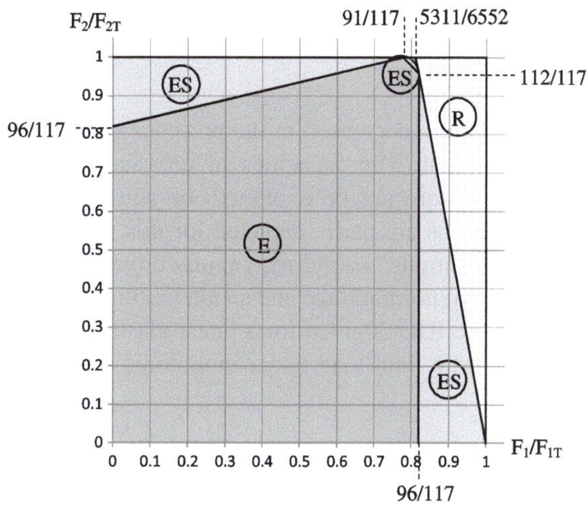

Fig. 2.28 RID of continuous beam with a constant force F_1 and cyclic force F_2 (both axes scaled by the respective plastic limit loads for F_1 and $F_2 = F$ alone)

support (the small triangle at the top of RID). A region of plastic shakedown does not exist. The elastic limit refers here to the advent of a plastic hinge in a section, not to incipient plastic straining in one of the outermost fibers of a section. Possibilities and limits of plastic system reserves are hereby demonstrated for this configuration of a two-field beam.

2.5.5 Interaction of Section Forces

In the previous section, plastic system reserves were the subject of consideration, but a look is now taken at plastic cross-section reserves.

Because the boundary between elastic or plastic shakedown on the one hand and ratcheting on the other is always reached at load levels below the plastic limit load with linear elastic-perfectly plastic material behavior, the plastic capacity of a

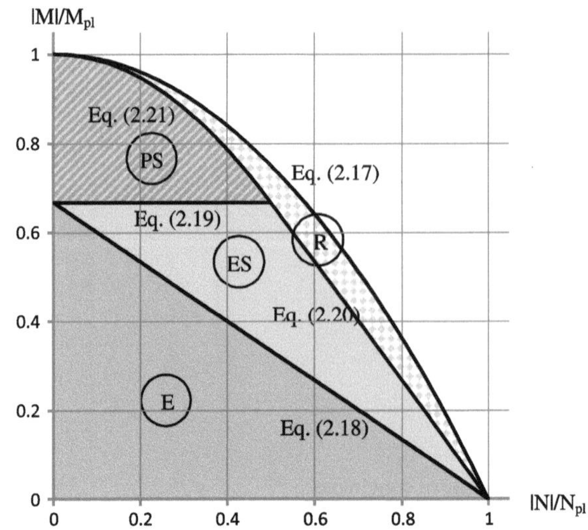

Fig. 2.29 Bearing capacity of a rectangular cross-section: RID of constant normal force and alternating bending moment

cross-section is decreased in the presence of even slightly varying load, as almost always present. The capacities of cross-sections derived for monotonic loading should therefore not be exploited, or you should at least be aware that existing safety margins are partly depleted for this.

As an example, a solid rectangular cross-section under force-controlled loading by a constant normal force and an alternating bending moment is considered. In [14] and [15] the RID is given for a linear elastic-perfectly plastic material (Fig. 2.29).

The plastic limit load under monotonic loading is given by

$$\frac{|M|}{M_{\mathrm{pl}}} + \left(\frac{N}{N_{\mathrm{pl}}}\right)^2 = 1 \tag{2.17}$$

and the elastic limit load by

$$\frac{3}{2}\frac{|M|}{M_{\mathrm{pl}}} + \frac{|N|}{N_{\mathrm{pl}}} = 1. \tag{2.18}$$

The region of elastic shakedown is bounded by

$$\frac{3}{2}\frac{|M|}{M_{\mathrm{pl}}} = 1 \quad \text{for} \quad 0 \le \frac{|N|}{N_{\mathrm{pl}}} \le 0.5 \tag{2.19}$$

$$\frac{3}{4}\frac{|M|}{M_{\mathrm{pl}}} + \frac{|N|}{N_{\mathrm{pl}}} = 1 \quad \text{for} \quad 0.5 \le \frac{|N|}{N_{\mathrm{pl}}} \le 1 \tag{2.20}$$

and the region of plastic shakedown by

$$\frac{|M|}{M_{\rm pl}} + \frac{4}{3}\left(\frac{N}{N_{\rm pl}}\right)^2 = 1. \tag{2.21}$$

The origin of Eq. (2.21) depicted in the RID of Fig. 2.29 possibly requires some explanation. For determining the upper boundary of the plastic shakedown region, the relation between normal force or bending moment and membrane strain ε_m or bending strain ε_b may be expressed by

$$\frac{N}{N_{\rm pl}} = \frac{\varepsilon_m}{\varepsilon_b} \tag{2.22}$$

$$\frac{M}{M_{\rm pl}} = 1 - \frac{1}{3}\left(\frac{f_y/E}{\varepsilon_b}\right)^2 - \left(\frac{\varepsilon_m}{\varepsilon_b}\right)^2. \tag{2.23}$$

Now use is made of the relationship between a constant normal force and displacement-controlled bending as represented by the Bree diagram (Fig. 2.22, left). The boundary between plastic shakedown and ratcheting, Eq. (2.13), can, because σ_t is the linearly elastic calculated bending stress, also be formulated by

$$\frac{\sigma_t}{f_y} = \frac{\varepsilon_b}{f_y/E} = \frac{1}{N/N_{\rm pl}} \geq 2. \tag{2.24}$$

Insertion of Eqs. (2.22) and (2.24) into Eq. (2.23) then yields Eq. (2.21).

Ratcheting takes place in the region between the shakedown states and the limit load, which manifests itself in an increase in the membrane stretching in the direction of the normal force in each load cycle. For a given normal force, the cross-section can only be loaded less for cyclic than for monotonic loading. The reduction is up to one-ninth of the plastic moment (at $N/N_{\rm pl} = 2/3$), or, in other terms, the load bearing capacity of the cross-section is reduced by up to one-third compared to monotonic loading (for $N/N_{\rm pl} \rightarrow 1$).

2.6 Myths About Ratcheting

Among working engineers and scientific publications, and even in codes and standards for the design of structures, a number of myths have been disseminated which can be corrected based on the findings in the previous sections.

Myth 1: Ratcheting Only Occurs at Very High Cyclic Loads

The underlying idea, nourished by design codes, is that ratcheting is a byproduct of LCF (low cycle fatigue: fatigue damage at small numbers of load changes and high strain ranges), and could thus only occur in conjunction with alternating plasticity, i.e., in terms often used by design codes such as the KTA rules [9] only when $S_n > 3\ S_m$.

This perception is not true as we can see from the ratcheting-interaction diagrams of the last sections. The two-bar model (Fig. 2.7) shows that ratcheting can occur far below the threshold of cyclic plasticity (i.e., when $\sigma_t/f_y < 2$) if the constant primary stress is large enough.

Myth 2: In Order For a Ratcheting Mechanism to Develop, a Force-Controlled Load or Primary Stress is Required as a Driving Force, Which Also Must be Present Constantly

This idea of ratcheting is invalidated through the three-bar model treated in Sect. 2.5.1. The ratcheting interaction diagram (Fig. 2.9) shows that, even without a primary stress (i.e., at $\sigma_P/f_y = 0$), ratcheting may occur.

Myth 3: In Order to Exclude Structural Ratcheting, all Loads on the Structure Need to be Analyzed only in Pairs

In fact, the consideration of the extremes of a pair of variable loads is sufficient in many cases to detect or exclude structural ratcheting. This includes the two-bar model (F alone, $F + T$ in the left bar), the Bree tube (p alone, $p + \Delta T$), and the continuous beam (F_1 alone, $F_1 + F_2$). In the three-bar model in Sect. 2.5.1, there are four actions: an external force F and a temperature peak in each of the three bars. These cause four different loading extremes (F alone, $F + T_1$, $F + T_2$, and $F + T_3$). However, none of the six possible pairs of these four states can constitute a ratcheting mechanism, but only the combined effect of all four states. So, a ratcheting analysis restricted to pairwise consideration of load conditions could not identify ratcheting correctly.

2.7 Residual Stresses

Under variable loads, the terms "residual stress" and "residual strain" play a major role. Thus, stresses and strains are those which are present in the structure when, after prior plasticizing, all loads are completely removed, assuming purely elastic behavior. So it is not that the point in time is considered where the variable portion of the loading vanishes, but when the structure is completely unloaded, including constant loads. In the absence of any loads, the residual stresses must satisfy self-equilibrium. They are therefore sometimes also referred to as eigenstresses.

A full discharge of all previously applied loads would often actually be associated with reversed plasticity because of the removal of constant loads, even in the case of elastic shakedown. However, this is not considered. Instead, the fictitious elastic stresses and strains are subtracted from the elastic-plastic state, with the consequence that the residual stresses ρ and the residual strains ε^* are often pure operands which are not really present in the structure after complete unloading (Fig. 2.30):

$$\rho_i = \sigma_i - \sigma_i^{\text{fel}} \tag{2.25}$$

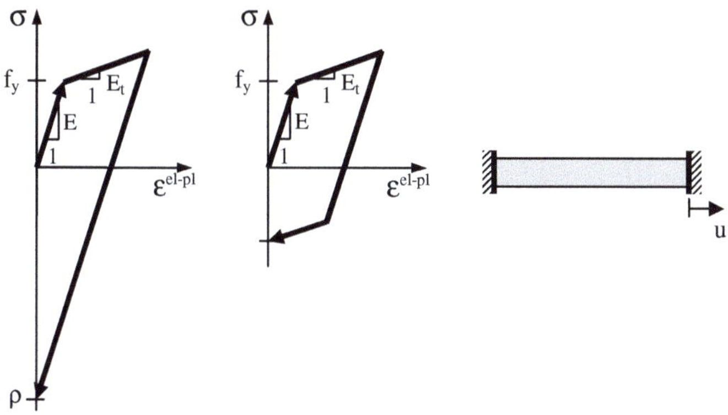

Fig. 2.30 *Left* residual stress for a tension bar under strain-controlled load with linear kinematic hardening; *right* stress actually remaining after unloading

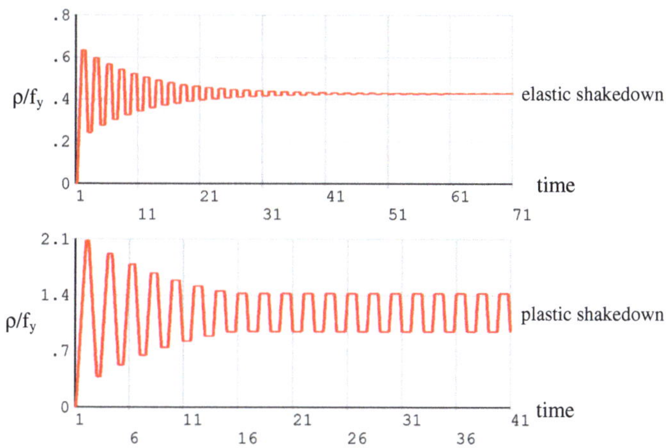

Fig. 2.31 Histogram of residual stresses for the left bar of the two-bar model with linear kinematic hardening (σ_P/f_y = 0.8; E_t/E = 0.05); *top* σ_t/f_y = 0.9; *bottom* σ_t/f_y = 2.5

$$\varepsilon_i^* = \varepsilon_i - \varepsilon_i^{fel}. \tag{2.26}$$

The evolution of residual stresses with the number of cycles is shown in Fig. 2.31 for the left bar of the two-bar model with linear kinematic hardening (see Sect. 2.3, Figs. 2.5 and 2.6). It is apparent that the residual stresses remain constant after reaching the state of elastic shakedown.

In Fig. 2.32 the stress–strain hysteresis for both bars is duplicated from Fig. 2.6, supplemented by marking the residual stresses at the time of load reversals in the two bars after the state of plastic shakedown has been reached (squares for times 19, 21, 23, etc., in the load histogram of Fig. 2.3, triangles for times 18, 20, 22, etc.).

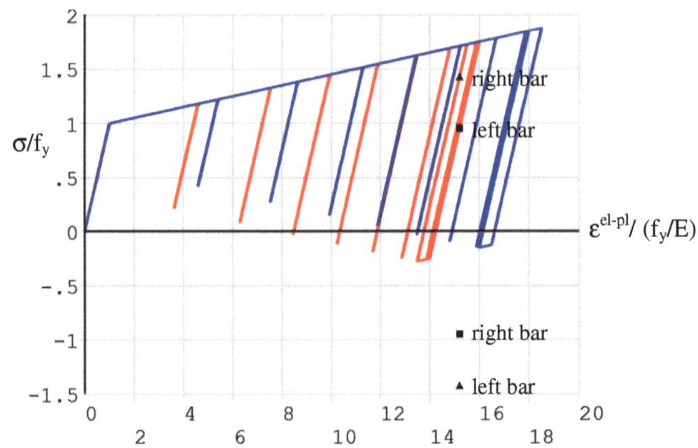

Fig. 2.32 Stress–strain diagram for the two-bar model as in Fig. 2.6, here complemented with residual stresses and residual strains at load reversals after the plastic shakedown state has been reached

The residual stress states do not apparently necessarily lie on the elastic-plastic material curve. The residual strains are at all times the same in this example after plastic shakedown has been reached, because the thermal load is then strain-controlled with the effect that the plastic strain enhancement factor K_e, defined as ratio of the elastic-plastic and the fictitiously elastically calculated strain, amounts to $K_e = 1$.

2.8 Service Life

2.8.1 Fatigue

Sufficient fatigue strength is always to be proven under variable load. There are a number of different concepts, differing in both the way of determining as well as assessing the stresses. Some are based on classical continuum mechanics (nominal stress concept, structure stress concept, local concept), others on fracture mechanics to calculate crack propagation. Some only make use of the elementary strength of materials to determine the stresses without taking account of geometric or metallurgical transitions or locally confined disturbances of the stress pattern at locations of load application. Others require the determination of local stresses with linear elastic or elastic-plastic material laws, usually adopting the finite element method.

Because this book is aimed at determining elastic-plastic stresses and strains, primarily those life proof concepts are of interest here, which can make use of them. This applies to the local concept, which is based on the elastic-plastic strain range used to determine the fatigue usage by means of strain-based fatigue curves (Wöhler curve, SN curve) that have been experimentally determined by uniaxial strain-controlled tests (see, for example, [17]). Here it is essentially to distinguish

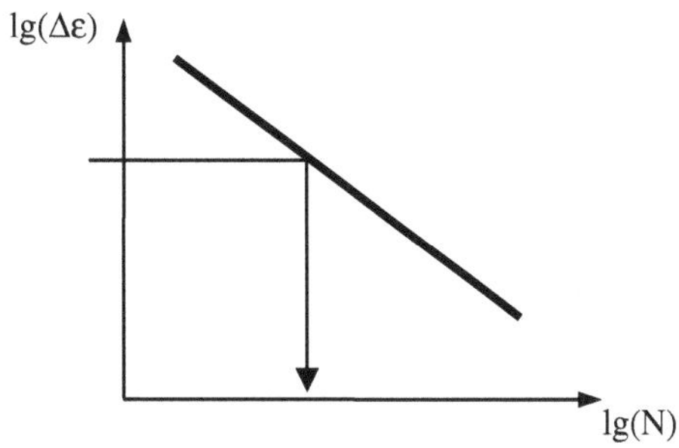

Fig. 2.33 Fatigue curve (schematic)

between the LCF (= low cycle fatigue, i.e. large strain ranges associated with portions of plastic strain; less than 20,000 cycles to failure), the HCF (= high cycle fatigue, less than 2,000,000 cycles to failure), and the endurance regime.

For fatigue analysis the elastic-plastic strain ranges are to be determined for all six different strain components of a load pair (in certain circumstances also the mean strains or mean stresses). For these, the equivalent strain range is formed, for which the number of allowable load cycles N can be read from the fatigue curve (Fig. 2.33).

The fatigue usage factor u then results from the number of load cycles n, either specified or known from in-service monitoring, and must not exceed the value of 1:

$$u = \frac{n}{N} \leq 1. \tag{2.27}$$

If a structure is subjected to different kinds or levels of loading, then several pairs of loading are formed by means of a counting procedure. Here, mostly the rainflow or the reservoir method is used. The number of occurrences of the elastic-plastic strain ranges of each of these pairs of loadings generates a partial usage factor from the fatigue curve. This in turn is summed up over all load pairs i according to the linear fatigue accumulation rule by Palmgren-Miner:

$$u = \sum_i \left(\frac{n}{N}\right)_i \leq 1. \tag{2.28}$$

2.8.2 Strain Accumulation

The life of a cyclically loaded structure operated in the elastic-plastic regime depends not only on the fatigue phenomenon but also on the phenomenon of ratcheting. Furthermore, the service life of a structure can be limited by other phenomena such as corrosion, wear as a result of contact friction, etc., which, however, are not considered here. Because of the one-way increase in strain and displacement per cycle, ratcheting is primarily associated with the risk of incremental collapse after an infinite number of cycles. This is an ideal imagination, similar to that of an instantaneous collapse by exceeding the plastic limit load, which is based on assumptions that are no longer met in the state of collapse, such as perfectly plastic material, small strains, and small displacements. In addition, the distortion may become so great in an active ratcheting mechanism, even at a finite number of load cycles, that failure occurs by exceeding the ductility of the material. Also, the serviceability of the structure may have already been lost because of excessive displacements.

As an indirect consequence of ratcheting, cross-sections might be so weakened after a certain number of cycles that the plastic limit load is exceeded, or a stable equilibrium state is no longer possible. Although ratcheting and fatigue are different phenomena, and ratcheting can also occur without significant fatigue, an interaction between the two is also possible. So the strain may alternate around a mean strain or mean stress enlarged by ratcheting with the consequence that the fatigue usage is increased, or the range of applicability of the fatigue curve may be exceeded because of excessive strains.

For this reason, in some regulations, such as the nuclear part of the ASME code, which serves as a basis for many standards worldwide (also [9]), proof is required that certain strain limits are not exceeded at the end of life. This differs depending on the proof format, i.e., whether the strains are determined by simplified or detailed elastic-plastic analyses. They also depend on the distribution of strains across the cross-section. So, different strain limits apply to membrane, linearized, and peak strains. Furthermore, the position of the point under consideration in the structure plays a role, depending on whether it is located in the base material or in a welded joint. Accordingly, the strain limits vary between 0.5 % for the membrane strain in welded joints and 5.0 % for the peak strains in the base material.

For example, in the German KTA safety standards [9] for components of the primary circuit of light water reactors, Sect. 17.13.2 "Simplified check with approximate formulas" under 7.13.2.3 "Proof by limiting strains" requires:

> "(5) The sum of all strain increments may not exceed 2 % at the end of life"

or Sect. 17.13.3 "General proof by elastic-plastic analysis" requires:

> "(3) The locally accumulated plastic tensile principal strain must not exceed to end of life at any point of a cross-section the following maximum values: 5.0 % in the base material, 2.5 % in welds."

At this point, it should be noted that ratcheting and fatigue proofs are always required if a variable load is present. In the respectively applicable design codes, the procedure for the fatigue assessment is usually described and for the proof against ratcheting usually not. Sometimes a ratcheting proof is required in design codes only if certain stress ranges are exceeded ($S_n > 3S_m$) and is thus coupled with the need for fatigue analyses, which is, however, theoretically not justified (see Sect. 2.6, Myth 1).

If the proof against ratcheting is performed by limiting the strains, the strain components accumulated through the load cycles at the end of the service life of the

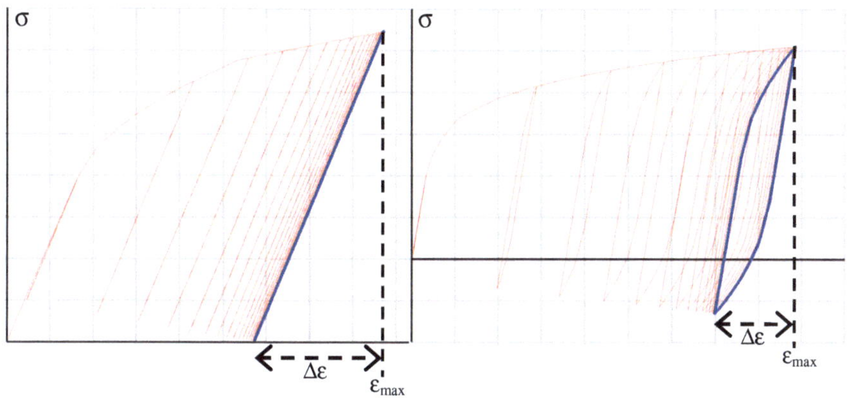

Fig. 2.34 Elastic-plastic strains required for proof of life; *left* elastic shakedown; *right* plastic shakedown; the cycle of shakedown is highlighted

structure are to be calculated; see the schematic diagram of Fig. 2.34 in a uniaxial stress state. In the state of elastic shakedown (Fig. 2.34, left) the accumulated strain ε_{max} is often decisive for the service life, because the strain range $\Delta\varepsilon$ is relatively small. In the state of plastic shakedown (Fig. 2.34, right) either the accumulated strain or the strain range may be life-determining.

It is not sufficient, as a substitute for limiting strains, to prove that the structure for a given load configuration shakes down, and not even that it shakes down elastically (regions ES or ES-FR in the ratcheting interaction diagram), because the state of shakedown may be associated with excessive strains. For the same reason, it is also irrelevant whether finite or infinite ratcheting is present, which is why a strict distinction between the two is omitted in the following.

This finding has implications for the choice of a material model for the calculation of elastic-plastic strains. So for this purpose models may also be used, which in principle cannot predict infinite ratcheting and tend to underestimate finite ratcheting, e.g., pure kinematic hardening without recovery terms. This deficiency should, however, be compensated either by reducing the maximum allowable strains or by appropriately modifying material parameters obtained by curve fitting, e.g., artificially reducing the tangent modulus E_t.

2.9 Analysis Methods

To obtain the elastic-plastic strains that are necessary for fatigue and ratcheting proof work, different methods are possible. Some provide only information on the range of strains for the fatigue check, and are not applicable to determining the accumulated strains for the ratcheting proof.

2.9.1 Fictitious Elastic Analysis and Correction Factors

This method calculates the stress range for each pair of loads through adoption of linear elastic material behavior, and the resulting equivalent fictitious elastic strain range is multiplied with a plastic strain enhancement factor that can be taken from design codes. In many cases the plastic strain enhancement factor K_e ("K_e-factor") of the ASME Code is being used, e.g., in [9]. A critical assessment of this factor can be found in [16]. Other proposals go back to Neuber [18], who has dealt with elastic-plastic strain concentration at sharp notches. A number of other proposals were developed by, amongst others, Hübel [16], Roche [19], and Seshadri [20], which partly entered into French and German design codes.

Generally, however, it is difficult to reflect adequately the complex elastic-plastic behavior depending on the material behavior, component geometry, load configuration, level of loading, etc., by a factor. This concept is historical and dates back to periods when the finite element method was not even available for linear calculations in engineering practice. From today's perspective, this factorial approach is for general use, apart from special configurations of component geometry and kinds of loading, to be considered not very well founded on continuum mechanics. It draws its justification mainly from the fact that a rich experience in the application of this concept is available in narrowly circumscribed design situations.

Figure 2.35 shows some solutions for the plastic strain enhancement factor K_e for linear hardening ($C/E = 0.1$). Shown in dashed lines are the material curves for different yield stresses or, because of the normalization, ultimately for different load levels, expressed using the fictitious elastic calculated stress σ^{fel}. The five colored lines show the dependence of the plastic strain enhancement factor K_e on the stress level because of stress redistribution: The curves for the limit behavior of a stress-controlled load (the stress is independent of the behavior of the material) and for the limit behavior of a strain-controlled load (the strain is independent of the behavior of the material) are horizontal or vertical.

Furthermore, the Neuber hyperbola is shown. It results from the requirement that the product of stress and strain is independent of the material behavior:

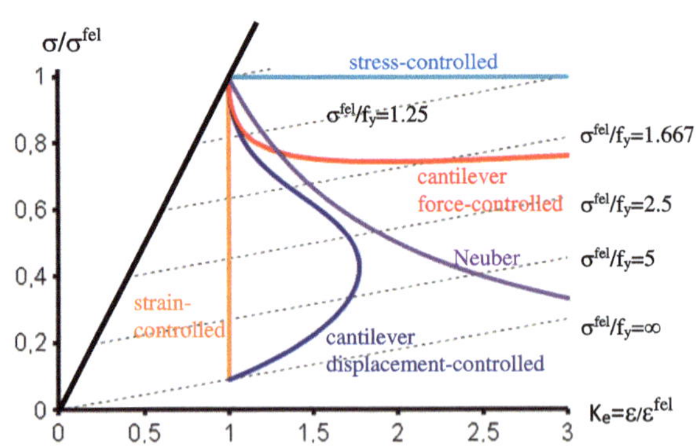

Fig. 2.35 Examples of plastic strain enhancement with linear kinematic hardening material ($C/E = 0.1$) under monotonic loading

$$\sigma \cdot \varepsilon(\sigma, f_y, C/E) = \text{const.} \neq f(C/E) \tag{2.29}$$

which therefore also applies to $C/E \rightarrow \infty$ and thus to elastic behavior:

$$\sigma \cdot \varepsilon(\sigma, f_y, C/E) = \sigma^{\text{fel}} \cdot \varepsilon^{\text{fel}}. \tag{2.30}$$

In addition the solution for a cantilever beam under force- or displacement-controlled concentrated load at its free end, already mentioned in Sect. 1.1, is obtained from [16] and accordingly processed in Fig. 2.35.

The fictitious elastic stresses can be obtained with relatively little effort. This is true even for transient thermal loading in which the stresses must be determined for many points in time in order to identify those two points in time associated with the maximum range of stress or strain by subsequent post-processing. However, the determination of the relevant plastic strain enhancement factor in the context of the post-processing of finite element analyses can be quite costly, such as when the factor is based on the linearization of nonlinear elastic stress distributions across the wall thickness [9]. This is because the linearizations must be made to very many points in time at many cross-sections of a structure for transient loading. Finally, the location of the maximum fictitious elastically calculated stress range is not necessarily identical to the location of the maximum plastic strain enhancement factor, nor to the location of the maximum product of stress range and plastic strain enhancement factor.

Possible effects of mean strains or mean stresses on the fatigue usage cannot be captured with such factors that are purely related to strain ranges.

For the estimation of cyclically accumulated strain, such correction factors applied to elastically calculated stresses are unknown, so that an extension of this factorial concept to ratcheting is not possible.

2.9.2 Incremental Elastic-Plastic Analysis

In the incremental elastic-plastic analysis a predetermined load histogram is calculated step by step. To do this, an elastic-plastic material model is required, which is suitable for the description of loading and unloading at multiaxial stress states with varying ratios of the stress components, and which is based on the characteristics treated in Sect. 1.2, e.g., the Besseling or the Chaboche model, etc. Apart from academic examples or uniaxial tension-compression tests for the identification of material parameters, this is possible only by means of implementing these models in finite element programs.

As can hereby be identified which portions of the volume (zones) of a structure are plastically and which are not, it is also called the theory of plastic zones – in contrast to the theory of plastic hinges or the yield line theory, which allow plastic deformation to occur only in discrete cross-sections, thus disregarding local plastic straining in sections that are only partly plastic, e.g., in notches. Furthermore, in application of the theory of plastic hinges and the yield line theory, usually only linear elastic-perfectly plastic or rigid plastic behavior is assumed. Hardening is thus disregarded, which makes sense for the calculation of load conditions produced by monotonic loading for some materials, but not for cyclically varying loads (cf. Sect. 1.2.5).

The term "incremental" refers not only to the fact that a load histogram needs to be divided into a sequence of monotonic loading changes, i.e., into a number of half cycles, but also that the flow law at Eq. (1.18) is formulated differentially. This can therefore, as already mentioned in Sect. 1.2.4, in general not be integrated in closed form, so that the change in load must be further divided into small sections (increments) within a half cycle and, because of the resulting path dependence of the loading process, for each of these intermediate states, a solution must be obtained. This can only be done iteratively because of the nonlinearity of the problem, because the finite element method is ultimately based on the creation and solution of linear systems of equations, and the linear system, which simulates the nonlinear problem adequately, without violating equilibrium conditions, must be found iteratively, usually with the help of Newton–Raphson iterations. With each change in load it is first assumed that this is connected with linearly elastic behavior. The stresses lying outside the yield surface (trial stress) are then projected to the yield surface (radial return method).

To capture this full path dependence of plastic structural behavior, a high computational effort is required. Each of l half cycles is to be subdivided into a number of s substeps, each calling for i equilibrium iterations. How many half cycles must be calculated until shakedown is achieved, how many substeps are required in each half cycle, and how many equilibrium iterations in each substep depend on the specific circumstances (component geometry, finite element mesh, load type, load level, material model, desired convergence tolerances, etc.) and varies greatly. To get, for example, the histograms of Fig. 2.34, a system of equations was set up and used to solve about $l \times s \times i = 36 \times 10 \times 2 = 720$ times, which corresponds to the numerical effort of performing 720 linear elastic analyses.

If the sequence of different loadings is not known by in-service monitoring, a conservative load histogram must be developed first, which can also be associated with high computational effort.

An incremental analysis is the only method by which it can be determined how many cycles are required to achieve approximately the shakedown state, or how many cycles are required to reach a certain strain level.

2.9.3 Twice-Yield Method

The estimation of elastic-plastic strain by means of factorial correction of fictitiously elastic calculated stress (Sect. 2.9.1) and performing incremental cyclic analyses with an elastic-plastic material model (Sect. 2.9.2) represent the extreme approaches to determine elastic-plastic strains. Although one requires only little computational effort but is usually associated with unsatisfactory quality of the results, with the other, depending on the choice of an appropriate material model, theoretically the best quality of results is achievable, but only with very high computational effort.

There is therefore a need for appropriate simplifications which lead to a compromise between quality of the results and computational effort required. It may be necessary to distinguish between methods that provide only an estimate of the strain range for use in a fatigue analysis and those which, in addition, also permit approximations for the accumulated strains (direct methods, Sect. 2.9.4).

One way to determine the strain range alone is the so-called twice-yield method of Kalnins [21]. After that, the strain range between the extremes of the two load conditions in the state of shakedown can approximately be determined by a monotonic incremental analysis by applying the load range as a monotonous load and calculating the structural response incrementally with an elastic-plastic material model. The material parameters have to be adapted to the cyclic stabilized stress–strain hysteresis. For a material without cyclic hardening and bilinear stress–strain curve, for example, the yield stress is doubled (hence the name), but the elastic modulus and the tangent modulus E_t is maintained.

This approach can lead to a good approximation for the strain range if the local and directional stress redistribution is not particularly pronounced, the temperature dependence of the material parameters plays no major role, and mean stress or mean strain effects need not be determined [22].

The twice-yield method is not applicable for the determination of the accumulated strain required for a ratcheting proof.

2.9.4 Direct Methods

So-called "direct methods" are intended to identify the state of shakedown without having to calculate a load histogram cycle by cycle incrementally. There is therefore no load histogram needed but only a load domain, which is multidimensional in general, containing all possible states and sequences of loads.

Direct methods are dealing with different questions:

- Which increase of the load, so what load factor, is possible in a specified load domain, just to guarantee shakedown?
- What stresses (strain, deformation, etc.) and which ranges is a structure experiencing in the shakedown state?

Different analysis methods are required to answer these questions. Some of these methods only allow the determination for which load intensity shakedown can be achieved, and what the nature of this shakedown state is (elastic or plastic shakedown), without having to determine the accumulated strains or strain ranges associated. If only two load states are involved in the cyclic loading, the result to be obtained corresponds to that of a ratcheting-interaction diagram. Their advantage is, however, to allow such statements for multiple load conditions, i.e., in a multi-dimensional load domain.

For this purpose use is made of so-called shakedown theorems:

- Either of the static shakedown theorem of Melan, known as the "lower bound theorem", providing a lower and thus conservative bound of the load factor.
- Or of the kinematic shakedown theorem of Koiter, known as the "upper bound theorem", providing an upper and thus non-conservative bound of the load factor.

The bounding character stems from the fact that, by analogy with the static and kinematic bounding theorems for plastic limit loads, solutions are sought that do not necessarily satisfy the field equations and the material law simultaneously. There are numerous references to both variants. Basics and numerous references are found in [4, 15, 23, 24], for example. Ultimately, the optimum fields of residual stresses or kinematic variables are sought, for which the structural behavior is formulated as a mathematical optimization problem [25, 26]. Mostly, linear elastic-perfectly plastic material behavior is assumed, but there are also extensions to capture unlimited linear, limited linear, or nonlinear kinematic hardening. Further extensions are related to temperature-dependent material data, dynamic loads, and the formulation of equilibrium conditions for the deformed system. The required computational effort for solving the optimization problem (mathematical programming) can be very high, depending on the finite element model because of the number of restrictions.

There are also proposals for direct methods in which, unlike the optimization problems mentioned above, the boundary value problem of structural mechanics is completely dissolved, i.e., all field equations (equilibrium and compatibility) and the yield condition are simultaneously fulfilled. Such methods have the advantage that they also provide information about deformations, displacements, etc., present in the state of shakedown, i.e., for all quantities that could otherwise be obtained by an incremental analysis of a load histogram. Because these quantities are required for a life assessment analysis in the sense of Sect. 2.8, such methods are, in principle, applicable both for a fatigue proof as well as for a ratcheting proof. Those methods are "direct" insofar as they directly aim at the shakedown state without performing a step-by-step analysis of a load histogram and therefore potentially require less numerical effort.

As the first limitation, it is already apparent that the path dependence of plastic behavior is lost. Likewise, it cannot be determined how many loading cycles are required until the shakedown state is reached. Also, the shape of the stress–strain hysteresis in the state of plastic shakedown cannot readily be obtained, which, however, is not necessarily a significant loss of information in strain-based fatigue analyses, because the dissipated energy does not enter there. As compensation for fully resolving the boundary value problem, other simplifications are introduced, which ultimately lead to a blurring of the results. A widely used simplification is to disregard the material hardening. So the results are estimates that differ from those obtained by an incremental analysis with a more realistic material model. This is the price that must be paid for this kind of simplified analysis method.

Amongst them are those methods based on an iterative sequence of linear elastic analyses with spatially varying Young's modulus, sometimes called EMAP (elastic modulus adjustment procedures). These include the Generalized Local Stress–Strain (GLOSS) Method of Seshadri [20, 27], the Elastic Compensation Method (ECM) of Mackenzie [28], and the Linear Matching Method (LMM) of Ponter and Chen [29, 30] (which still wasn't termed LMM in the first-mentioned publication). The GLOSS method can approximate the strain range in the state of shakedown, taking into account the material hardening, but cannot be used to predict accumulated strains. The ECM supplies the load factors in the shakedown state for non-hardening material, but neither strain ranges nor accumulated strains. The LMM applies only to non-hardening materials and supplies the strain ranges in the state of shakedown and the increment of strain per cycle when ratcheting occurs, which is a constant value because of the lack of hardening.

As direct methods, not belonging to the EMAP family, the Large Time Increment Method (LATIN) of Ladevèze [31, 32] and the Residual Stress Decomposition Method (RSDM) of Spiliopoulos and Panagiotou [33, 34] should be mentioned here.

A more detailed discussion of the methods listed, for example in relation to the achievable accuracy of the results, the required computational effort, the ability to model temperature dependent material behavior, taking into account multi-dimensional load domains, etc., is beyond the scope of this book by far.

Instead, another direct method, known as Zarka's method in the literature, and which serves as basis for the Simplified Theory of Plastic Zones, is presented in the following chapters in detail. As is seen there, a solution is sought iteratively through a series of successive elastic analyses. The theoretical basis and the procedure for doing this, however, differ greatly from the methods mentioned above. The aim is to estimate strain ranges and accumulated strains in the shakedown state, taking into account the hardening of the material.

References

1. Miller, D.R.: Thermal-stress ratchet mechanism in pressure vessels. ASME J. Basic Eng. **81**, 190–196 (1959)
2. Mulcahy, T.M.: An assessment of kinematic hardening thermal ratcheting. Trans. ASME J. Eng. Mater. Technol. **96**(3), 214–221 (1974)
3. Jiang, W., Leckie, F.A.: A direct method for the shakedown analysis of structures under sustained and cyclic loads. J. Appl. Mech. **59**, 251–260 (1992)
4. Ponter, A.R.S.: Shakedown and Ratchetting Below the Creep Range, CEC Report EUR8702 EN. European Commission, Brussels (1983)
5. ANSYS Release 14.5, ANSYS Inc. Canonsburg, USA (2012)
6. Owen, R.J., Prakash, A., Zienkiewicz, O.C.: Finite Element Analysis of Non-Linear Composite Materials by Use of Overlay Systems, Computers and Structures, vol. 4, pp. 1251–1267. Pergamon Press, New York (1974)
7. Wolters, J., Majumdar, S.: A three-bar Model for Ratcheting of Fusion Reactor First Wall. Argonne National Laboratory, Argonne, Illinois (1994)

8. Hübel, H.: Basic conditions for material and structural ratcheting. Nucl. Eng. Des. **162**, 55–65 (1996)
9. Sicherheitstechnische Regel des KTA, KTA 3201.2. Komponenten des Primärkreises von Leichtwasserreaktoren, Teil 2: Auslegung, Konstruktion und Berechnung. Fassung 6/96 (including correction from BAnz. Nr. 129, 13.07.2000). Office of the KTA c/o Bundesamt für Strahlenschutz, Salzgitter (2000)
10. Bree, J.: Elastic-plastic behaviour of thin tubes subjected to internal pressure and intermittent high-heat fluxes with application to fast-nuclear-reactor fuel elements. J. Strain Analysis **2**(3), 226–238 (1967)
11. Hill, R.: The Mathematical Theory of Plasticity, pp 292–294. Oxford University Press, London (1950)
12. Sartory, W.K.: Structural Design for Elevated Temperature Environments – Creep. Ratchet, Fatigue, and Fracture Effect of peak thermal strain on simplified ratchetting analysis procedures, ASME Proceedings, PVP **163**, 31–38 (1989)
13. Burth, K., Brocks, W.: Plastizität: Grundlagen und Anwendungen für Ingenieure. Vieweg, Braunschweig/Wiesbaden (1992)
14. Hübel, H.: Bemerkungen zur Ausnutzung plastischer Querschnitts- und Systemreserven. STAHLBAU **72**(12), 844–852 (2003)
15. Sawczuk, A.: shakedown analysis of elastic-plastic structures. Nucl. Eng. Des. **28**, 121–136 (1974)
16. Hübel, H.: Plastische Dehnungserhöhungsfaktoren in Regelwerken und Vorschlag zur Etablierung angemessenerer Faktoren. Gesamthochschule Kassel, Institut für Mechanik, Mitteilung Nr. 4 (Dissertation) (1985)
17. Haibach, E.: Betriebsfestigkeit. Verfahren und Daten zur Bauteilberechnung. Springer, Berlin/Heidelberg (2006)
18. Neuber, H.: Theory of stress concentration for shear strained prismatical bodies with arbitrary, nonlinear stress-strain law. Trans. ASME, J. Appl. Mech. **28**(4), 544–550 (1961)
19. Roche, R.L.: Practical procedure for stress classification. Int. J. Pres. Ves. Piping **37**, 27–44 (1989)
20. Seshadri, R.: The Generalized Local Stress Strain (GLOSS) Analysis – Theory and Applications. Trans. ASME J. Pressure Vessel Technol. **113**, 219–227 (1991)
21. Kalnins, A.: Fatigue analysis of pressure vessels with twice-yield plastic FEA. ASME PVP **419**, 43–52 (2001)
22. Hübel, H., et al.: Performance study of the simplified theory of plastic zones and the Twice-Yield method for the fatigue check. Int. J. Press. Vessels Pip. **116**, 10–19 (2014). doi:10.1016/j.ijpvp.2014.01.003
23. Ponter, A.R.S., Karadeniz, S., Carter, K.F.: The computation of shakedown limits for structural components subjected to variable thermal loading – Brussels diagrams, CEC Report EUR 12686 EN. European Commission, Brussels (1990)
24. König, J.A., Maier, G.: Shakedown Analysis of Elastoplastic Structures: A Review of Recent Developments. Nucl. Eng. Des. **66**, 81–95 (1981)
25. Heitzer, M., Staat, M.: FEM-computation of load carrying capacity of highly loaded passive components by direct methods. Nucl. Eng. Des. **193**, 349–358 (1999)
26. Staat, M., Heitzer, M.: LISA – a European project for FEM-based limit and shakedown analysis. Nucl. Eng. Des. **206**, 151–166 (2001). doi:10.1016/S0029-5493(00)00415-5
27. Seshadri, R.: residual stress estimation and shakedown evaluation using GLOSS analysis. J. Press. Vessel Technol. **116**(3), 290–294 (1994). doi:10.1115/1.2929590
28. Mackenzie, D., Boyle, J.T., Hamilton, R.: The elastic compensation method for limit and shakedown analysis: a review. Trans. IMechE J. Strain Anal. Eng. Des. **35**(3), 171–188 (2000)
29. Ponter, A.R.S., Carter, K.F.: Shakedown state simulation techniques based on linear elastic solutions. Comput. Methods Appl. Mech. Eng. **140**, 259–279 (1997)
30. Chen, H.: Linear matching method for design limits in plasticity, computers. Materials Continua. Tech. Sci. Press **20**(2), 159–183 (2010)

31. Ladevèze, P.: Nonlinear Computational Structural Mechanics – New Approaches and Non-Incremental Methods of Calculation. Springer, New York (1999)
32. Maier, G., Comi, C., Corigliani, A., Perego, U., Hübel, H.: Bounds and estimates on inelastic deformations, Commission of the European Communities, contract RA1-0162-I and RA1-0168-D, Report EUR 16555 EN. European Commission, Brussels (1992)
33. Spiliopoulos, K.V., Panagiotou, K.D.: A direct method to predict cyclic steady states of elastoplastic structures. Comput. Methods Appl. Mech. Eng. **223–224**, 186–198 (2012)
34. Spiliopoulos, K.V., Panagiotou, K.D.: The residual stress decomposition method (RSDM): a novel direct method to predict cyclic elastoplastic states. In: Spiliopoulos, K., Weichert, D. (Eds.) Direct Methods for Limit States in Structures and Materials, pp 139–155. Springer Science + Business Media, Dordrecht (2014). doi:10.1007/978-94-007-6827-7

3 STPZ at Monotonic Loading

STPZ is, as already mentioned, intended primarily to estimate the stresses and strains under variable load. The basic idea of the STPZ, as founded on Zarka's method, can, however, be more easily presented for monotonic loading. A bilinear stress–strain diagram, i.e., linear kinematic hardening, is assumed first, as well as constant material parameters that do not vary in the course of a thermal loading process. In Chap. 4 an extension to cyclic loading is presented, temperature-dependent material parameters are introduced in Chap. 5, and multilinear stress–strain relations in Chap. 7.

3.1 Transformed Internal Variable

3.1.1 Basic Assumptions for the Material Law

In the following, the assumptions for time-independent (inviscid) material behavior are adopted as presented in Chap. 1:

- Additivity of elastic and plastic deformation Eq. (1.15)
- Isotropy and linearity of the elastic behavior Eq. (1.3)
- Incompressibility of the plastic portions of deformation Eq. (1.16)
- Mises yield surface Eq. (1.12)
- Normality rule for the direction of the plastic portions of the deformation increments Eq. (1.17)
- Linear kinematic hardening Eqs. (1.20)–(1.25)
- Constancy of the material data (yield stress f_y, elastic parameters E and ν, hardening modules E_t and C), so that they remain constant during a thermal loading process, for example.

H. Hübel, *Simplified Theory of Plastic Zones*,
DOI 10.1007/978-3-319-29875-7_3

3.1.2 Reformulation of the Material Law

On this basis, Zarka, partly with co-workers, has initially proposed rephrasing of the material law (see, for example, [1–6]) by subtracting the deviatoric part of the residual stress ρ_i (Sect. 2.7) from the internal variable, the backstress ξ_i introduced in Sect. 1.2.5. The resulting quantity is termed "transformed internal variable" (TIV) Y_i, sometimes referred to in the literature as "modified backstress", with the defining equation

$$Y_i = \xi_i - \rho_i' . \tag{3.1}$$

Because the backstress ξ_i is deviatoric, Y_i is also deviatoric, so that

$$Y_x + Y_y + Y_z = 0. \tag{3.2}$$

At a location of the structure, which does not plasticize itself, hardening ξ_i is not effective and Eq. (3.1) reduces to

$$Y_i = -\rho_i'. \tag{3.3}$$

If there is no plastic straining at any location of the structure, so residual stresses do not exist anywhere, the TIVs disappear throughout the entire structure and take on the zero vector 0_i anywhere:

$$Y_i = 0_i . \tag{3.4}$$

Using the defining equation for the residual stresses, Eq. (2.25), the Mises yield condition of Eq. (1.22) can be reformulated to

$$g\left(\sigma_i^{\prime\text{fel}} - Y_i\right) = f_y. \tag{3.5}$$

If restricted to principal stresses, the yield condition of Eq. (1.23) becomes

$$\sqrt{\frac{3}{2}}\sqrt{\left(\sigma_1^{\prime\text{fel}} - Y_1\right)^2 + \left(\sigma_2^{\prime\text{fel}} - Y_2\right)^2 + \left(\sigma_3^{\prime\text{fel}} - Y_3\right)^2} = f_y. \tag{3.6}$$

This formulation of the yield condition can be visualized in the space of the TIV as a circle with radius f_y and the center in the fictitiously elastic calculated deviatoric stress $\sigma_i^{\prime\text{fel}}$. From Fig. 3.1 it can be seen that just a translation with respect to the deviatoric stress space has taken place. Equation (3.6) could also be interpreted as a circle with center Y_i in the space of the fictitiously elastic calculated deviatoric stress, but this would not turn out to be helpful.

With the introduction of the TIV Y_i and the corresponding modification of the flow law, Eq. (3.6) merely contains a single unknown vector, namely Y_i, because the fictitiously elastic calculated stresses can be viewed as known, because their

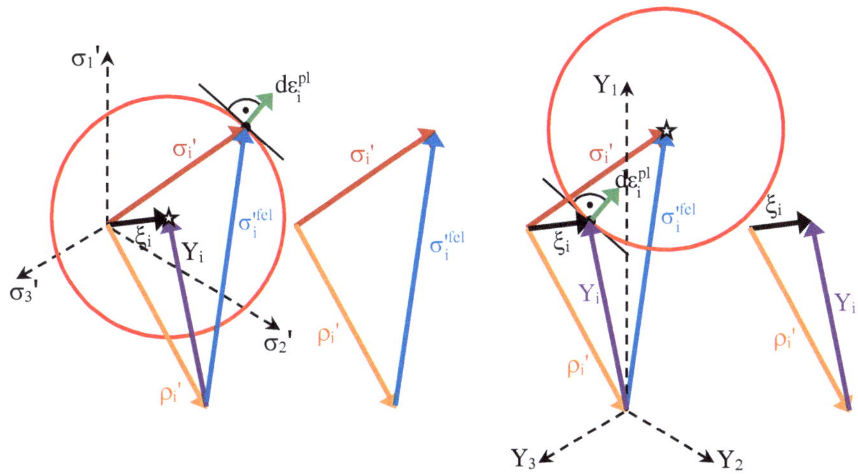

Fig. 3.1 Mises yield surface with geometric interpretation of Eqs. (2.25) and (3.1); *left* in the space of the deviatoric stresses; *right* in the space of the TIV

determination does not pose special difficulties and can therefore always be obtained with relatively little computational effort. The crucial difference in the consideration of the TIV-space compared to the deviatoric stress space is that in the space of deviatoric stresses the positions of the Mises circle (center ξ_i) and of the stress state on its edge (σ_i') are initially completely unknown, whereas in the space of the TIV the location of the Mises circle (center $\sigma_i'^{\mathrm{fel}}$) is known a priori. As is seen later, in addition, the position of Y_i on its edge may at least be reasonably estimated.

Also noteworthy is that the yield surface moves in a purely elastic process in the space of the TIV ($\mathrm{d}\sigma_i'^{\mathrm{fel}} \neq 0$), but not in the space of the deviatoric stresses ($\mathrm{d}\xi_i = 0$).

The direction of the increments of the plastic strain from Eq. (1.21) is given by

$$\mathrm{d}\varepsilon_i^{\mathrm{pl}} = -\frac{3}{2}\frac{\mathrm{d}\varepsilon_{\mathrm{v}}^{\mathrm{pl}}}{f_y}\left(Y_i - \sigma_i'^{\mathrm{fel}}\right). \tag{3.7}$$

Although it is directed outward in the deviatoric stress space from the edge of the flow surface, it is an inward normal to the flow surface in the space of the TIV (Fig. 3.1).

During the increment of plastic deformation and because of the hardening law of Eq. (1.24), the center of the yield surface in the deviatoric stress space develops normal to the yield surface but the TIVs do not. As is clear from the first derivative of the defining Eq. (3.1), the direction of $\mathrm{d}Y_i$ depends instead on the direction of the deviatoric residual stress increment $\mathrm{d}\rho_i'$.

After introducing the additivity theorem Eq. (1.15) in the formulation for the residual strain ε^* in Eq. (2.26), and substituting the elastic portion $\varepsilon^{\mathrm{el}}$ of the elastic-plastic strain, and the fictitiously elastic calculated strain ε^{fel} by the elastic material law Eq. (1.3), we obtain with the aid of Eq. (2.25)

$$\varepsilon_i^* = E_{ij}^{-1} \rho_j + \varepsilon_i^{\text{pl}}. \tag{3.8}$$

If, as required, the hardening parameter C of linear kinematic hardening remains constant during the loading process, the defining equation for the TIV, Eq. (3.1), can also be introduced in Eq. (1.26). Rearranging for the plastic strains, Eq. (3.8) becomes

$$\varepsilon_i^* = E_{ij}^{-1} \rho_j + \frac{3}{2C} \left(\rho_i^{'} + Y_i \right). \tag{3.9}$$

It should be emphasized once again that a finite relation between the backstress and the plastic strain was required (see Eq. (1.26)), i.e., that a differential formulation, such as in the Chaboche model in the form of Eq. (1.29), cannot be used.

At a location of the structure where no plastic strains occur, because of Eq. (3.3) it is immediately apparent that

$$\varepsilon_i^* = E_{ij}^{-1} \rho_j. \tag{3.10}$$

For other locations, i.e., those with plastic strain, we can use the operator L_{ij} introduced in Eq. (1.9) to convert a vector into its deviatoric share and we can rewrite Eq. (3.9) as

$$\varepsilon_i^* = \left(E_{ij}^{-1} + \frac{3}{2C} L_{ij} \right) \rho_j + \frac{3}{2C} Y_i. \tag{3.11}$$

In this form the material law for the residual state can be interpreted as a linear elastic material law with a modified elasticity matrix E_{ij}^* and initial strains $\varepsilon_{i,0}$:

$$\varepsilon_i^* = \left(E_{ij}^* \right)^{-1} \rho_j + \varepsilon_{i,0}, \tag{3.12}$$

in which

$$\left(E_{ij}^* \right)^{-1} = \left(E_{ij}^{-1} + \frac{3}{2C} L_{ij} \right) \tag{3.13}$$

$$\varepsilon_{i,0} = \frac{3}{2C} Y_i. \tag{3.14}$$

The difference between the formally similar looking Eqs. (3.8) and (3.12) can be illustrated in the stress–strain curve for a uniaxial stress state (Fig. 3.2).

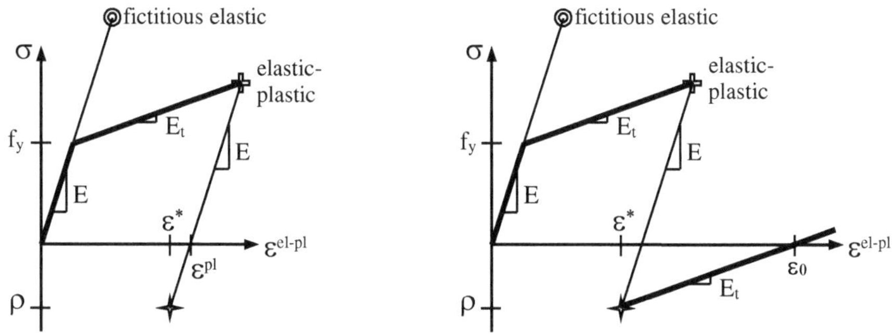

Fig. 3.2 Residual state of uniaxial stress; *left* according to Eq. (3.8); *right* according to Eq. (3.12)

It should be noted that these initial strains $\varepsilon_{i,0}$ are deviatoric. Comparing the matrices on the left and the right side of Eq. (3.13):

$$\frac{1}{E^*}\begin{pmatrix} 1 & -\nu^* & -\nu^* & 0 & 0 & 0 \\ -\nu^* & 1 & -\nu^* & 0 & 0 & 0 \\ -\nu^* & -\nu^* & 1 & 0 & 0 & 0 \\ 0 & 0 & 0 & 1+\nu^* & 0 & 0 \\ 0 & 0 & 0 & 0 & 1+\nu^* & 0 \\ 0 & 0 & 0 & 0 & 0 & 1+\nu^* \end{pmatrix}$$

$$=\frac{1}{E}\begin{pmatrix} 1 & -\nu & -\nu & 0 & 0 & 0 \\ -\nu & 1 & -\nu & 0 & 0 & 0 \\ -\nu & -\nu & 1 & 0 & 0 & 0 \\ 0 & 0 & 0 & 1+\nu & 0 & 0 \\ 0 & 0 & 0 & 0 & 1+\nu & 0 \\ 0 & 0 & 0 & 0 & 0 & 1+\nu \end{pmatrix}$$

$$+\ \frac{3}{2C}\begin{pmatrix} \frac{2}{3} & -\frac{1}{3} & -\frac{1}{3} & 0 & 0 & 0 \\ -\frac{1}{3} & \frac{2}{3} & -\frac{1}{3} & 0 & 0 & 0 \\ -\frac{1}{3} & -\frac{1}{3} & \frac{2}{3} & 0 & 0 & 0 \\ 0 & 0 & 0 & 1 & 0 & 0 \\ 0 & 0 & 0 & 0 & 1 & 0 \\ 0 & 0 & 0 & 0 & 0 & 1 \end{pmatrix}$$

and expressing C by E and E_t according to Eq. (1.25), it is evident that

$$E^* = E_t \tag{3.15}$$

$$\nu^* = \frac{1}{2} - \frac{E_t}{E}\left(\frac{1}{2} - \nu\right). \tag{3.16}$$

The modified elastic material law for the residual state thus depends only on the actual elastic and plastic material parameters E, ν, and E_t, but not from the level of loading or of the TIV.

It should be mentioned that a transition to non-hardening material ($E_t = C = 0$) is problematic, because the modified loading in Eq. (3.14) cannot be defined, the Young's modulus of the modified elastic material law becomes 0, and the Poisson's ratio 0.5 (i.e., elastic incompressible).

If a structure that occupies the volume V is divided into two parts, V_p where plastic straining occurs and the complementary part V_e that remains elastic, so

$$V = V_e \cup V_p \quad ; \quad V_e \cap V_p = \emptyset, \tag{3.17}$$

then for a location of the structure with the coordinate vector $\underline{x}$ the modified elastic material can be summarized to read

$$\varepsilon_i^* = \left(E_{ij}^*\right)^{-1} \rho_j + \varepsilon_{i,0} \quad \forall \underline{x} \in V \tag{3.18}$$

with $\varepsilon_{i,0}$ from Y_i according to Eq. (3.14). Although in the following sections, because of the unknown backstress ξ_i in the defining equation of Y_i, appropriate procedures for the determination or estimation of Y_i and thus of $\varepsilon_{i,0}$ in the plastic zone V_p have yet to be found, Y_i in V_e is directly known from Eq. (3.3):

$$\varepsilon_{i,0} = \frac{3}{2C} Y_i \quad \text{with} \quad Y_i = -\rho_i' \quad \forall \underline{x} \in V_e. \tag{3.19}$$

As an alternative to Eq. (3.18) with Eq. (3.19) we can also formulate

$$\varepsilon_i^* = \begin{cases} \left(E_{ij}^*\right)^{-1} \rho_j + \varepsilon_{i,0} & \forall \underline{x} \in V_p \\ E_{ij}^{-1} \rho_j & \forall \underline{x} \in V_e \end{cases}. \tag{3.20}$$

In the elastic zone V_e, the modified elastic material law can therefore be defined either with the actual elastic material parameters E and ν without initial strains, or with the modified elastic material parameters E^* and ν^* with initial strains according to Eq. (3.19). In many applications the formulation of Eq. (3.20) turned out as more advantageous than Eq. (3.18).

Sometimes, however, it is also useful to make use of the inverted form of Eq. (3.20):

$$\rho_i = \begin{cases} E^*_{ij}\,\varepsilon^*_j + \sigma_{i,0} & \forall \underline{x} \in V_p \\ E_{ij}\,\varepsilon^*_j & \forall \underline{x} \in V_e \end{cases}, \tag{3.21}$$

or the inversion of Eq. (3.18) with Eq. (3.19), where the deviatoric initial stresses $\sigma_{i,0}$ are given by

$$\sigma_{i,0} = -\frac{3}{2C} E^*_{ij}\, Y_j\,. \tag{3.22}$$

Because the TIV Y is of deviatoric nature, the term $E^*_{ij} Y_j$ can be written even shorter, and Eq. (3.22) becomes

$$\sigma_{i,0} = -\frac{3}{2C} \frac{E^*}{1+\nu^*} Y_i. \tag{3.23}$$

The reformulation of the modified elastic material law with initial stresses instead of initial strains provides benefits if initial strains cannot, but only initial stresses can be applied in analysis software. It sometimes also permits the treatment of linear elastic-perfectly plastic material behavior, as the term E_t/C included in the term E^*_{ij}/C is not indeterminate for the limiting case $E_t \to 0$, but approaches unity. This was, for example, used in [7, pp. 104–108], where a cantilever with solid rectangular cross-section was investigated under displacement-controlled loading with a modification of Zarka's method for non-hardening material.

3.1.3 Modified Elastic Analysis

Because the residual state is, by definition, characterized in that all the previously applied loads are removed so that external forces do not act any more and all displacement-controlled loadings such as temperature loads, support displacement, etc., are set to zero (see Sect. (2.7)), the initial strains of Eq. (3.14) can be considered as the only load for the determination of the residual state. They are therefore referred to as modified loading and depend not only on the Young's modulus E and the hardening modulus E_t but also on the TIV Y_i, and therefore the level of the real load.

If the geometry of the plastic zone of the structure (V_p) and at any point therein the TIV Y_i are known, it is now possible to perform a so-called modified elastic analysis (MEA). This involves a linear elastic analysis with the modified elastic material parameters E^* and ν^* in V_p, and E and ν in V_e to calculate the residual stresses and residual strains produced by the modified loading of the structure, which consists only of the initial strains $\varepsilon_{i,0}$. As a result, a statically admissible stress field is obtained which satisfies self-equilibrium, as well as a kinematically admissible strain field with geometric constraints set to zero.

After superposition of the residual state with the fictitious elastic state, determined with the actual elastic material parameters E and ν and the actual load of the structure, the elastic-plastic state is then obtained by definition. Thus, for all stress components, strain components, displacements u, section forces F, bearing forces A, etc.,

$$\sigma_i = \sigma_i^{\text{fel}} + \sigma_i^{\text{MEA}} \quad \forall \underline{x} \in V \quad ; \quad \sigma_i^{\text{MEA}} = \rho_i \tag{3.24}$$

$$\varepsilon_i^{\text{el-pl}} = \varepsilon_i^{\text{fel}} + \varepsilon_i^{\text{MEA}} \quad \forall \underline{x} \in V \quad ; \quad \varepsilon_i^{\text{MEA}} = \varepsilon_i^* \tag{3.25}$$

$$u_i = u_i^{\text{fel}} + u_i^{\text{MEA}} \quad \forall \underline{x} \in V \tag{3.26}$$

$$F_i = F_i^{\text{fel}} + F_i^{\text{MEA}} \tag{3.27}$$

$$A_i = A_i^{\text{fel}} + A_i^{\text{MEA}}. \tag{3.28}$$

The elastic shares of strain in V_p are then obtained from the stresses using the elastic material law, and the plastic strain components, for example, from Eq. (1.15):

$$\varepsilon_i^{\text{el}} = E_{ij}^{-1}\,\sigma_j \quad ; \quad \varepsilon_i^{\text{pl}} = \varepsilon_i^{\text{el-pl}} - \varepsilon_i^{\text{el}}. \tag{3.29}$$

The result thus obtained is exact in the framework of the assumptions made in Sect. 3.1.1 concerning the material law, i.e., identical to the result obtained by a conventional elastic-plastic analysis in accordance with the theory of plastic zones, because so far no approximations have been introduced, but instead only an equivalent reformulation of the material law has been carried out.

The superposition in Eqs. (3.24)–(3.28) is true only if the same constraints are defined for the MEA, the fictitious elastic analysis, and, thus, also for the elastic-plastic behavior to be calculated. Movements of the bearings, possibly acting as the actual load of the structure, are, however, to be set to zero in the MEA. This makes the analysis of structures problematic where the bearing conditions are not known a priori, because they depend on the plastic behavior of the structure. Such situations may occur, for example, in contact problems.

Through the reformulation of the material law, the calculation of elastic-plastic behavior is ultimately led back to the problem to detect the geometry of the plastic zone V_p and to determine the TIV Y_i at any point in V_p. Herein lies the core problem of the Zarka method and the STPZ, and here both differ.

In the following sections, procedures are suggested to obtain, if necessary, iteratively, approximations for V_p and Y_i. This allows approximate solutions to be obtained for the elastic-plastic behavior of a structure. The computational effort, required beyond the routine fictitious elastic analyses, consists of local calculations, such as the evaluation of stresses to estimate the plastic zone and the initial strains, the superposition of the results, and performing linear elastic analyses (the MEAs).

The approximate solutions of Zarka's method and the STPZ always meet the static and kinematic conditions (equilibrium and compatibility) exactly, unless other approximations are exercised, such as the discretization of a continuous structure when adopting the finite element method. This is because, with both the fictitious elastic and the modified elastic analyses, the respective boundary value problems are solved exactly. Thus, classical methods for error estimates, based on energy considerations, are not readily applicable here.

3.1.4 Uniaxial Stress State

The initial strains and initial stresses determined according to Eq. (3.14) or Eq. (3.22) must be applied as loading in the MEA. Even with uniaxial or biaxial stress states, usually none of the three direct components of the initial strains and initial stresses is zero because of their deviatoric character. In a finite element environment, where uniaxial and plane stress conditions are calculated with specially reformulated constitutive laws, the initial strains or initial stresses cannot be specified in all directions. Instead, the equations listed in Sect. 3.1.2 are to be modified as follows for uniaxial stress states (for plane stress conditions see Sect. 3.1.5).

In Euler–Bernoulli beams or trusses, which exhibit only a uniaxial state of stress, the stress vector of Eq. (1.1) reduces to

$$\sigma_i = \begin{pmatrix} \sigma_x \\ 0 \\ 0 \\ 0 \\ 0 \\ 0 \end{pmatrix}. \tag{3.30}$$

The linear elastic material law of Eq. (1.2) connects the stress component σ_x with the elastic strain in the x-direction giving

$$\varepsilon_x^{\mathrm{el}} = \frac{\sigma_x}{E}. \tag{3.31}$$

Of course, lateral strains are also present in the y- and z-directions. The x-component of the deviatoric fictitious elastic stress amounts to

$$\sigma_x^{\prime\mathrm{fel}} = \frac{2}{3}\sigma_x^{\mathrm{fel}}. \tag{3.32}$$

Because

$$Y_y = Y_z = -\frac{1}{2}Y_x, \tag{3.33}$$

Eq. (3.6) becomes

$$\mathrm{abs}\left(\sigma_x^{\mathrm{fel}} - \frac{3}{2} Y_x\right) = f_y. \tag{3.34}$$

Because the residual stresses in the y- and z-directions must also disappear, Eq. (3.11) becomes

$$\varepsilon_x^* = \frac{1}{E^*} \rho_x + \frac{3}{2C} Y_x \tag{3.35}$$

and Eq. (3.14)

$$\varepsilon_{x,0} = \frac{3}{2C} Y_x. \tag{3.36}$$

If the component Y_x is replaced by the total value of Y (cf. Fig. 1.2) then

$$Y = \frac{3}{2} Y_x \tag{3.37}$$

and if the index x is dropped in the uniaxial stress state, we get the notation for the yield criterion:

$$\mathrm{abs}\left(\sigma^{\mathrm{fel}} - Y\right) = f_y, \tag{3.38}$$

for the modified elastic material law:

$$\varepsilon^* = \frac{1}{E^*} \rho + \varepsilon_0, \tag{3.39}$$

for the initial strain:

$$\varepsilon_0 = \frac{1}{C} Y, \tag{3.40}$$

and the initial stress:

$$\sigma_0 = -\frac{1}{C} E^* Y. \tag{3.41}$$

E^* is unchanged, given by Eq. (3.15).

3.1.5 Plane Stress State

Under plane stress conditions the direct stress in the z-direction and two shear stresses disappear, so that Eq. (1.1) can be reduced to three components:

$$\sigma_i = \begin{pmatrix} \sigma_x \\ \sigma_y \\ \tau_{xy} \end{pmatrix}. \tag{3.42}$$

The linear elastic material law, Eq. (1.3), links these stress components with the three strain components:

$$\varepsilon_i = \begin{pmatrix} \varepsilon_x \\ \varepsilon_y \\ \gamma_{xy}/2 \end{pmatrix}$$

so that

$$\varepsilon_i^{el} = E_{ij}^{-1}\sigma_j \quad ; \quad i,j = 1\ldots3 \tag{3.44}$$

$$E_{ij} = \frac{E}{1-\nu^2}\begin{pmatrix} 1 & \nu & 0 \\ \nu & 1 & 0 \\ 0 & 0 & 1-\nu \end{pmatrix} \tag{3.45}$$

$$E_{ij}^{-1} = \frac{1}{E}\begin{pmatrix} 1 & -\nu & 0 \\ -\nu & 1 & 0 \\ 0 & 0 & 1+\nu \end{pmatrix}. \tag{3.46}$$

Furthermore, it applies to the third direction because of the lateral deformation:

$$\varepsilon_z^{el} = -\frac{\nu}{E}\left(\sigma_x + \sigma_y\right). \tag{3.47}$$

The modified elastic material law for the plane stress condition is formally the same as Eq. (3.12), but taking into account the fact that E_{ij}^* is now a 3 × 3 matrix, whereas E^* and ν^* are unchanged from Eqs. (3.15) and (3.16):

$$\varepsilon_i^* = \left(E_{ij}^*\right)^{-1}\rho_j + \varepsilon_{i,0} \quad ; \quad i,j = 1\ldots3. \tag{3.48}$$

For the three in-plane initial strains $\varepsilon_{i,0}$ in Eq. (3.48), simply the corresponding components of the six spatial initial strains of Eq. (3.14) can be used, so that $\varepsilon_{z,0}$ is first ignored:

$$\varepsilon_{x,0}^{2D} = \varepsilon_{x,0}^{3D} \quad ; \quad \varepsilon_{y,0}^{2D} = \varepsilon_{y,0}^{3D} \quad ; \quad \gamma_{xy,0}^{2D} = \gamma_{xy,0}^{3D}. \tag{3.49}$$

However, if the modified elastic material law from Eq. (3.21) is to be used, i.e., with specification of initial stresses instead of initial strains, then the two-dimensional initial stresses in the x- and y-directions are now required which cause the same effect as the three-dimensional initial stresses in the x-, y-, and z-directions from Eq. (3.22):

$$\sigma_{x,0}^{2D} = \frac{\sigma_{x,0}^{3D} + \nu^* \sigma_{y,0}^{3D}}{1 - \nu^*} \tag{3.50}$$

$$\sigma_{y,0}^{2D} = \frac{\nu^* \sigma_{x,0}^{3D} + \sigma_{y,0}^{3D}}{1 - \nu^*} \tag{3.51}$$

$$\tau_{z,0}^{2D} = \tau_{z,0}^{3D}. \tag{3.52}$$

Of course,

$$\sigma_{z,0}^{2D} = 0. \tag{3.53}$$

When performing an MEA, in both cases, i.e., when setting either initial strains or initial stresses, all three residual stress components and all three components of residual strains related to them via the modified elastic material law, Eq. (3.48), are determined correctly, but not the strain in the z-direction. Because of lateral strain effects we first get

$$\varepsilon_z^* = -\frac{\nu^*}{E^*}\left(\rho_x + \rho_y\right). \tag{3.54}$$

The share of the initial strain to actually be applied in the z-direction is still not correctly captured however. Equation (3.54) is therefore to be expanded by a term reflecting a free change of thickness without causing stresses:

$$\varepsilon_z^* = -\frac{\nu^*}{E^*}\left(\rho_x + \rho_y\right) + \varepsilon_{z,0}\frac{1 - 2\nu^*}{1 - \nu^*}. \tag{3.55}$$

In an FE environment, initial strains in the z-direction causing stresses produced by Poisson's effect in the x-y-plane of a plane stress condition cannot be specified; the last term in Eq. (3.55) can usually be simulated by a free thermal expansion. Introducing an anisotropic modified coefficient of thermal expansion that equals 0 in the x- and y-directions and 1 in the z-direction, a modified temperature

$$T^* = \varepsilon_{z,0}\frac{1 - 2\nu^*}{1 - \nu^*} \tag{3.56}$$

is to be applied to obtain the correct strain in the thickness direction.

3.2 Plastic Zone

After completion of the fictitious elastic analysis, no information is currently available concerning the residual stresses. Thus, initially no better information is available for estimating the plastic zone V_p than to assume that those locations in the structure, in which the fictitious elastic calculated equivalent stress exceeds the yield stress, are identical to the plastic zone:

$$\sigma_{\mathrm{v}}^{\mathrm{fel}} = \sqrt{\frac{3}{2}}\sqrt{\left(\sigma_1^{\prime\mathrm{fel}}\right)^2 + \left(\sigma_2^{\prime\mathrm{fel}}\right)^2 + \left(\sigma_3^{\prime\mathrm{fel}}\right)^2} \tag{3.57}$$

$$V_p = \left\{ \underline{x} \mid \sigma_{\mathrm{v}}^{\mathrm{fel}} \geq f_y \right\}. \tag{3.58}$$

Correspondingly, the remaining part of the structure is then attributable to V_e:

$$V_e = \left\{ \underline{x} \mid \sigma_{\mathrm{v}}^{\mathrm{fel}} < f_y \right\}. \tag{3.59}$$

Performing an MEA, using this assignment and adopting values of the TIV Y_i in V_p we still need to know (see the next sections), residual stresses are obtained which lead, after superposition with the fictitious elastic stresses according to Eq. (3.24), to a first estimation of the elastic-plastic stress components. We refer to them as $\sigma_i^{(1)}$. They can be used to form the equivalent stress:

$$\sigma_{\mathrm{v}}^{(1)} = \sqrt{\frac{3}{2}}\sqrt{\left(\sigma_1^{\prime(1)}\right)^2 + \left(\sigma_2^{\prime(1)}\right)^2 + \left(\sigma_3^{\prime(1)}\right)^2}. \tag{3.60}$$

On this basis, the estimate of the plastic zone can be improved:

$$V_p = \left\{ \underline{x} \mid \sigma_{\mathrm{v}}^{(1)} \geq f_y \right\} \tag{3.61}$$

$$V_e = \left\{ \underline{x} \mid \sigma_{\mathrm{v}}^{(1)} < f_y \right\}. \tag{3.62}$$

A new MEA probably delivers somewhat different residual stresses, leading to a slightly different estimate of the stress components in the elastic-plastic state, $\sigma_i^{(2)}$, and the equivalent stress

$$\sigma_{\mathrm{v}}^{(2)} = \sqrt{\frac{3}{2}}\sqrt{\left(\sigma_1^{\prime(2)}\right)^2 + \left(\sigma_2^{\prime(2)}\right)^2 + \left(\sigma_3^{\prime(2)}\right)^2}, \tag{3.63}$$

and then to a new estimate of V_p, etc., resulting in an iterative process for estimating the plastic zone. For the nth iteration step:

$$V_p^{(n)} = \left\{ \underline{x} \mid \sigma_{\mathrm{v}}^{(n-1)} \geq f_y \right\} \tag{3.64}$$

$$V_e^{(n)} = \left\{ \underline{x} \mid \sigma_{\mathrm{v}}^{(n-1)} < f_y \right\}. \tag{3.65}$$

So the estimation of the geometry of the plastic zone can be continuously improved. Whether this sequence converges, and if so, at what rate, i.e., how many MEAs are required, also depends on an appropriate estimation of the TIV, which is made in the next sections. Examples to be dealt with later illustrate this.

It is worth noting at this point that, in contrast to Zarka's method in [1], such regions of the structure that were assigned to V_p in one iteration according to the above described procedure of the STPZ, may be reassigned to V_e in the next iteration.

3.3 Transformed Internal Variable at Radial Loading

In certain cases the TIV Y_i can be known a priori, namely at so-called radial loading, which is characterized by a strictly radial development of the stress with respect to the yield surface in the deviatoric stress space during a loading process. This is the case if no directional stress redistribution occurs, i.e., if the ratio of all stress components to each other during the loading process remains constant (Sect. 2.1). This was also a prerequisite for the elementary integration of the differential flow law Eq. (1.18) to arrive at Eq. (1.19). As the simplest example of this, the uniaxial stress state is considered.

With such a radial loading, all six vectors depicted in Fig. 3.1 are collinear, so they differ only by a scalar factor. In the case of active plasticizing at a location $\underline{x} \in V_p$, this means that, on the one hand, Y_i has the same direction as the fictitious elastic calculated deviatoric stress $\sigma_i'^{\mathrm{fel}}$, known beforehand, and, on the other hand, it must also be located on the edge of the yield surface. Thus immediately arises

$$Y_i = \sigma_i'^{\mathrm{fel}} \left(1 - \frac{f_y}{\left| \sigma_i'^{\mathrm{fel}} \right|} \right). \tag{3.66}$$

Here, $\left| \sigma_i'^{\mathrm{fel}} \right|$ is the length of the fictitious elastic calculated stress vector and thus their Mises equivalent stress, Eq. (3.57).

Specifically for the uniaxial state of stress (Sect. 3.1.4) we get

$$Y = \sigma^{\mathrm{fel}} - f_y \, \mathrm{sgn}\left(\sigma^{\mathrm{fel}} \right). \tag{3.67}$$

Because the TIVs given by Eqs. (3.66) and (3.67) are not approximations but are precisely known, they cannot and need not be iteratively improved during a sequence of MEAs. Consequently, a sequence of MEAs is related only to iteratively improving the estimation of the plastic zone V_p in the case of radial loading.

3.4 Examples of Radial Loading

Before non-radial loading is considered, examples of radial loading are treated first. Because only the geometry of the plastic zone must be determined iteratively, whereas the TIVs are exactly known, it can be expected that the iterative result approaches the exact result.

3.4.1 Tension Bar with Step-like Change in Cross-Section

A bar with step-like transition from the cross-sectional area A_1 (length l_1) to a smaller area A_2 (length l_2), but the same material in both parts, is subjected to displacement-controlled tension u (Fig. 3.3).

The notch stresses occurring at the transition of the cross-sections are not considered, and the transverse strains at the bearings are not constrained so that only the elementary stresses σ_1 and σ_2 as well as the strains ε_1 and ε_2 in the longitudinal direction need to be considered. The indices 1 and 2 of the stresses and strains here are not related to the direction, but to the left and right part of the structure.

Because of equilibrium we have

$$N_1 = N_2 \quad ; \quad N_1 = A_1\sigma_1 \quad ; \quad N_2 = A_2\sigma_2, \tag{3.68}$$

and the kinematic relations are

$$u = \Delta l_1 + \Delta l_2 \quad ; \quad \Delta l_1 = l_1\varepsilon_1 \quad ; \quad \Delta l_2 = l_2\varepsilon_2. \tag{3.69}$$

The fictitious elastic solution is obtained by inserting the material law

$$\sigma_1^{\mathrm{fel}} = E\varepsilon_1^{\mathrm{fel}} \quad ; \quad \sigma_2^{\mathrm{fel}} = E\varepsilon_2^{\mathrm{fel}} \tag{3.70}$$

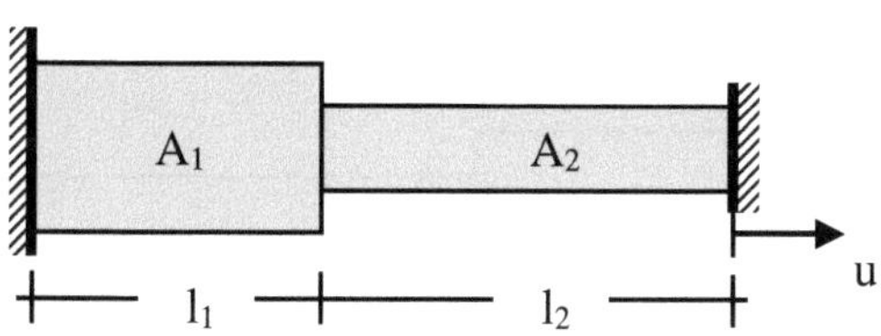

Fig. 3.3 Bar with step-like change in cross-section subjected to displacement-controlled tension

in Eqs. (3.68) and (3.69), so that

$$\sigma_1^{\text{fel}} = \frac{uE}{l_2 \frac{A_1}{A_2} + l_1} \quad ; \quad \sigma_2^{\text{fel}} = \frac{uE}{l_2 + l_1 \frac{A_2}{A_1}}. \tag{3.71}$$

If the geometry parameters, the Young's modulus E, the load level u, and the yield strength f_y are selected such that $\sigma_1^{\text{fel}} < f_y$ and $\sigma_2^{\text{fel}} > f_y$, then part 1 is assigned to the subvolume V_e according to Eq. (3.59), and part 2 to V_p according to Eq. (3.58):

$$\sigma_1^{\text{fel}} \leq f_y \Rightarrow \text{part } 1 \rightarrow V_e \tag{3.72}$$

$$\sigma_2^{\text{fel}} > f_y \Rightarrow \text{part } 2 \rightarrow V_p. \tag{3.73}$$

According to Eq. (3.67), the TIV in part 2 is

$$Y_2 = \sigma_2^{\text{fel}} - f_y \tag{3.74}$$

and the initial strains according to Eq. (3.14), where, because of the limitation to a uniaxial instead of the deviatoric approach, the factor 3/2 can be dropped (Sect. 3.1.4):

$$\varepsilon_{2,0} = \frac{1}{C}\left(\sigma_2^{\text{fel}} - f_y\right). \tag{3.75}$$

Through an MEA of the configuration shown in Fig. 3.4 ($u = 0$, $E^* = E$ in part 1, $E^* = E_t$ in part 2), the residual stresses are to be calculated.

From the equilibrium condition of Eq. (3.68) we get

$$A_1 \rho_1 = A_2 \rho_2, \tag{3.76}$$

Eq. (3.69) becomes

$$l_1 \varepsilon_1^* + l_2 \varepsilon_2^* = 0, \tag{3.77}$$

and the material law of Eq. (3.20)

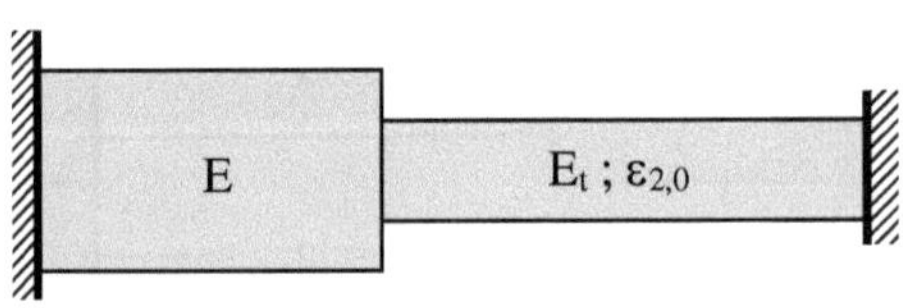

Fig. 3.4 System for the first MEA

$$\varepsilon_1^* = \frac{\rho_1}{E} \quad ; \quad \varepsilon_2^* = \frac{\rho_2}{E_t} + \frac{1}{C}\left(\sigma_2^{\text{fel}} - f_y\right), \tag{3.78}$$

and hence the solution is

$$\rho_1 = \rho_2 \frac{A_2}{A_1} \quad ; \quad \rho_2 = -\frac{\left(1 - \frac{E_t}{E}\right)\left(\sigma_2^{\text{fel}} - f_y\right)}{1 + \frac{E_t}{E}\frac{l_1}{l_2}\frac{A_2}{A_1}}. \tag{3.79}$$

Superposition with the fictitious elastic solution according to Eq. (3.24) provides for the elastic-plastic state after the first MEA:

$$\sigma_1^{(1)} = \sigma_1^{\text{fel}} - \frac{\left(1 - \frac{E_t}{E}\right)\left(\sigma_1^{\text{fel}} - f_y \frac{A_2}{A_1}\right)}{1 + \frac{E_t}{E}\frac{l_1}{l_2}\frac{A_2}{A_1}} \quad ; \quad \sigma_2^{(1)} = \sigma_2^{\text{fel}} - \frac{\left(1 - \frac{E_t}{E}\right)\left(\sigma_2^{\text{fel}} - f_y\right)}{1 + \frac{E_t}{E}\frac{l_1}{l_2}\frac{A_2}{A_1}}. \tag{3.80}$$

A comparison with the yield stress now reveals whether the assignment made in Eq. (3.72) of part 1 to remain elastic was correct. If so, Eq. (3.80) already represents the correct solution for the stresses:

$$\sigma_1^{(1)} \leq f_y \Rightarrow \text{part } 1 \rightarrow V_e \quad \Rightarrow \quad \sigma_1 = \sigma_1^{(1)} \quad ; \quad \sigma_2 = \sigma_2^{(1)}. \tag{3.81}$$

The corresponding elastic-plastic strains can then be determined according to Eq. (3.25), after the residual stresses from Eq. (3.79) were used in the material law Eq. (3.78).

If, however, the assignment of part 1 as remaining elastic turns out to be false, then another MEA is to be performed, now assuming that not only part 2 but also part 1 is going plastic:

$$\sigma_1^{(1)} > f_y \Rightarrow \text{part } 1 \rightarrow V_p. \tag{3.82}$$

Although Y_2 and $\varepsilon_{2,0}$ remain unchanged as given by Eqs. (3.74) and (3.75), initial strains are now also prescribed in part 1:

$$Y_1 = \sigma_1^{\text{fel}} - f_y \tag{3.83}$$

$$\varepsilon_{1,0} = \frac{1}{C}\left(\sigma_1^{\text{fel}} - f_y\right). \tag{3.84}$$

In addition, the modified Young's modulus in part 1 is no longer E, but E_t (Fig. 3.5).

The result of the second MEA is

$$\rho_1 = \rho_2 \frac{A_2}{A_1} \quad ; \quad \rho_2 = -\left(1 - \frac{E_t}{E}\right)\left(\sigma_2^{\text{fel}} - f_y \frac{1 + \frac{l_1}{l_2}}{1 + \frac{l_1}{l_2}\frac{A_2}{A_1}}\right). \tag{3.85}$$

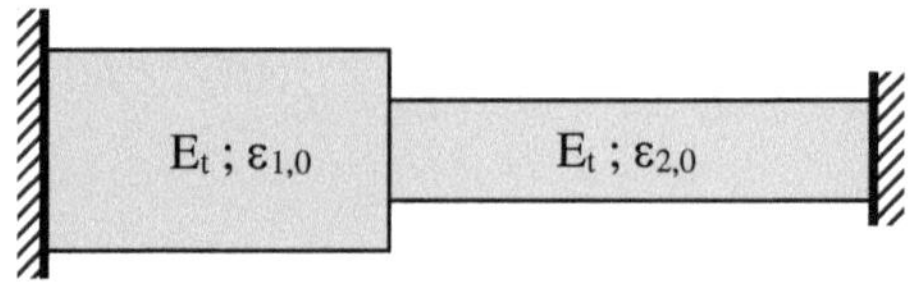

Fig. 3.5 System for a possibly required second MEA, or for the first MEA if $\sigma_1^{\mathrm{fel}} > f_y$

The same result would have been reached with the first MEA for each load level u, which instead of Eq. (3.72) leads to $\sigma_1^{\mathrm{fel}} > f_y$.

Anyway, the correct end result is achieved at the latest after the second MEA because the plastic zone is then correctly identified. If the structure is plasticized at all, then there exist only two possibilities because of its discrete nature: either part 2 alone is plastic, or both parts are plastic.

3.4.2 Bending Beam with Sandwich Cross-Section

A beam, clamped at one end and simply supported at the other end is subjected to a movement of the bearing w_0, so that virtually a cantilever beam is created which is subjected to a displacement-controlled load (Fig. 3.6). The cross-section consists of a sandwich cross-section, which can be thought of as an I-profile with infinitely thin web, where the two flanges are so thin and so far apart (distance 2h) that only membrane stresses occur in them. Stress redistribution across the section is also excluded.

Because the membrane stresses in the two flanges have the same absolute values, it is sufficient to consider only the upper flange in the following calculations, in which only positive stresses occur. If restricted to pure bending deformations without shear, the distribution of the fictitious elastic calculated axial stresses is given by

$$\sigma_{(x)}^{\mathrm{fel}} = 3E \frac{w_0}{L} \frac{h}{L} \frac{x}{L}. \tag{3.86}$$

According to Eqs. (3.57) and (3.58), we get the classification of V_p and V_e:

$$V_p = \{ x \mid x \geq x_{pl} \} \tag{3.87}$$

$$V_e = \{ x \mid x < x_{pl} \}, \tag{3.88}$$

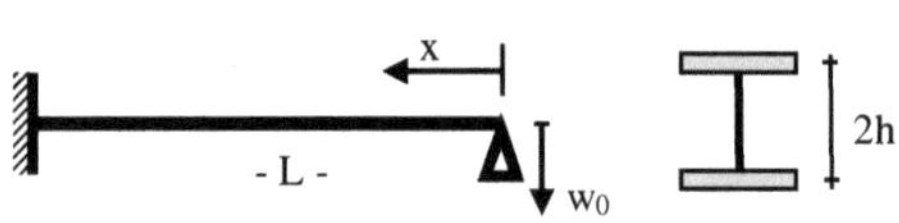

Fig. 3.6 Bending beam with sandwich section under displacement-controlled loading

where the boundary between the elastic and the plastic zone initially is given by

$$\frac{x_{pl}}{L} = \frac{f_y}{\sigma^{\text{fel}}_{(x=L)}}. \tag{3.89}$$

Because of the uniaxial stress state, the TIV and the initial strain in V_p can again be written according to Eq. (3.67) and Sect. 3.1.4 as

$$Y_{(x)} = \sigma^{\text{fel}}_{(x)} - f_y \tag{3.90}$$

$$\varepsilon_{0(x)} = \frac{1}{C}\left(\sigma^{\text{fel}}_{(x)} - f_y\right). \tag{3.91}$$

The material law for the MEA is

$$\varepsilon^*_{(x \le xpl)} = \frac{\rho_{(x)}}{E} \quad ; \quad \varepsilon^*_{(x \ge xpl)} = \frac{\rho_{(x)}}{E_t} + \frac{1}{C}\left(\sigma^{\text{fel}}_{(x)} - f_y\right). \tag{3.92}$$

Because of the spatial dependence of the material law, the bending differential equation ($w_{(x)}$ is the deflection, ()‘ = d()/dx the derivative in the axial direction)

$$w''_{(x)} = \frac{\varepsilon^*_{(x)}}{h} \tag{3.93}$$

must be integrated separately for V_e and V_p. Because the section moments of the residual state must be distributed linearly for equilibrium reasons, and the residual stresses are independent of the cross-sectional coordinates because of the sandwich cross-section, the residual stresses in the x-direction must be linearly distributed. Integrating each of

$$w''_{(x \le xpl)} = \frac{\rho_{(x=L)}}{hE}\frac{x}{L} \quad \text{in } V_e \tag{3.94}$$

$$w''_{(x \ge xpl)} = \left(\frac{\rho_{(x=L)}}{hE_t} + \frac{\sigma^{\text{fel}}_{(x=L)}}{hC}\right)\frac{x}{L} - \frac{f_y}{hC} \quad \text{in } V_p \tag{3.95}$$

twice, we get four integration constants. Because of hardening, a kink must not develop at the clamped end, in contrast to the plastic hinge theory, i.e., the rotation must vanish there. So five boundary and transitional conditions are available:

$$w_{(x=0)} = w_0 \quad ; \quad w_{(x=L)} = w'_{(x=L)} = 0 \tag{3.96}$$

$$w_{(x=xpl,Ve)} = w_{(x=xpl,Vp)} \quad ; \quad w'_{(x=xpl,Ve)} = w'_{(x=xpl,Vp)}. \tag{3.97}$$

They are sufficient to determine the integration constants and the residual stresses at the clamped end as a function of the fictitious elastic stress at this location as well as the position of the transition between V_e and V_p:

$$\frac{\rho_{(x=L)}}{f_y} = \left(1 - \frac{E_t}{E}\right) \frac{\frac{3}{2}\left[1 - \left(\frac{x_{pl}}{L}\right)^2\right] - \frac{\sigma^{\text{fel}}_{(x=L)}}{f_y}\left[1 - \left(\frac{x_{pl}}{L}\right)^3\right]}{\frac{E_t}{E}\left(\frac{x_{pl}}{L}\right)^3 + \left[1 - \left(\frac{x_{pl}}{L}\right)^3\right]}. \tag{3.98}$$

Because no use has yet been made of Eq. (3.89), Eq. (3.98) also applies for more upcoming iterations. The stresses of the elastic-plastic state are obtained from Eq. (3.24). For the first iteration, Eq. (3.98) is to be inserted there and we get

$$\frac{\sigma^{(1)}_{(x=L)}}{f_y} = \frac{\frac{3}{2} - \frac{3}{2}\left(\frac{f_y}{\sigma^{\text{fel}}_{(x=L)}}\right)^2 + \left(\frac{C}{E_t} - 1\right)\left(\frac{\sigma^{\text{fel}}_{(x=L)}}{f_y}\right)}{\frac{C}{E_t} - \left(\frac{f_y}{\sigma^{\text{fel}}_{(x=L)}}\right)^3}. \tag{3.99}$$

The corresponding elastic-plastic strains at the clamped end after the first MEA are given by

$$\frac{\varepsilon^{\text{el-pl}(1)}_{(x=L)}}{f_y/E} = 1 + \frac{E}{E_t}\left(\frac{\sigma^{(1)}_{(x=L)}}{f_y} - 1\right) \tag{3.100}$$

or, in the form of the plastic strain enhancement factor K_e,

$$K_e^{(1)} = \frac{\varepsilon^{\text{el-pl}(1)}_{(x=L)}}{\sigma^{\text{fel}}_{(x=L)}/E} = \frac{1 + \frac{E}{E_t}\left(\frac{\sigma^{(1)}_{(x=L)}}{f_y} - 1\right)}{\frac{\sigma^{\text{fel}}_{(x=L)}}{f_y}}. \tag{3.101}$$

Adopting Eqs. (3.61) and (3.62), it can now be checked whether the estimation of the plastic zone carried out by Eq. (3.89) was true. Using the stresses calculated for the elastic-plastic state based on the first MEA, an improved estimation can now be made by

$$\frac{x_{pl}}{L} = \frac{f_y}{\sigma^{(1)}_{(x=L)}}. \tag{3.102}$$

Inserting in Eq. (3.98) provides new residual stresses for the second MEA and, after superposition with the fictitious elastic stresses, a new approximation for the stress, $\sigma^{(2)}_{(x=L)}$, and correspondingly for $\varepsilon^{\text{el-pl}(2)}_{(x=L)}$ and $K_e^{(2)}$.

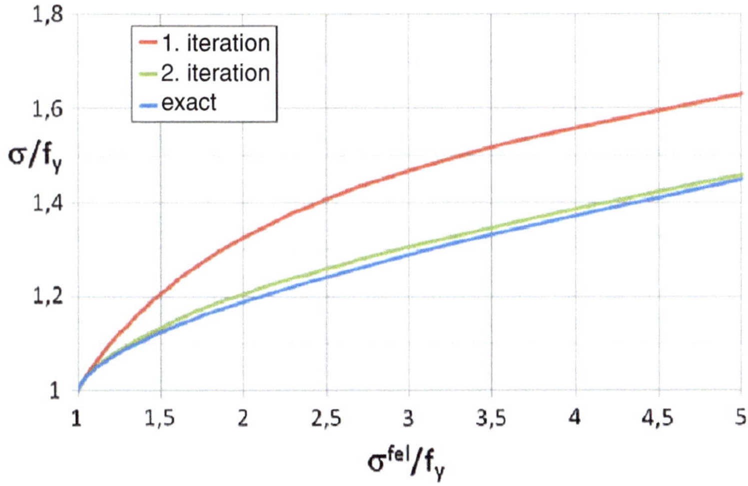

Fig. 3.7 Stress for the bending beam with sandwich section subjected to displacement-controlled load for $E_t/E = 0.05$

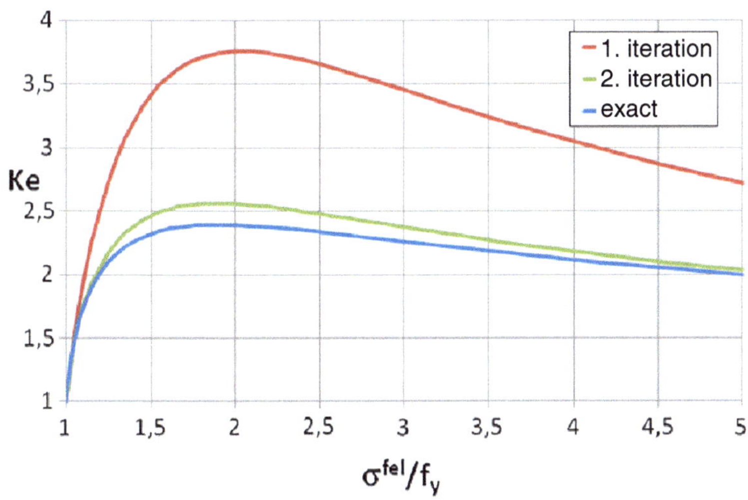

Fig. 3.8 Plastic strain enhancement factor K_e for the bending beam with sandwich section subjected to displacement-controlled load for $E_t/E = 0.05$

In Figs. 3.7 and 3.8 the stresses and the factor K_e are shown for $E_t/E = 0.05$ for the first two iterations, and for comparison also the exact solution, as a function of the load level, represented by the fictitious elastic stress at the clamped end.

The exact solution can be obtained analytically by double integration of the bending differential equation

$$w''_{(x)} = \frac{\varepsilon^{\text{el-pl}}_{(x)}}{h}, \tag{3.103}$$

where the actual elastic-plastic material law is to be inserted in the numerator. The result is preferably given in inverse form:

$$\frac{\sigma^{\text{fel}}_{(x=L)}}{f_y} = \frac{-\frac{3}{2} + \frac{1}{2}\left(\frac{f_y}{\sigma^{\text{exact}}_{(x=L)}}\right)^2 + \left(1 + \frac{C}{E}\right)\left(\frac{\sigma^{\text{exact}}_{(x=L)}}{f_y}\right)}{\frac{C}{E}} \tag{3.104}$$

$$\frac{\varepsilon^{\text{el-pl(exact)}}_{(x=L)}}{f_y/E} = 1 + \frac{E}{E_t}\left(\frac{\sigma^{\text{exact}}_{(x=L)}}{f_y} - 1\right) \tag{3.105}$$

$$K_e^{\text{exact}} = \frac{\varepsilon^{\text{el-pl(exact)}}_{(x=L)}}{\sigma^{\text{fel}}_{(x=L)}/E} = \frac{1 + \frac{E}{E_t}\left(\frac{\sigma^{\text{exact}}_{(x=L)}}{f_y} - 1\right)}{\frac{\sigma^{\text{fel}}_{(x=L)}}{f_y}}. \tag{3.106}$$

In Figs. 3.7 and 3.8 it can be seen that even after the second MEA the exact result is not yet reached, although already a fairly good approximation. If necessary, further improvements through more iterations would be possible.

The quality of the approximation by the STPZ depends on the hardening modulus and the load level. The more pronounced the hardening, the less MEAs are required for a good approximation. As a measure for the error involved in a particular iteration, the portion of the total volume could be used in this example that is treated as V_p but turns out to be V_e at the end of the same iteration, or vice versa. Although such an error measure has the advantage of not requiring any information not already available in an iteration step, it is not suited to a generalization. In general, the volume fraction, in which the yield condition is violated, is not an appropriate measure of an error estimate, because this error naturally occurs in areas that do not belong to the highly stressed areas of a structure.

In the previous example, the tension bar with step-like change in cross-section in Sect. 3.4.1, the plastic zone V_p was accurately found after a few MEAs, though in some ranges of the load level with intermediate underestimate of their extent. In the present example of a bending beam with sandwich section, however, V_p is systematically overestimated because x_{pl} is not found exactly, but approaches the exact solution from below.

Therefore, it is noted that in the STPZ the plastic zone is determined iteratively and, at least for continuous structures, has only an approximate character.

3.5 Transformed Internal Variable at Non-Radial Loading

If the loading is not radial, the TIV can, instead of Eq. (3.66), be estimated by

$$Y_i = \sigma_i^{\prime \mathrm{fel}} - \sigma_i^{\prime}\left(\frac{f_y}{\sigma_\mathrm{v}}\right) \quad \forall \underline{x} \in V_p. \tag{3.107}$$

The fact that this is merely an approximation can be taken from Fig. 3.1. Equation (3.107) can be interpreted as a projection of the negative deviatoric residual stresses on the yield surface in the space of TIV (Fig. 3.1, right), similar to the radial return for projecting the trial stress on the flow surface in an incremental analysis. By this approximation, neither the equilibrium nor the compatibility conditions nor the yield condition are violated, but possibly the flow law Eq. (1.18). If the yield surface is not a Mises circle, the projection of the negative deviatoric residual stress on the yield surface cannot be formulated as easily as in Eq. (3.107). Instead, the minimum distance to the yield surface is to be found, which, because of its convexity, turns out to be an optimization problem to be solved using the methods of convex analysis.

The quality of the approximation Eq. (3.107) is crucial for the quality of Zarka's method and the STPZ. As Eq. (3.107) cannot be deduced from principles of mechanics, it must be regarded as heuristic, and the quality of its approximation is to be checked in the application environment as plausible and useful. Accordingly, intervention is possible at this point to be able to introduce, where appropriate, better estimations by a modification of Eq. (3.107). In general, Eq. (3.107) has so far proved to be a good approximation, provided directional redistribution is not too pronounced.

If the stresses of the elastic-plastic state resulting from the previous MEA are used for $\sigma_i^{\prime}$ and σ_v, then not only is the plastic zone V_p improved iteratively, but also the estimation of the TIV in V_p. For the nth iteration, Eq. (3.107) then becomes

$$Y_i^{(n)} = \sigma_i^{\prime \mathrm{fel}} - \sigma_i^{\prime (n-1)}\left(\frac{f_y}{\sigma_\mathrm{v}^{(n-1)}}\right) \quad \forall \underline{x} \in V_p^{(n)}. \tag{3.108}$$

Prior to the first iteration ($n = 1$), no residual stresses have been determined, so that

$$\rho_i^{\prime (0)} = 0_i \quad \rightarrow \quad \sigma_i^{\prime (0)} = \sigma_i^{\prime \mathrm{fel}} \tag{3.109}$$

and Eq. (3.108) can be led back to Eq. (3.66). This corresponds to a projection of the origin of the coordinate system to the yield surface in the space of the TIV.

With the iteration of the plastic zone and the iteration of the TIV, we get the workflow shown in Fig. 3.9.

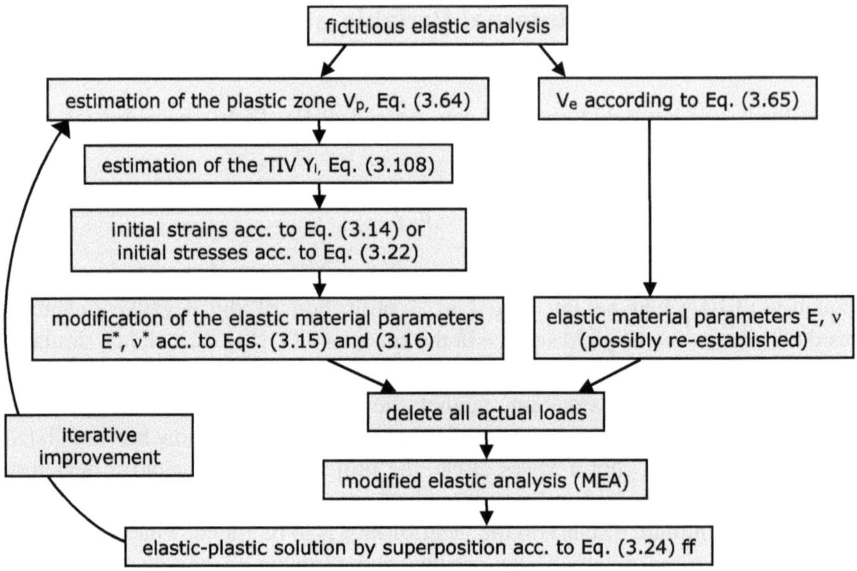

Fig. 3.9 Workflow for the iterative application of the STPZ at monotonic loading in the case of linear kinematic hardening with temperature-independent material data

3.6 Examples of Non-Radial Loading

Whereas in Sect. 3.4 only configurations were considered in which the plastic zone V_p had to be determined iteratively, but not the TIV Y_i, a configuration is considered in Sect. 3.6.1 in which only the estimation of the TIV Y_i, but not of V_p, is subjected to an iterative improvement. After that, examples are treated where both the TIV and V_p are determined iteratively. However, in the case of non-radial loading, hand calculations are hardly possible. These examples are therefore analyzed using the finite element program ANSYS [8], in which the STPZ was implemented by means of a user subroutine. Solutions obtained with ANSYS by conventional incremental calculations, regarded as more or less exact solutions, are also given for comparison purposes.

In Sects. 3.6.1 and 3.6.2 a test for STPZ is carried out using simple examples to find what impact the projection of residual stresses on the yield surface in the space of the TIV has on the approximation quality of the STPZ in the presence of directional stress redistribution.

For all calculations it is assumed that the displacements and strains are so small that the displacement-strain relations may be linearized and the equilibrium conditions may be formulated on the undeformed system.

3.6.1 Directional Stress Redistribution Produced by Elastic Incompressibility

In Sect. 2.1 a homogeneously stressed volume of material was considered as an example of directional stress redistribution, simultaneously subjected to a plane stress condition ($\sigma_z = 0$), a plane strain condition ($\varepsilon_y = 0$), and a displacement-controlled load (displacement u) in the third direction. If unit lengths are chosen, the latter directly corresponds to the strain ε_x (Fig. 3.10).

3.6.1.1 Fictitious Elastic Analysis

Because of the boundary conditions ($\sigma_z = 0$, $\varepsilon_y = 0$), the fictitious elastic solution arises directly from the linear elastic material law, Eq. (1.3), for the load ε_x:

$$\sigma_i^{\text{fel}} = \begin{pmatrix} 1 \\ \nu \\ 0 \end{pmatrix} \frac{E\,\varepsilon_x}{1-\nu^2} \quad \rightarrow \quad \sigma_i^{\prime\text{fel}} = \frac{1}{3} \begin{pmatrix} 2-\nu \\ -1+2\nu \\ -1-\nu \end{pmatrix} \frac{E\,\varepsilon_x}{1-\nu^2} \tag{3.110}$$

$$\varepsilon_i^{\text{fel}} = \begin{pmatrix} 1 \\ 0 \\ -\frac{\nu}{1-\nu} \end{pmatrix} \varepsilon_x \tag{3.111}$$

and the equivalent stress from Eq. (1.13):

$$\sigma_{\text{v}}^{\text{fel}} = \frac{\sqrt{1-\nu+\nu^2}}{1-\nu^2} E\,\varepsilon_x. \tag{3.112}$$

3.6.1.2 Incremental Elastic-Plastic Analysis

For the incremental calculation with ANSYS, a Poisson's ratio $\nu = 0$ and a hardening modulus $E_t/E = 0.05$ are chosen. Figure 3.11 shows the evolution of the stress state and of the TIV as trajectories in the deviatoric stress space or in the Y-space up to a load level equal to five times the elastic limit load. In this example, the TIV leaves the origin of the coordinate system only with incipient plasticization, which is caused by the homogeneous state of stress, allowing no stress redistribution to occur.

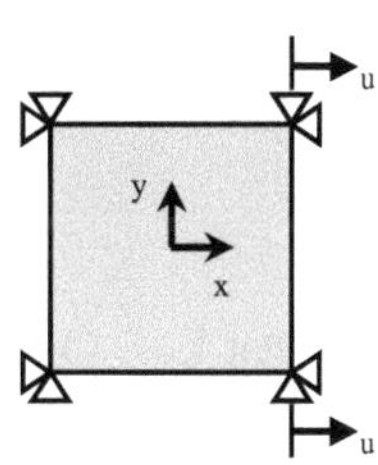

Fig. 3.10 Volume element with plane stress and plane strain, subjected to a displacement u

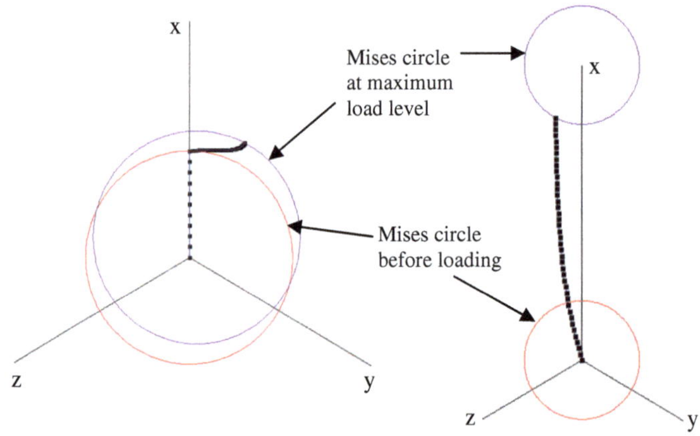

Fig. 3.11 Volume element with plane stress and plane strain; *left* trajectories of the stress state in the deviatoric stress space; *right* trajectories of the TIV in the *Y*-space according to an incremental elastic-plastic analysis

3.6.1.3 STPZ

Similar to the stress state in the fictitious elastic state, the residual stress state for the *n*th MEA can be obtained from the modified elastic material law, Eq. (3.20), with the boundary conditions $\rho_z = 0$, $\varepsilon_x^* = \varepsilon_y^* = 0$, provided that the load level of ε_x is sufficiently high to cause plasticization, i.e., if the entire element is to be considered as plastic zone V_p:

$$\rho_i^{(n)} = -\frac{3}{2}\frac{1-\frac{E_t}{E}}{1-(\nu^*)^2}\begin{pmatrix} Y_x^{(n)} + \nu^* Y_y^{(n)} \\ \nu^* Y_x^{(n)} + Y_y^{(n)} \\ 0 \end{pmatrix} \tag{3.113}$$

$$\varepsilon_i^{*(n)} = \frac{3}{2}\frac{1-\frac{E_t}{E}}{E_t}\frac{1-2\nu^*}{1-\nu^*}\begin{pmatrix} 0 \\ 0 \\ Y_z^{(n)} \end{pmatrix}. \tag{3.114}$$

Here, the TIV $Y_i^{(n)}$ is to be used as estimated according to Eq. (3.108). For the first iteration ($n = 1$), $\sigma_i^{'(0)} = \sigma_i^{'\text{fel}}$ is still valid, and therefore

$$Y_i^{(1)} = \frac{1}{3}\left(\frac{E\,\varepsilon_x}{1-\nu^2} - \frac{f_y}{\sqrt{1-\nu+\nu^2}}\right)\begin{pmatrix} 2-\nu \\ -1+2\nu \\ -1-\nu \end{pmatrix}, \tag{3.115}$$

so that Eqs. (3.113) and (3.114) yield

$$\rho_i^{(1)} = -\frac{1}{2}\frac{1-\frac{E_t}{E}}{1-(\nu^*)^2}\left(\frac{E\,\varepsilon_x}{1-\nu^2} - \frac{f_y}{\sqrt{1-\nu+\nu^2}}\right)\begin{pmatrix}(2-\nu)+\nu^*(-1+2\nu)\\ \nu^*(2-\nu)+(-1+2\nu)\\ 0\end{pmatrix} \tag{3.116}$$

$$\varepsilon_i^{*(1)} = \frac{1}{2}\frac{1-\frac{E_t}{E}}{E_t}\frac{(1-2\nu^*)}{(1-\nu^*)}\left(\frac{E\,\varepsilon_x}{1-\nu^2} - \frac{f_y}{\sqrt{1-\nu+\nu^2}}\right)\begin{pmatrix}0\\ 0\\ -1-\nu\end{pmatrix}. \tag{3.117}$$

The elastic-plastic stress results from superposition with the fictitious elastic stress according to Eq. (3.24):

$$\sigma_i^{(1)} = \sigma_i^{\text{fel}} + \rho_i^{(1)}. \tag{3.118}$$

Because the plastic zone is independent of the load level in this example, the residual stress and residual strain, and thus also the elastic-plastic stress, depend only linearly on the load level in the first iteration.

If the material is elastic incompressible ($\nu = 0.5$), the exact result is already achieved with this first iteration, so that an iterative improvement is neither possible nor necessary. This is because the loading is radial in this case. When $\nu \neq 0.5$, however, the exact result has not been reached with the first iteration. An iterative improvement is achieved by forming the equivalent stress and the deviatoric stresses from the stresses given by Eq. (3.118), and obtaining $Y_i^{(2)}$ by inserting them in Eq. (3.108). We then get $\rho_i^{(2)}$ from Eq. (3.113) and $\sigma_i^{(2)}$ from Eq. (3.24). More iterations are performed in the same way. The result approaches the "exact" result closer and closer but never quite reaches it, revealing the approximate character of the STPZ at non-radial loading. The largest deviations from the incrementally calculated "exact" solution arise in non-hardening material ($E_t = 0$), where the result after the first iteration cannot be further iteratively improved.

For Poisson's ratio $\nu = 0$, the directional redistribution is most pronounced. Figure 3.12 gives an impression of the approximate character and the improvements through the iteration of the TIV.

For a Poisson's ratio that is characteristic of steel, $\nu = 0.3$, the approximation of the STPZ to the exact solution is much faster, as shown in Table 3.1 for a fivefold elastic limit load at $E_t/E = 0.05$. After five MEAs the error of the STPZ is less than 1 %.

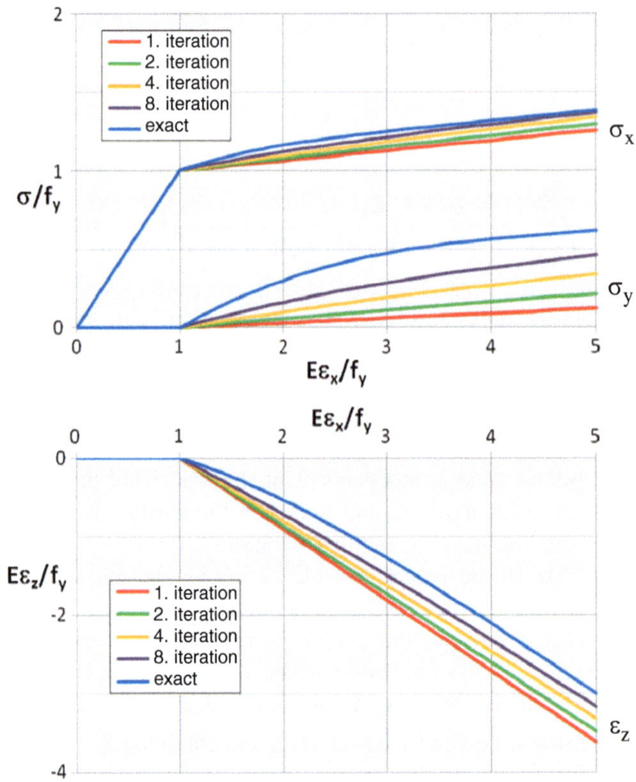

Fig. 3.12 Volume element with plane stress and plane strain; stresses σ_x and σ_y (*top*) and strain ε_z (*bottom*) for several MEAs produced by loading by ε_x ($\nu = 0$, $E_t/E = 0.05$)

Table 3.1 Development of the strain in the *z*-direction with the MEAs of the STPZ in comparison with the result obtained by an incremental analysis

MEA	$E/f_y{*}\varepsilon_z$
1	−4.374
2	−4.353
3	−4.338
4	−4.327
5	−4.319
6	−4.313
7	−4.308
8	−4.305
Incremental	−4.279

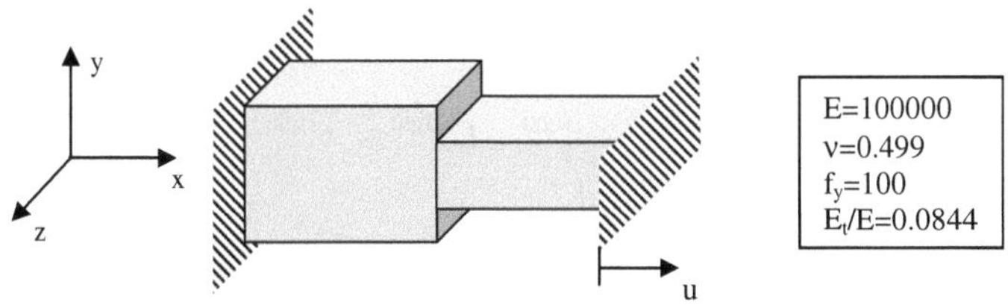

Fig. 3.13 Two-element model with generalized plane strain in the z-direction

3.6.2 Directional Stress Redistribution Despite Elastic Incompressibility

3.6.2.1 Geometry, Material, Load

Although a homogeneously multiaxial stressed element was considered in the previous section in order to examine the directional redistribution caused by different Poisson's ratios in the elastic ($\nu^{el} < 0.5$) and in the plastic range ($\nu^{pl} = 0.5$), another cause for directional stress redistribution is considered now that occurs, as a result of inhomogeneous multiaxial loading, even with elastic incompressibility ($\nu^{el} = 0.5$). As far as numerical solutions are found, such as by employing the "exact" theory of plastic zones, the Poisson's ratio of 0.5 is replaced by 0.499 for numerical reasons.

The tension bar with step-like change in cross-section treated in Sect. 3.4.1 serves as a basis. This is modified by introducing a so-called generalized plane strain condition perpendicular to the plane of the sketch, ensuring that, in contrast to the plane strain condition ($\varepsilon_z = 0$), both parts have the same, but yet unknown, strain in the z-direction ($\varepsilon_z \neq 0$). Both parts are chosen for the numerical calculations as equally long (unit length). Although they have the same thickness in the z-direction, their cross-sectional thicknesses in y-direction are in the ratio 2:1 (see Fig. 3.13).

3.6.2.2 Fictitious Elastic Analysis

In both parts there is a homogeneous stress state, so that eight unknown quantities exist: in the left as well as in the right part, the stresses and the strains in the x- and the z-directions. In the y-direction the stresses are zero, but not the strains.

The eight equations required for their determination are:

$$\text{equilibrium in } x - \text{direction}: \; \sigma_x^{\text{right}} = 2\sigma_x^{\text{left}} \tag{3.119}$$

$$\text{equilibrium in } z - \text{direction}: \; \sigma_z^{\text{right}} = -2\sigma_z^{\text{left}} \tag{3.120}$$

$$\text{generalized plane strain}: \; \varepsilon_z^{\text{right}} = \varepsilon_z^{\text{left}} \tag{3.121}$$

$$\text{kinematics}: \; \varepsilon_x^{\text{right}} + \varepsilon_x^{\text{left}} = u \,\text{because of unit lengths} \tag{3.122}$$

linear elastic material law:

$$E\varepsilon_x^{\text{right}} = \sigma_x^{\text{right}} - \nu\sigma_z^{\text{right}}\;;\; E\varepsilon_z^{\text{right}} = \sigma_z^{\text{right}} - \nu\sigma_x^{\text{right}} \tag{3.123}$$

$$E\varepsilon_x^{\text{left}} = \sigma_x^{\text{left}} - \nu\sigma_z^{\text{left}}\;;\; E\varepsilon_z^{\text{left}} = \sigma_z^{\text{left}} - \nu\sigma_x^{\text{left}} \tag{3.124}$$

So we get the fictitious elastic stresses, for example in the higher stressed right part:

$$\sigma_x^{\text{right}} = \frac{6Eu}{9-\nu^2}\;;\; \sigma_z^{\text{right}} = \frac{2\nu Eu}{9-\nu^2}\;;\; \sigma_{\text{v}}^{\text{right}} = \frac{2Eu}{9-\nu^2}\sqrt{9-3\nu+\nu^2}. \tag{3.125}$$

3.6.2.3 Incremental Elastic-Plastic Analysis

The incremental elastic-plastic calculation provides the evolution of the stress and the TIV, shown in Fig. 3.14 as trajectories in the deviatoric stress space and the TIV-space, respectively, until the load level $u = 0.01$ is reached, which corresponds to the 6.363-fold elastic limit load.

The second kink of the stress path in the deviatoric stress space indicates that the left part (and thus the entire structure) has just become plastic. In this state, the directional redistribution is obviously most pronounced.

The corresponding histogram of the stress components is shown in Fig. 3.15.

3.6.2.4 STPZ

The application of the STPZ follows the workflow shown in Fig. 3.9. Because a multiaxial stress state is present with non-radial loading, both the plastic zone and the TIV are to be determined iteratively according to Sect. 3.5. The plastic zone, consisting only of the right element, is already correctly identified after the first iteration, so that further iterations are used only to improve the estimates of the TIV.

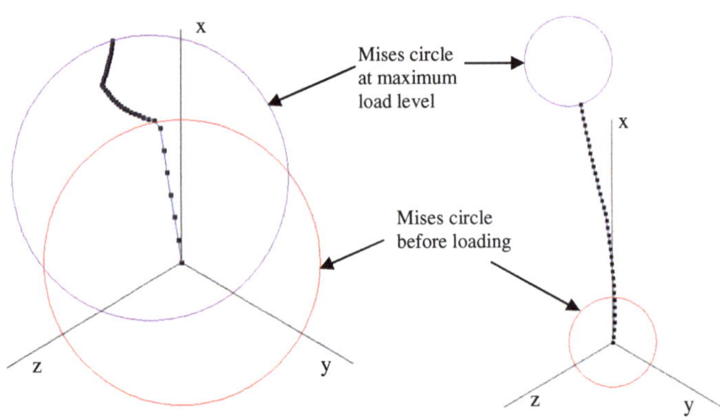

Fig. 3.14 Right part of the two-element-model; *left* trajectories of the stress in the deviatoric stress space; *right* trajectories of the TIV in the *Y*-space as determined by an incremental analysis

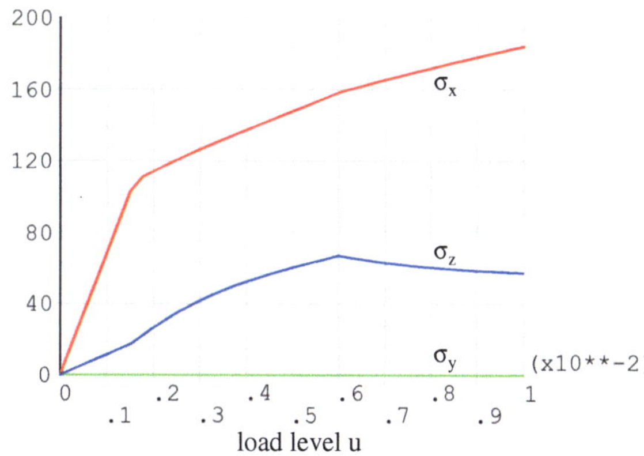

Fig. 3.15 Right part of the two-element model: stress histogram as determined by an incremental analysis

Table 3.2 Development of the stresses and strains in the right element with the MEAs of the STPZ compared to the incremental solution for $u = 0.006$

MEA	σ_x	σ_z	ε_x [%]	ε_z [%]
1	161.60	42.21	0.5930	−0.0671
2	155.77	48.21	0.5101	−0.0630
3	156.75	53.15	0.5084	−0.0657
4	157.24	56.20	0.5074	−0.0673
5	157.49	58.09	0.5068	−0.0683
6	157.62	59.27	0.5064	−0.0690
7	157.70	60.01	0.5062	−0.0693
8	157.75	60.47	0.5060	−0.0696
Exact	158.12	66.72	0.5043	−0.0728

The development of stresses and strains with the MEAs are listed in Table 3.2 for the right element at the load level of $u = 0.006$.

After a few MEAs, a good approximation for the stresses and strains in the most highly stressed element has been reached, at least in the direction of the maximum stress component. The quality of stresses and strains in the z-direction is lower.

3.6.3 Disc with a Hole

3.6.3.1 Geometry, Material, Load

A rectangular disc (size 180 × 360; here and in the following, physical quantities without units are to be understood in any arbitrary consistent set of units) with concentric hole (diameter 70) is subjected to a stress-controlled load ($p = 76$) on its narrow sides (Fig. 3.16). Because of the condition of plane stress, the stresses perpendicular to the disc vanish. For reasons of symmetry, only the first quadrant needs to be considered. The elastic and the plastic material parameters for linear kinematic hardening are shown in Fig. 3.16.

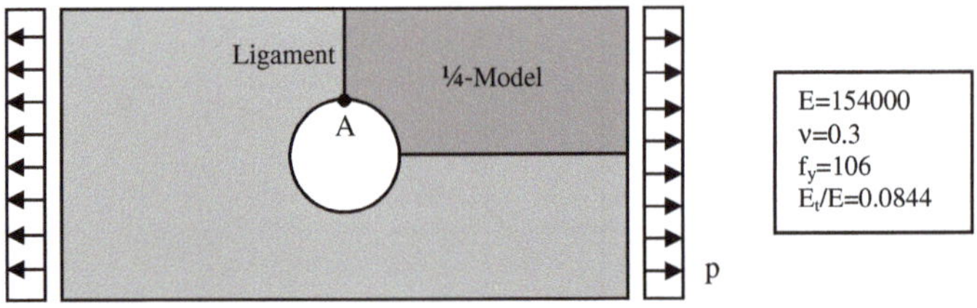

Fig. 3.16 Double symmetric disc with a hole (plane stress state) under monotonic loading

3.6.3.2 Fictitious Elastic Analysis

The fictitious elastic calculation shows that the maximum equivalent stress in point *A* is 279.1 and the load level is thus 2.633 times the elastic limit load.

3.6.3.3 Incremental Elastic-Plastic Analysis

The distribution of the equivalent stress obtained with an elastic-plastic incremental analysis is shown in Fig. 3.17. The plastic zone includes the colors light green, yellow, and red. One recognizes two disjoint plastic zones.

This calculation, too, shows that the term "exact" for the solution employing the theory of plastic zones and thus the measure of the quality of the approximation by the STPZ is questionable. The results depend, for example, on the chosen element type, the chosen finite element mesh, the convergence tolerance, and the selected number of load increments to capture the dependence on the loading path. As a result, it turns out that the area in which the yield limit is exceeded is not exactly the same as the area in which the plastic straining occurs. Despite the uniaxial stress state at point *A*, the plastic strain components in the other two directions are not exactly the same there.

Figure 3.18 shows the development of the TIV at the highest stressed point in the notch root of the ligament (point *A*). The stress state is uniaxial there, so the TIV components Y_i depend linearly on the fictitious elastic calculated stress according to

Fig. 3.17 Disc with a hole: equivalent stress obtained by an incremental elastic-plastic analysis applying the "exact" theory of plastic zones

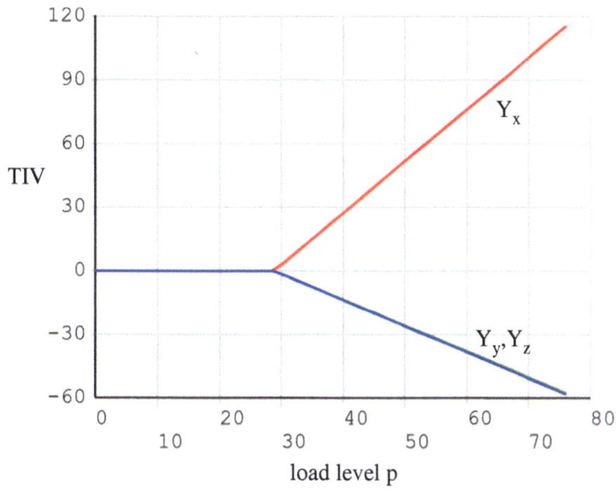

Fig. 3.18 Disc with a hole: development of the three TIV-components at point *A* with increasing load level

Eq. (3.66) and thus also linearly on the load level. For other parts of the plastic zone, however, this is not true.

3.6.3.4 STPZ

The application of the STPZ follows the workflow shown in Fig. 3.9. Because the stress state is multiaxial with non-radial loading, both the plastic zone and the TIV are to be determined iteratively according to Sect. 3.5.

Figure 3.19 shows the distribution of the equivalent stress after one, two, and four MEAs. The differences between the MEAs are getting smaller, so that after four MEAs only minor changes occur. A comparison with Fig. 3.17 reveals that the geometry of the flow zone, and the distribution of the equivalent stress after four MEAs hardly deviates from the "exact" analysis.

Figure 3.20 represents the development of the elastic-plastic axial strain distribution along the ligament.

After the second MEA, i.e., together with the fictitious elastic analysis after a total of three linear elastic calculations, a very good approximation for the strain at the most highly stressed point is achieved. And after four MEAs, the distribution of stresses and the strain is determined over the entire structure in a very good approximation.

3.6.4 Thick-Walled Cylinder Under Internal Pressure

3.6.4.1 Geometry, Material, Load

A thick-walled cylinder (internal radius r_i = 1,000, external radius r_a = 2,000) is subjected to internal pressure (p = 450). The ends of the cylinder are closed, so that the axial force $N = \pi p r_i^2$ arises through the internal pressure acting on the end caps.

Consider a pipe section which is sufficiently far away from the bending disturbances caused by end caps, so that they need not be considered. As a result,

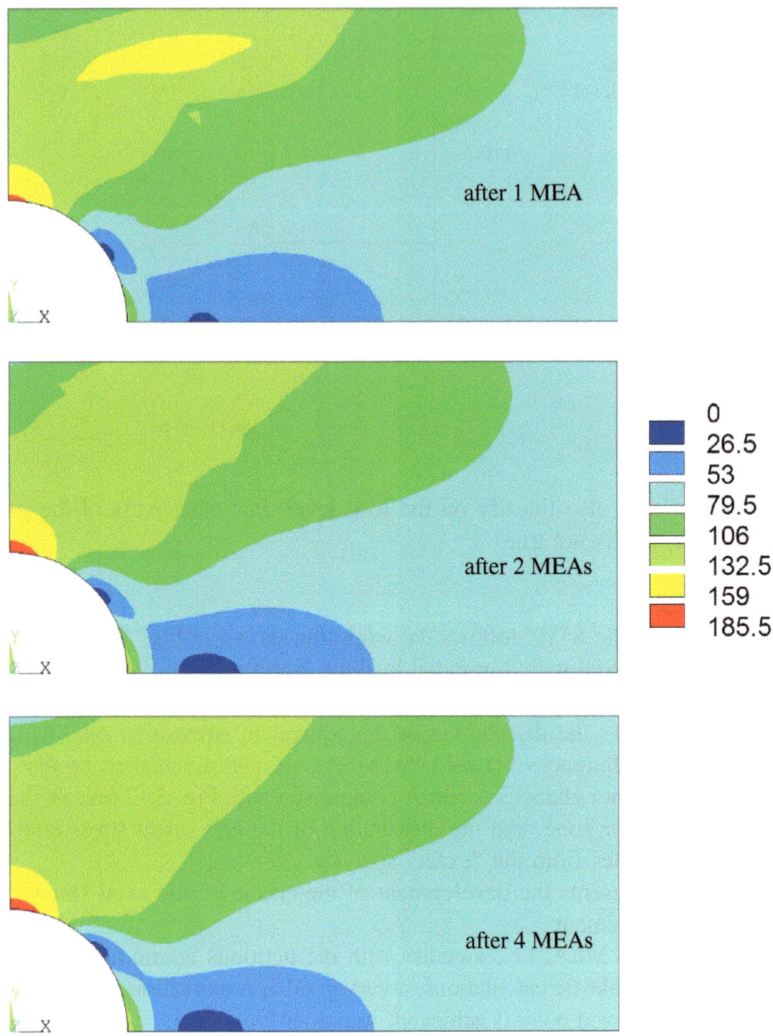

Fig. 3.19 Disc with a hole: equivalent stress according to the STPZ after one, two, and four MEAs

stresses are constant along the axial direction. Therefore, only a cross-sectional slice is to be modeled with a single finite element layer in the axial direction, one side of which is supported in the axial direction, and the nodes at the opposite side are coupled to each other so that they all have the same axial displacement (see Fig. 3.21).

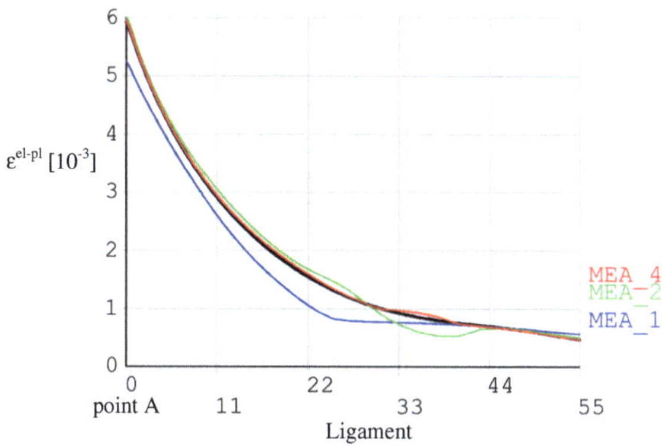

Fig. 3.20 Disc with a hole: distribution of axial strain along the ligament: comparison between "exact" incremental analysis (*black curve*) and various MEAs of the STPZ

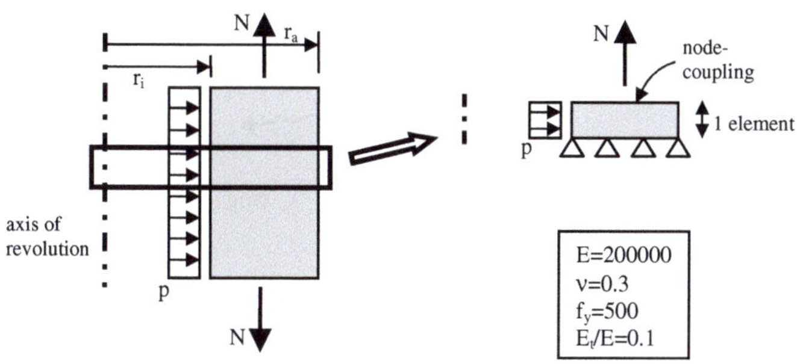

Fig. 3.21 Thick-walled cylinder under internal pressure

3.6.4.2 Fictitious Elastic Analysis

The stress state is three-dimensional in the entire structure except on the outside surface. The most highly stressed location is the inside surface. For the load level of 2.077 times the elastic limit load, the distribution of stresses across the wall thickness is shown in Fig. 3.22.

3.6.4.3 Incremental Elastic-Plastic Analysis

After exceeding the elastic limit load, the boundary between the plastic and the elastic zones moves further outwards with increasing load. At a load level of 2.077 times the elastic limit load, 95.0 % of the wall thickness is plasticized.

Figure 3.23 shows how the stress and the TIV on the inner surface evolve with increasing load according to an incremental analysis with ANSYS. On the inner surface, the TIV leaves the origin only with incipient plasticization. At other locations, however, this is not necessarily the case, because residual stresses may develop there because of local stress redistribution, even though the location under consideration is itself not plasticizing.

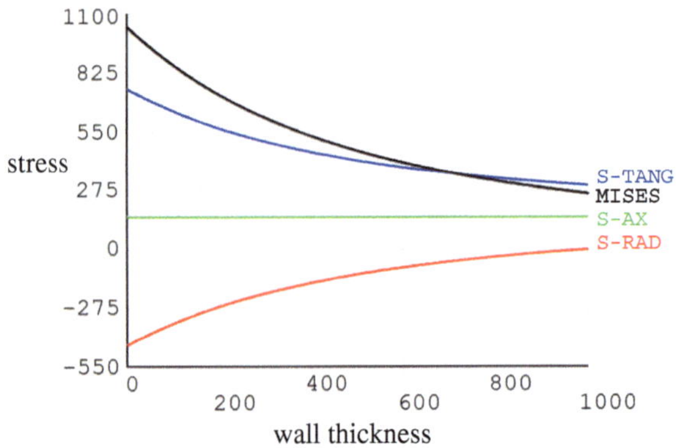

Fig. 3.22 Thick-walled cylinder: distribution of the fictitious elastic stresses along the wall thickness for a load level of 2.077 times the elastic limit load

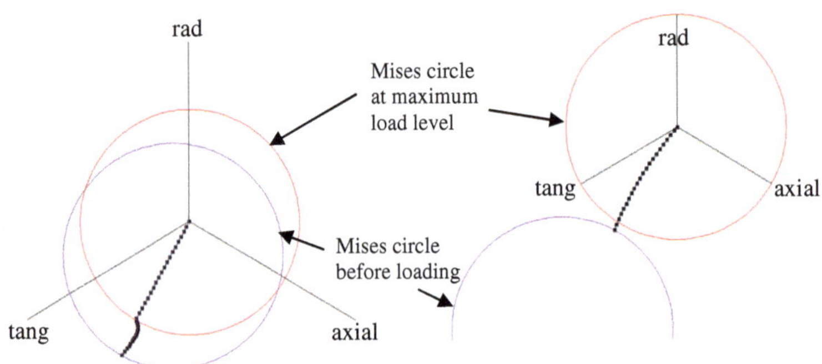

Fig. 3.23 Thick-walled cylinder; *left* trajectories of the stress on the inner surface in the deviatoric stress space; *right* trajectories of the TIV in the *Y*-space as determined by an incremental analysis up to a load level of 2.077 times the elastic limit load

In Fig. 3.24 the profile of the stresses and strains across the wall thickness is depicted.

3.6.4.4 STPZ

Figure 3.25 illustrates the profiles of stresses and strain components for various MEAs with the STPZ. It can be seen how the plastic zone V_p is searched for in the first MEAs. After three to four MEAs, hardly any difference to the incremental solution (Fig. 3.24) can be detected with bare eyes.

The examples of monotonic loading dealt with in Sects. 3.6.1–3.6.4 should allow a first insight into the way the STPZ works, and the achievable quality of results at non-radial loading, and in particular the "hardship" of directional stress redistribution. The computational effort required for the application of the STPZ is still not necessarily significantly less than for conventional incremental calculations

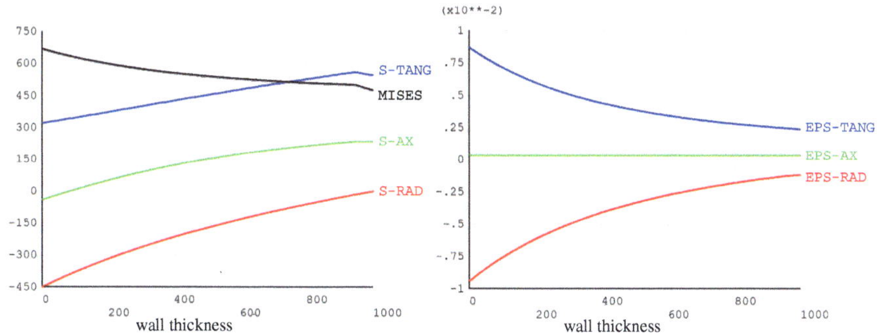

Fig. 3.24 Thick-walled cylinder: profile of the stresses (*left*) and strain components (*right*) across the wall thickness as determined by an incremental analysis for a load level of 2.077 times the elastic limit load

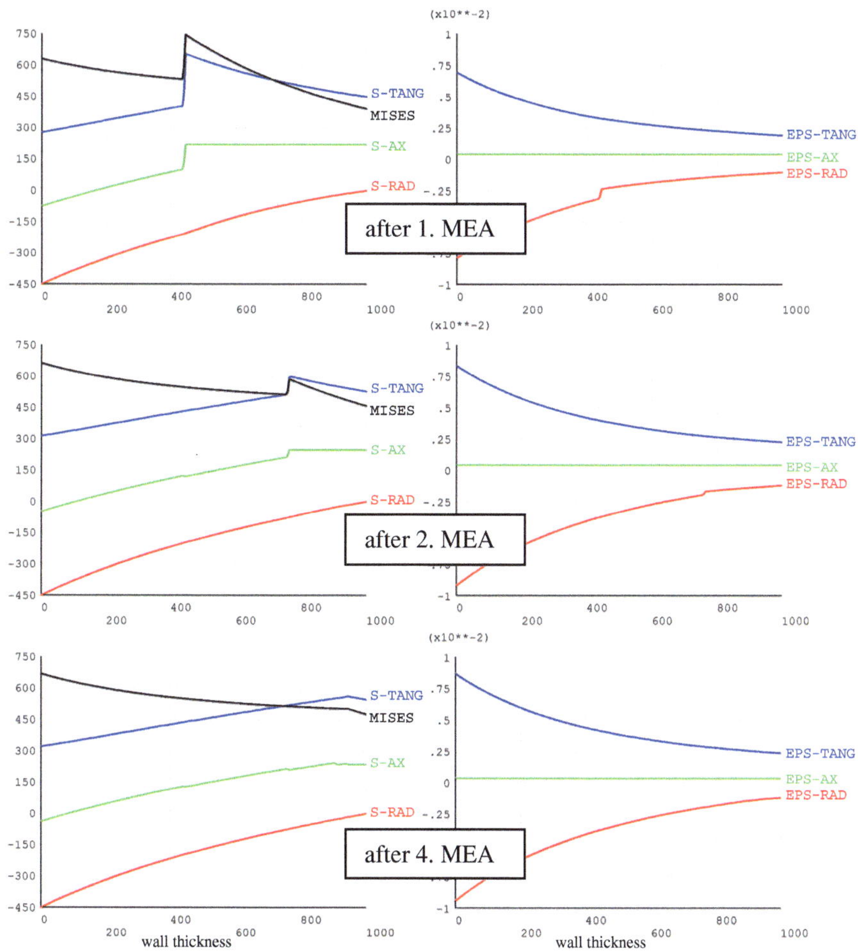

Fig. 3.25 Thick-walled cylinder: profile of the stresses (*left*) and strain components (*right*) across the wall thickness for various MEAs of the STPZ

with the "exact" theory of plastic zones employing the Newton–Raphson iterations usually used there. This advantage arises primarily under cyclic loading.

References

1. Zarka, J., Frelat, J., Inglebert, G., Kasmai Navidi, P.: A new Approach to Inelastic Analyses of Structures. Martinus Nijhoff Publishers, Dordrecht/Boston/Lancaster (1988) and (greatly expanded) 1990
2. Zarka, J., Inglebert, G., Engel, J.J.: On a Simplified Inelastic Analysis of Structures. Nucl. Eng. Des. **57**, 333–368 (1980)
3. Zarka, J.: On Simplified Mathematical Modelling of Cyclic Behaviour. Res Mechanica **4**, 203–223 (1982)
4. Zarka, J., Casier, J.: Elastic-Plastic Response of a Structure to Cyclic Loading: Practical Rules. In: Nemet-Nasser, S. (ed.) Mechanics Today, vol. 6, pp. 93–198. Pergamon Press, New York (1981)
5. Zarka, J.: Direct Analysis of Elastic-Plastic Structures with 'Overlay' Materials during Cyclic Loading. Int. J. Numerical Methods in Engineering **15**, 225–235 (1980)
6. Zarka, J.: Sur l'étude du comportement global des matériaux soumis à un chargement cyclique. Journal de Mécanique appliquée **3**(3), 291–326 (1979)
7. Hübel, H.: Ermittlung realistischer Ke-Faktoren (Plastifizierungsfaktoren) als Grundlage für die Präzisierung des kerntechnischen Regelwerkes hinsichtlich der Ermüdungsanalyse. Project SR2221 for the Federal Ministry for environment, nature conservation and nuclear safety (BMU-1997-481). ISSN 0724-3316 (1997)
8. ANSYS Release 14.5, ANSYS Inc. Canonsburg, USA (2012)

STPZ at Cyclic Loading

4

For structures subjected to cyclic loading above the elastic limit, the phenomena addressed in Chap. 2 are important for their service life, in particular the deformations accumulated as a result of a possible ratcheting mechanism, as well as elastic-plastic strain ranges.

With unlimited kinematic hardening, shakedown is guaranteed to occur because of cyclic loading, either elastic or plastic shakedown. The nature of the state of shakedown can be found adopting the STPZ, and all desired quantities in the state of shakedown such as maximum strain, mean strain, strain range, displacements, etc., can be determined approximately with relatively little effort.

This chapter describes how the elastic-plastic strain range in the state of plastic shakedown and the accumulated strains in the state of elastic and plastic shakedown can be determined with the STPZ based on Zarka's method (see also [1]). First, it is assumed that the material properties remain constant in the course of cyclic loading process. In Chap. 5 it is described how a possible temperature dependency can be taken into account. For the time being, linearity of the hardening law is assumed. In Chap. 7 an expansion to multilinear kinematic hardening is carried out.

4.1 Nature of the Shakedown State

In the following, cyclic loading is considered that is mainly attributed to two extreme loading conditions, hereafter referred to as the "minimum" and "maximum" loading condition ($L_{\min}$ and $L_{\max}$), between which all loads vary proportionally to each other:

$$L(t) = L_{\min} + \lambda(t)(L_{\max} - L_{\min}) \,, \tag{4.1}$$

wherein $\lambda(t)$ is an arbitrary scalar function, monotonically increasing from 0 to 1 and monotonically decreasing from 1 to 0. If the load can be described by a single time-dependent parameter, it is also called a one-parameter loading. The assignment, which load condition is said to be the "minimal" and which should be the

H. Hübel, *Simplified Theory of Plastic Zones*,
DOI 10.1007/978-3-319-29875-7_4

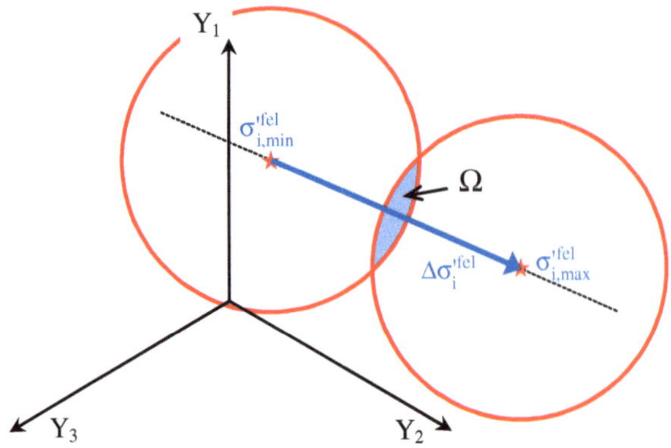

Fig. 4.1 Mises circles at cyclic loading in the space of the TIV

"maximum", is arbitrary, so that no statement about the amounts and sign of the loads or the corresponding stresses is associated.

The range of the fictitious elastic calculated stresses is then

$$\Delta\sigma_i^{\mathrm{fel}} = \sigma_{i,\max}^{\mathrm{fel}} - \sigma_{i,\min}^{\mathrm{fel}} \,. \tag{4.2}$$

Section 4.8 addresses how accumulated strains can be determined if more than two loading conditions must be considered.

If there is unlimited kinematic hardening, the nature of the state of shakedown, developing after an indeterminate number of loading cycles, can be detected easily by purely elastic calculations. If the fictitious elastic equivalent stress range between all possible load conditions is not larger than twice the yield stress at each location of the structure, then elastic shakedown (ES) occurs, otherwise plastic shakedown (PS). In the case of a one-parameter loading we thus have

$$\Delta\sigma_{\mathrm{v}(\underline{x})}^{\mathrm{fel}} \leq 2f_y \quad \forall \underline{x} \in V \rightarrow \mathrm{ES} \tag{4.3}$$

$$\Delta\sigma_{\mathrm{v}(\underline{x})}^{\mathrm{fel}} > 2f_y \quad \exists \underline{x} \in V \rightarrow \mathrm{PS}. \tag{4.4}$$

The $2f_y$-boundary between ES and PS is already known from the ratcheting interaction diagrams in Sects. 2.4 and 2.5. Transferred to the space of the TIV, this means that when the Mises circles centered in the fictitious elastic calculated stresses at the maximum and the minimum load conditions form an intersection area Ω that is not void at some location of the structure, then it comes to ES, otherwise to PS (Fig. 4.1).

4.2 Strain Range at Plastic Shakedown

To calculate the stress or strain ranges with the STPZ, the ranges of the residual stresses $\Delta\rho_i$ and the residual strains $\Delta\varepsilon_i^*$ are computed with the help of MEAs, analogous to the procedures described in Chap. 3. For this purpose, the plastic subvolume V_p and the ranges of the initial strains $\Delta\varepsilon_{i,0}$ or the initial stresses $\Delta\sigma_{i,0}$ are required. The modified elastic material law, Eqs. (3.20) and (3.21), then turn into

$$\Delta\varepsilon_i^* = \begin{cases} \left(E_{ij}^*\right)^{-1}\Delta\rho_j + \Delta\varepsilon_{i,0} & \forall \underline{x} \in V_p \\ E_{ij}^{-1}\Delta\rho_j & \forall \underline{x} \in V_e \end{cases} \tag{4.5}$$

$$\Delta\rho_i = \begin{cases} E_{ij}^*\,\Delta\varepsilon_j^* + \Delta\sigma_{i,0} & \forall \underline{x} \in V_p \\ E_{ij}\Delta\varepsilon_j^* & \forall \underline{x} \in V_e \end{cases} \tag{4.6}$$

and the modified loading in terms of initial strains or initial stresses given by Eqs. (3.14) and (3.22) becomes

$$\Delta\varepsilon_{i,0} = \frac{3}{2C}\Delta Y_i \tag{4.7}$$

$$\Delta\sigma_{i,0} = -\frac{3}{2C}E_{ij}^*\Delta Y_j = -\frac{3}{2C}\frac{E^*}{1+\nu^*}\Delta Y_i\,. \tag{4.8}$$

For the elastic zone V_e, again the alternative formulation analogous to Eqs. (3.18) and (3.19) is also possible:

$$\Delta\varepsilon_i^* = \left(E_{ij}^*\right)^{-1}\Delta\rho_j + \Delta\varepsilon_{i,0} \quad \text{with} \quad \Delta\varepsilon_{i,0} = \frac{3}{2C}\Delta Y_i \quad \text{with} \quad \Delta Y_i = -\Delta\rho_i' \\ \forall \underline{x} \in V_e\,. \tag{4.9}$$

The modified elastic material parameters E^* and ν^* remain as in Eqs. (3.15) and (3.16):

$$E^* = E_t \tag{4.10}$$

$$\nu^* = \frac{1}{2} - \frac{E_t}{E}\left(\frac{1}{2} - \nu\right). \tag{4.11}$$

The plastic zone V_p or its complement V_e are defined for the nth MEA, analogous to Eq. (3.64) or Eq. (3.65) by

$$V_p^{(n)} = \left\{\underline{x}\middle|\Delta\sigma_{\mathrm{v}}^{(n-1)} \geq 2f_y\right\} \tag{4.12}$$

$$V_e^{(n)} = \left\{\underline{x}\middle|\Delta\sigma_{\mathrm{v}}^{(n-1)} < 2f_y\right\}, \tag{4.13}$$

where $\Delta\sigma_v$ is the equivalent stress of the components of the stress range, obtained according to the superposition law analogous to Eq. (3.24):

$$\Delta\sigma_i = \Delta\sigma_i^{\text{fel}} + \Delta\rho_i \quad \forall \underline{x} \in V. \tag{4.14}$$

For n = 1 the residual stress range has not yet been calculated so

$$\Delta\rho_i = 0_i \quad \forall \underline{x} \in V \tag{4.15}$$

$$\Delta\sigma_v^{(0)} = \Delta\sigma_v^{\text{fel}} \quad \forall \underline{x} \in V. \tag{4.16}$$

The workflow in Fig. 3.9 is applicable, the necessary changes having been made, to estimate the strain range.

To estimate the TIV-range ΔY_i, Zarka, Inglebert, and co-workers have developed proposals for a lower and an upper bound, for example in [2, Chap. 7], [3, Appendix 3], [4, 5]. The current author had some very good experiences with these proposals, but also, for some applications, less good, such as a power plant nozzle, which was subjected to a thermal transient.

For this reason, a different approach is suggested here. Use is made of the Masing behavior, observed in many materials, stating that the cyclic behavior is similar to the monotonic behavior, so that the cyclic stress–strain diagram can be obtained simply from the monotonous by doubling the stress and strain. With linear kinematic hardening this means that the tangent modulus during cyclic and monotonic plasticization is the same and the range of the yield stress is twice as large as the monotonic yield stress (see Fig. 1.8).

This means that the range of the TIV under cyclic loading can be estimated in a similar manner to that of the TIV for monotonic loading in Sect. 3.5, namely by projecting the negative deviatoric residual stress range at the Mises circle with a radius twice the yield strength, which has its center in the fictitious elastic deviatoric stress range (see Fig. 4.2). It should be noted that the two Mises circles on the left in

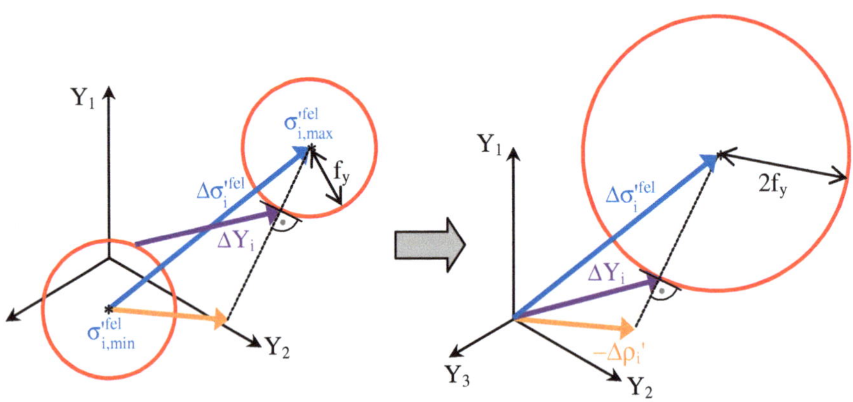

Fig. 4.2 Estimation of the TIV-range at cyclic loading in the space of the TIV

Fig. 4.2 need not necessarily be disjoint. They must just be disjoint at least at one location somewhere in the structure for plastic shakedown to occur.

This approach is similar to the twice-yield method mentioned in Sect. 2.9.3. Equation (3.108) thus becomes

$$\Delta Y_i^{(n)} = \Delta\sigma_i^{\prime\text{fel}} - \Delta\sigma_i^{\prime(n-1)}\left(\frac{2f_y}{\Delta\sigma_\text{v}^{(n-1)}}\right) \quad \forall \underline{x} \in V_p^{(n)} \tag{4.17}$$

or, in the special case of a uniaxial stress state, Eq. (3.67) now becomes

$$\Delta Y = \Delta\sigma^\text{fel} - 2f_y\,\text{sgn}(\Delta\sigma^\text{fel}) \quad \forall \underline{x} \in V_p^{(n)}\,. \tag{4.18}$$

The quality of this approximation is already reported in [6] for strain-based fatigue analyses of several practical examples. For this purpose, the knowledge of the strain range is sufficient without having resolved the stress–strain hysteresis, which would be required only when an energy-based concept of fatigue damage were adopted.

4.3 Examples of Strain Ranges at Plastic Shakedown

Because the procedure for estimating the strain range in cyclic loading is rather similar to estimating the strains in monotonic loading, the results for the examples in Sects. 3.4 and 3.6 can be used if the stresses and strains are replaced by their range values and the yield stress is doubled. Other examples are considered in the following sections. Some of these examples are pursued later in calculating accumulated strains. Some of these examples are also picked up after extension of the STPZ to account for temperature-dependent yield stress and of multilinear hardening in the following chapters.

4.3.1 Two-Bar Model

In Sect. 2.2, the two-bar model has already been presented as an introductory example of the phenomenon ratcheting (Fig. 4.3). Now, the range of stresses and strains is to be determined by the STPZ in the state of plastic shakedown. Both bars experience only a uniaxial stress state. The range of the TIV in longitudinal direction is thus given by Eq. (4.16) and is precisely known a priori. Because both bars have the same material properties and the same geometry, the stress and strain ranges in both bars are equal, so that they both plasticize cyclically in the state of plastic shakedown. Thus, the plastic zone V_p is exactly known in advance.

The fictitious elastic stress range in the left bar is

$$\Delta\sigma_\text{left}^\text{fel} = -\sigma_t = -{}^1\!/\!_2 E\alpha_T T\,. \tag{4.19}$$

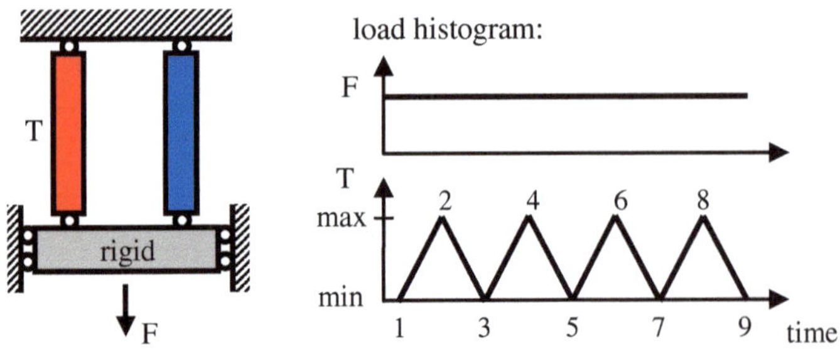

Fig. 4.3 Two-bar model under cyclic loading

and in the right bar

$$\Delta\sigma_{\text{right}}^{\text{fel}} = +\sigma_t = +{}^1\!/\!_2 E\alpha_T T\,. \tag{4.20}$$

The ranges of the TIV result from Eq. (4.18) and are

$$\Delta Y_{\text{left}} = \Delta\sigma_{\text{left}}^{\text{fel}} + 2f_y \tag{4.21}$$

$$\Delta Y_{\text{right}} = \Delta\sigma_{\text{right}}^{\text{fel}} - 2f_y, \tag{4.22}$$

so that, adapting Eq. (4.7) to uniaxial stress conditions (whereby the factor 3/2 is omitted, as described in Sect. 3.1.4), the initial strains become

$$\Delta\varepsilon_{\text{left},0} = \frac{1}{C}\Delta Y_{\text{left}}; \quad \Delta\varepsilon_{\text{right},0} = \frac{1}{C}\Delta Y_{\text{right}}\,. \tag{4.23}$$

With the modified elastic material law, Eq. (4.5), adjusted for the uniaxial stress state, we get

$$\Delta\varepsilon_{\text{left}}^* = \frac{1}{E_t}\Delta\rho_{\text{left}} + \frac{1}{C}\left(\Delta\sigma_{\text{left}}^{\text{fel}} + 2f_y\right) \tag{4.24}$$

$$\Delta\varepsilon_{\text{right}}^* = \frac{1}{E_t}\Delta\rho_{\text{right}} + \frac{1}{C}\left(\Delta\sigma_{\text{right}}^{\text{fel}} - 2f_y\right). \tag{4.25}$$

With the equilibrium condition

$$\Delta\rho_{\text{right}} = -\Delta\rho_{\text{left}} \tag{4.26}$$

and the kinematic condition

$$\Delta\varepsilon_{\text{left}}^* = \Delta\varepsilon_{\text{right}}^* \tag{4.27}$$

we obtain as a result of the first MEA:

$$\Delta\rho_{\text{left}} = -\left(1 - \frac{E_t}{E}\right)\left(\Delta\sigma_{\text{left}}^{\text{fel}} + 2f_y\right) \tag{4.28}$$

$$\Delta\varepsilon_{\text{left}}^{*} = 0\,. \tag{4.29}$$

The elastic-plastic strain range is obtained by superposition of the residual strain range of Eq. (4.29) with the fictitious elastic strain range. Thus, valid for both bars:

$$\Delta\varepsilon^{\text{el-pl}} = \Delta\varepsilon^{\text{fel}}\,. \tag{4.30}$$

Therefore, the loading is strain-controlled with respect to range values, i.e., in the terminology of the plastic strain enhancement factor:

$$K_e = 1, \tag{4.31}$$

regardless of material data and the load level. An iterative improvement is not possible, because both the TIV and the plastic zone V_p were a priori precisely known. It is also unnecessary because Eq. (4.30) represents the exact result, which can also be seen in Fig. 2.6 for the parameter constellation exemplarily investigated there.

So the STPZ provides the exact result for the stress and strain range after a single MEA.

4.3.2 Multiaxial Ratcheting

In Sect. 2.5.2, multiaxial ratcheting was treated. System and load histogram are again given in Fig. 4.4. If unit lengths are chosen for the edges, the force F directly corresponds to the stress σ_1 and the displacement u directly to the strain ε_2.

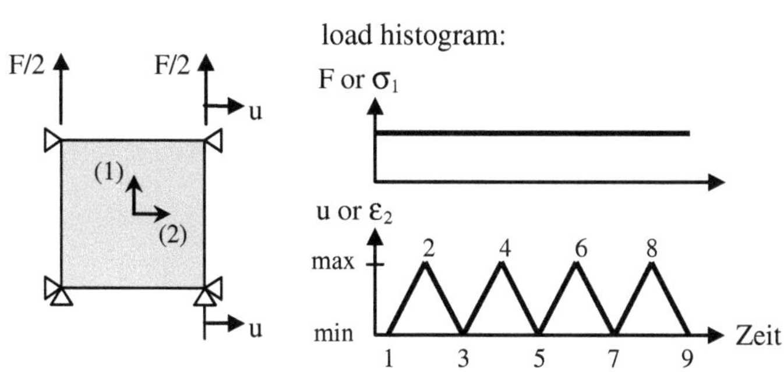

Fig. 4.4 Multiaxial ratcheting: one element with plane stress state

Again, the cyclic portion of the load leads to a uniaxial stress state. However, now the strain components are also determined in the other two directions, which is why the multiaxial formulation is maintained.

The ranges of the fictitious elastic stress components are

$$\Delta\sigma_i^{\mathrm{fel}} = \begin{pmatrix} 0 \\ \sigma_t \\ 0 \end{pmatrix}; \quad \sigma_t = E\varepsilon_2 \tag{4.32}$$

and their deviatoric components

$$\Delta\sigma_i^{\prime\mathrm{fel}} = \frac{1}{3}E\varepsilon_2 \begin{pmatrix} -1 \\ +2 \\ -1 \end{pmatrix}. \tag{4.33}$$

The corresponding equivalent stress is

$$\Delta\sigma_{\mathrm{v}}^{\mathrm{fel}} = E\varepsilon_2 \,. \tag{4.34}$$

The ranges of the fictitious elastic strains are

$$\Delta\varepsilon_i^{\mathrm{fel}} = \varepsilon_2 \begin{pmatrix} -\nu \\ 1 \\ -\nu \end{pmatrix}. \tag{4.35}$$

Because the element is in a homogeneous stress state, the whole element represents the plastic zone V_p, provided that $\Delta\sigma_{\mathrm{v}}^{\mathrm{fel}} > 2f_y$. From Eq. (4.17) the ranges of the TIV are obtained for the first MEA ($n = 1 \rightarrow \Delta\sigma_i^{\prime(0)} = \Delta\sigma_i^{\prime\mathrm{fel}}$):

$$\Delta Y_i^{(1)} = \frac{1}{3}\begin{pmatrix} -1 \\ +2 \\ -1 \end{pmatrix}\left(E\varepsilon_2 - 2f_y\right). \tag{4.36}$$

With the static and kinematic conditions

$$\Delta\rho_i = \begin{pmatrix} 0 \\ \Delta\rho_2 \\ 0 \end{pmatrix} \tag{4.37}$$

$$\Delta\varepsilon_i^* = \begin{pmatrix} \Delta\varepsilon_1^* \\ 0 \\ \Delta\varepsilon_3^* \end{pmatrix}, \tag{4.38}$$

the modified elastic material law of Eq. (4.5) leads to the system of equations

$$\begin{pmatrix} \Delta\varepsilon_1^{*(1)} \\ 0 \\ \Delta\varepsilon_3^{*(1)} \end{pmatrix} = \frac{1}{E_t} \begin{pmatrix} 1 & -\nu^* & -\nu^* \\ -\nu^* & 1 & -\nu^* \\ -\nu^* & -\nu^* & 1 \end{pmatrix} \begin{pmatrix} 0 \\ \Delta\rho_2^{(1)} \\ 0 \end{pmatrix} + \frac{E\varepsilon_2 - 2f_y}{2C} \begin{pmatrix} -1 \\ +2 \\ -1 \end{pmatrix} \tag{4.39}$$

with the solution for the first MEA

$$\Delta\rho_2^{(1)} = -\left(1 - \frac{E_t}{E}\right)(E\varepsilon_2 - 2f_y) \tag{4.40}$$

$$\Delta\varepsilon_1^{*(1)} = \Delta\varepsilon_3^{*(1)} = -\left(1 - \frac{E_t}{E}\right)\left(\varepsilon_2 - \frac{2f_y}{E}\right)\left(\frac{1}{2} - \nu\right). \tag{4.41}$$

The elastic-plastic stress and strain ranges are obtained analogously to Eq. (4.14) by superposition of the fictitious elastic and the modified elastic ranges

$$\Delta\sigma_i^{(1)} = \begin{pmatrix} 0 \\ 1 \\ 0 \end{pmatrix} \left[E_t\varepsilon_2 + \left(1 - \frac{E_t}{E}\right)2f_y\right] \tag{4.42}$$

$$\Delta\varepsilon_i^{(1)} = \begin{pmatrix} \varepsilon_2\left[-\frac{1}{2} + \left(\frac{1}{2} - \nu\right)\frac{E_t}{E}\right] + \left(\frac{1}{2} - \nu\right)\left(1 - \frac{E_t}{E}\right)\frac{2f_y}{E} \\ \varepsilon_2 \\ \varepsilon_2\left[-\frac{1}{2} + \left(\frac{1}{2} - \nu\right)\frac{E_t}{E}\right] + \left(\frac{1}{2} - \nu\right)\left(1 - \frac{E_t}{E}\right)\frac{2f_y}{E} \end{pmatrix}. \tag{4.43}$$

An iterative improvement is impossible and unnecessary as Eqs. (4.42) and (4.43) represent the exact result. Because of the uniaxial stress state and because the plastic zone was known a priori, the exact result of the stress and strain ranges is obtained in all coordinate directions by a single MEA with the STPZ, as already shown in the example of Sect. 4.3.1.

4.3.3 Bree Tube

The Bree tube has already been described in Sect. 2.5.3. It is a thin-walled tube under constant internal pressure, which is subjected to a cyclic radial temperature gradient with linear temperature distribution over the wall thickness. The difference between the temperature on the inner and outer surfaces is ΔT. Here the replacement model with uniaxial stress state shown in Fig. 2.21 is considered (Fig. 4.5).

In each cross-section of the bar the same behavior is present. The stresses are thus independent of the coordinate in longitudinal direction.

The stress state is uniaxial. If at maximum load cooling takes place at the top of the bar ($y = t/2$), the free thermal expansion strains are completely reversed by

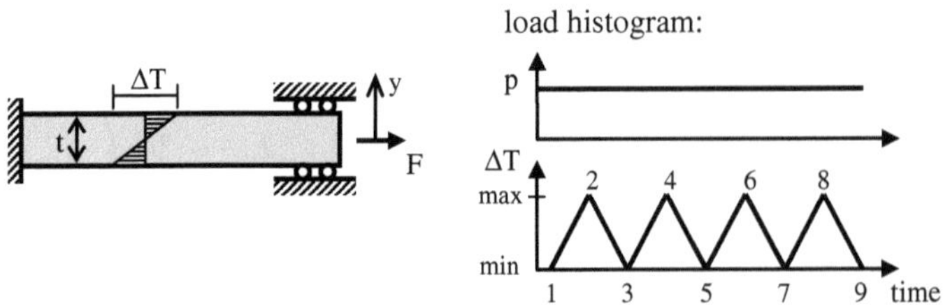

Fig. 4.5 Bree tube: replacement model with uniaxial stress state

tensile stresses. Because the thermal load here thus acts strain-controlled, the fictitious elastic calculated range of the axial stress is

$$\Delta\sigma^{\mathrm{fel}}_{(y)} = \frac{y}{t/2}\sigma_t; \quad \sigma_t = E\alpha_T\frac{\Delta T}{2}. \tag{4.44}$$

To achieve plastic shakedown as limiting state, it is required that

$$\Delta\sigma^{\mathrm{fel}}_{(y=t/2)} = \sigma_t > 2f_y . \tag{4.45}$$

Because of the uniaxial stress state and the strain-controlled loading, the subdivision of the cross-section in V_e and V_p is known a priori:

$$V_p = \left\{ y \middle| \Delta\sigma^{\mathrm{fel}}_{(y)} \geq 2f_y \quad \vee \quad \Delta\sigma^{\mathrm{fel}}_{(y)} \leq -2f_y \right\}. \tag{4.46}$$

The boundaries are marked by

$$\frac{y_{\mathrm{pl,above}}}{t/2} = +\frac{2f_y}{\sigma_t}; \quad \frac{y_{\mathrm{pl,below}}}{t/2} = -\frac{2f_y}{\sigma_t}. \tag{4.47}$$

According to Eq. (4.16), the range of the TIV is

$$\Delta Y_{(y>y_{\mathrm{pl,above}})} = \Delta\sigma^{\mathrm{fel}}_{(y)} - 2f_y \tag{4.48}$$

$$\Delta Y_{(y<y_{\mathrm{pl,below}})} = \Delta\sigma^{\mathrm{fel}}_{(y)} + 2f_y . \tag{4.49}$$

The initial strains in the longitudinal direction arise from Eq. (4.7) (without the pre-factor 3/2 because of the uniaxial approach, see Sect. 3.1.4) in V_p:

$$\Delta\varepsilon_{0(y>y_{\mathrm{pl,above}})} = \frac{1}{C}\left(\Delta\sigma^{\mathrm{fel}}_{(y)} - 2f_y\right) \tag{4.50}$$

$$\Delta\varepsilon_{0(y<y_{\mathrm{pl,below}})} = \frac{1}{C}\left(\Delta\sigma^{\mathrm{fel}}_{(y)} + 2f_y\right) \tag{4.51}$$

and therefore the material law for the MEA in the three different cross-sectional zones becomes

$$\Delta\varepsilon^*_{(y>y_{\text{pl,above}})} = \frac{\Delta\rho_{(y)}}{E_t} + \frac{1}{C}\left(\Delta\sigma^{\text{fel}}_{(y)} - 2f_y\right) \tag{4.52}$$

$$\Delta\varepsilon^*_{(y_{\text{pl,below}} \leq y \leq y_{\text{pl,above}})} = \frac{\Delta\rho_{(y)}}{E} \tag{4.53}$$

$$\Delta\varepsilon^*_{(y<y_{\text{pl,below}})} = \frac{\Delta\rho_{(y)}}{E_t} + \frac{1}{C}\left(\Delta\sigma^{\text{fel}}_{(y)} + 2f_y\right). \tag{4.54}$$

On the one hand, the residual strain range $\Delta\varepsilon^*$ must comply with the Bernoulli hypothesis that plane sections remain plane, so that it must be a linear function of y. On the other hand neither a curvature of the bar nor a cyclic change in length is admitted, so that

$$\Delta\varepsilon^*_{(y)} = 0 \tag{4.55}$$

must apply.

Equation (4.55) was actually foreseeable from the outset, because the cyclic portion of the load is strain-controlled, i.e., the elastic-plastic strain range is identical with the fictitious elastic strain range.

By inserting in Eqs. (4.52)–(4.54) we obtain the residual stress ranges in all three zones of the cross-section, calling forth ranges of residual bending moments but no ranges of residual axial forces. This is then also the exact solution.

4.3.4 Thick-Walled Cylinder Loaded by a Temperature Transient

Whereas the temperature distribution over the wall thickness of a thin-walled tube has been specified directly in the previous example, a thicker tube is now considered ($r_{\text{outside}}/r_{\text{inside}} = 1.5$). A fluid is moving through, whose temperature changes cyclically (Fig. 4.6 and Table 4.1), causing unsteady nonlinear temperature profiles across the wall thickness as a result of convective heat transfer on the inner surface (heat transfer coefficient h_f) and heat conduction through the wall. The outer surface of the tube is perfectly insulated. A location at sufficient distance from any bending perturbation is considered. There is also no axial temperature gradient. Geometry and load are axisymmetric.

Both the thermal analysis and the fictitious elastic and the modified elastic analyses are performed with ANSYS.

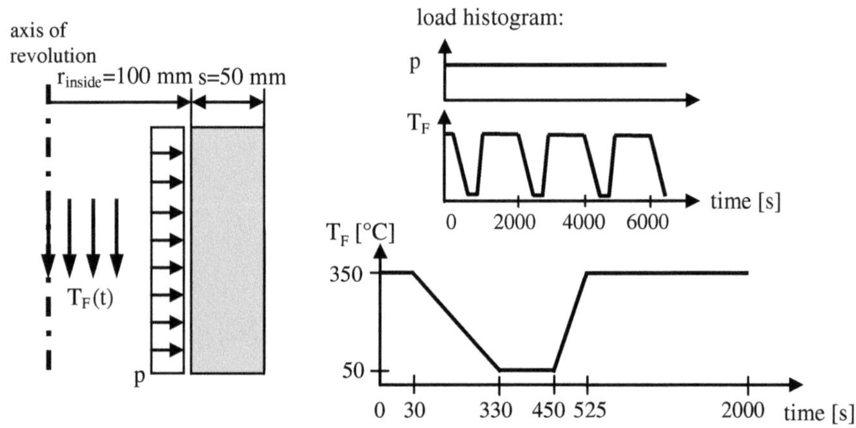

Fig. 4.6 Thick-walled tube under thermal transient

Table 4.1 Material data

Material data	
Density ρ	7.93×10^{-9} N s^2/mm^4
Thermal conductivity k	15 N/(s K)
Specific heat c	0.5×10^9 mm^2/(K s^2)
Heat transfer coefficient h_f	30 N/(K s mm)
E	200,000 N/mm^2
ν	0.3
α_T	1.6×10^{-5}/K
f_y	100 N/mm^2
E_t/E	0.05

At the end of each loading cycle, i.e., every 2,000 s, a complete temperature balance is enforced so that everywhere is 350 °C. Because no axial temperature gradient is created and possible end effects such as bending are disregarded, the analysis of an axisymmetric cross-sectional slice and thus one element layer in the axial direction, but many elements in the radial direction, is sufficient.

Figure 4.7 shows the results of the thermal analysis in the form of time histories of the temperature on the inner and outer surfaces as well as in the form of temperature distributions across the wall thickness at selected times. The temperature on the inner surface follows the fluid temperature with only a slight delay.

The greatest stresses occur shortly after each end of the down- and the up-transients. As "minimal" load condition, the time 330 s is chosen and as "maximum" load condition, 525 s. For these points in time, the stress components of a fictitious elastic calculation are displayed in Fig. 4.8. The tangential and axial stresses differ only slightly, and the radial stresses are small. There are no shear stresses.

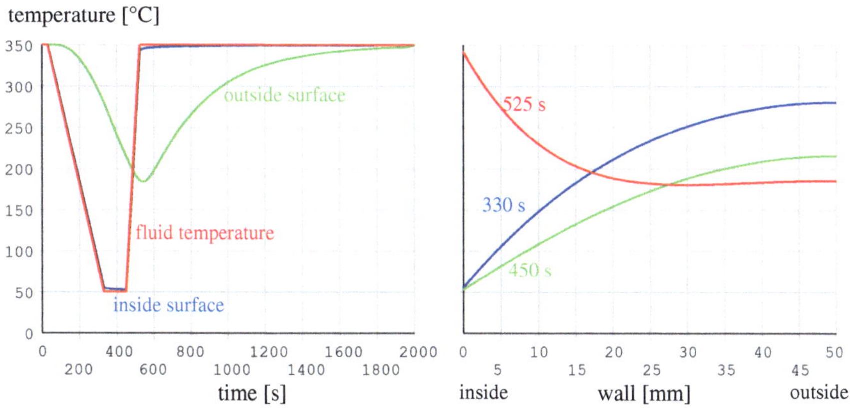

Fig. 4.7 Thermal analysis; *left* temperature histograms of fluid, inside surface, and outside surface; *right* temperature profiles across the wall at different times

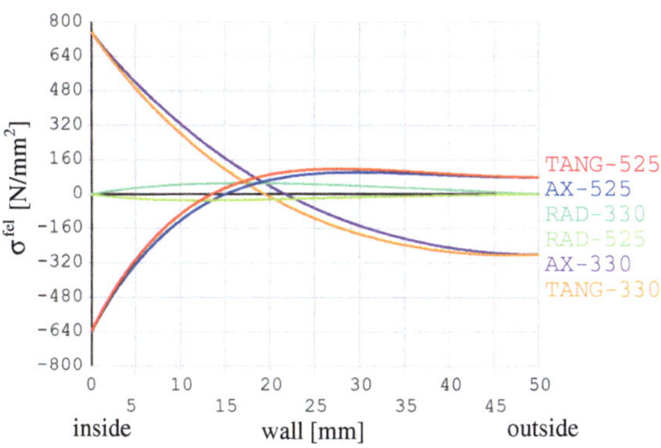

Fig. 4.8 Fictitious elastic analysis: distribution of stress components across the wall because of sole thermal loading at the minimum (330 s) and the maximum (525 s) load conditions

From the resulting ranges of the stress components, the equivalent stress ranges are formed, and thus, according to Eq. (4.12), the plastic zone V_p is estimated for the first MEA, and the associated initial strains via Eqs. (4.17) and (4.7) (see Fig. 4.9).

After performing the first MEA, an iterative improvement of the estimation of both the plastic zone V_p and the TIV in V_p is made. As a result, the elastic-plastic strain components for the first two MEAs are shown in Fig. 4.10. In further MEAs only minor changes are observed. After two MEAs the results already represent a good approximation, as is recognizable by comparison with the solution of an incremental analysis for several cycles of the thermal transient, each with numerous load steps, until shakedown is achieved (Fig. 4.11).

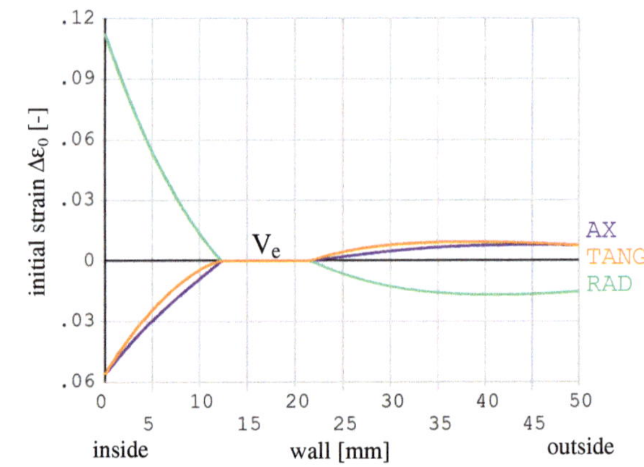

Fig. 4.9 Distribution of the range of initial strain components across the wall of the first MEA of the STPZ

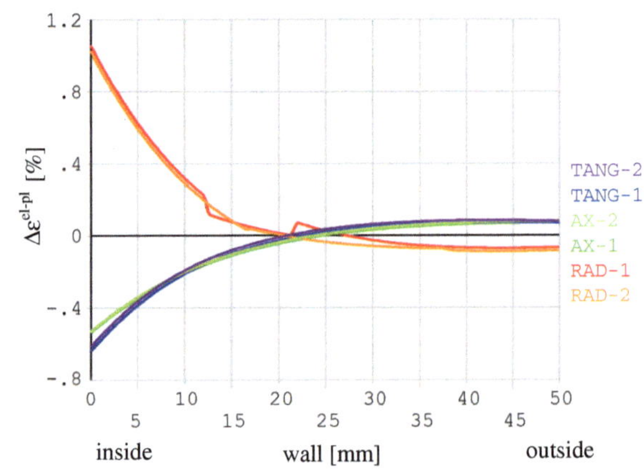

Fig. 4.10 Distribution of the ranges of the elastic-plastic strain components across the wall for different MEAs of the STPZ

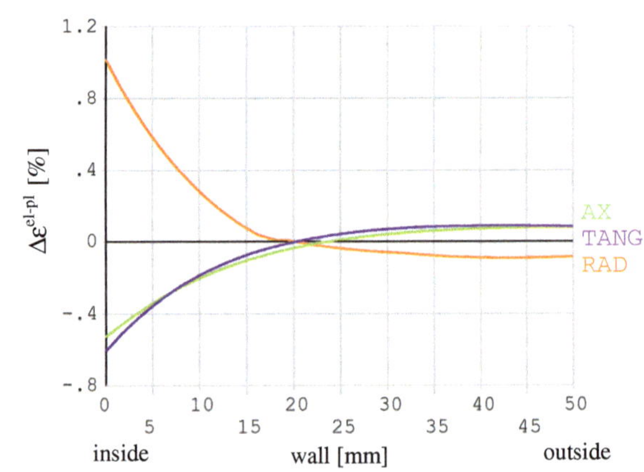

Fig. 4.11 Distribution of the ranges of the elastic-plastic strain components across the wall in the state of plastic shakedown (incremental analysis)

4.4 Accumulated Strains in the State of Elastic Shakedown

The condition for elastic shakedown (ES) has already been given by Eq. (4.3):

$$\Delta\sigma_v^{\text{fel}} \leq 2f_y \quad \forall \underline{x} \in V \rightarrow \text{ES}\,. \tag{4.56}$$

The modified elastic analyses (MEAs) for determining the residual stresses and strains in the state of elastic shakedown, especially the accumulated strains and associated stresses, are performed analogous to the procedure described in Sect. 3.1. For this purpose, the geometry of the plastic zone V_p and the initial strains to be applied there as modified loading are required.

4.4.1 Identification of the Elastic and the Plastic Subvolumes

Plastic strains that have emerged to reach the state of elastic shakedown are frozen during further load cycles, so that then only elastic changes of stress and strain occur. Consequently, the residual stresses ρ_i are as constant throughout the entire structure as the backstress ξ_i and hence the TIV Y_i (see Eq. (3.1)). As already noted in Sect. 4.1, this means that at any point of the structure an intersection area Ω must exist in the TIV-space, which is formed by the two yield surfaces belonging to the minimum and to the maximum load and is not empty (Fig. 4.1). Because of the constancy of the TIV, Y_i is definitely inside or on the edge of this intersection area. If plastic straining occurs at the location under consideration before the state of shakedown has been reached, then Y_i is located on the edge of Ω, otherwise in the interior of Ω. According to Eq. (3.3), the latter is the case when

$$Y_i = -\rho_i^{'}\,. \tag{4.57}$$

Based on these findings, the following separation of the volume V in the part remaining elastic, V_e, and the plastic zone V_p can be estimated in the nth iteration step. If the negative of the deviatoric residual stress determined in the previous iteration step

$$Y_i^{*(n)} = -\rho_i^{'(n-1)} \tag{4.58}$$

is located within Ω, the location of the structure is purely elastic and is thus allocated to V_e in the current nth iteration step. This is the case, if the Mises equivalent stress of the elastic-plastic stress components is less than the yield stress at both load conditions. Otherwise, this location belongs to V_p:

$$V_p^{(n)} = \left\{ \underline{x} \middle| \sigma_{\text{v,min}}^{(n-1)} \geq f_y \quad \vee \quad \sigma_{\text{v,max}}^{(n-1)} \geq f_y \right\} \tag{4.59}$$

$$V_e^{(n)} = \left\{ \underline{x} \middle| \sigma_{\mathrm{v,min}}^{(n-1)} < f_y \quad \wedge \quad \sigma_{\mathrm{v,max}}^{(n-1)} < f_y \right\}. \tag{4.60}$$

Provided the TIVs are known in V_p, initial strains can be determined and an MEA can be performed as described in Sects. 3.1.2 and 3.1.3 for monotonic loading.

4.4.2 Estimation of the Transformed Internal Variable

If the negative deviatoric residual stress is outside of Ω, it is obvious how to estimate the TIV, very similar to the monotonic loading (Sect. 3.5) or the strain range at cyclic loading (Sect. 4.2), by projection of the negative deviatoric residual stress, i.e., Y^*, onto the yield surface. More specifically, the projection has to be done on the intersection area Ω formed by the two yield surfaces at minimum and maximum load (see Fig. 4.12). In the first iteration step, there are still no residual stresses, so

$$Y_i^* = 0_i, \tag{4.61}$$

i.e., the origin of the coordinate system is to be projected on Ω. However, this applies only if the origin is not in itself in Ω, because the considered location is then in V_e.

How this projection is carried out in detail depends on the position of Y^* with respect to the regions ω_1–ω_4 of Fig. 4.12. In region ω_1, Y^* is to be projected on the yield surface that belongs to the maximum load, in region ω_2 on the yield surface of the minimal load, and in regions ω_3 and ω_4 on the nearest vertex of Ω.

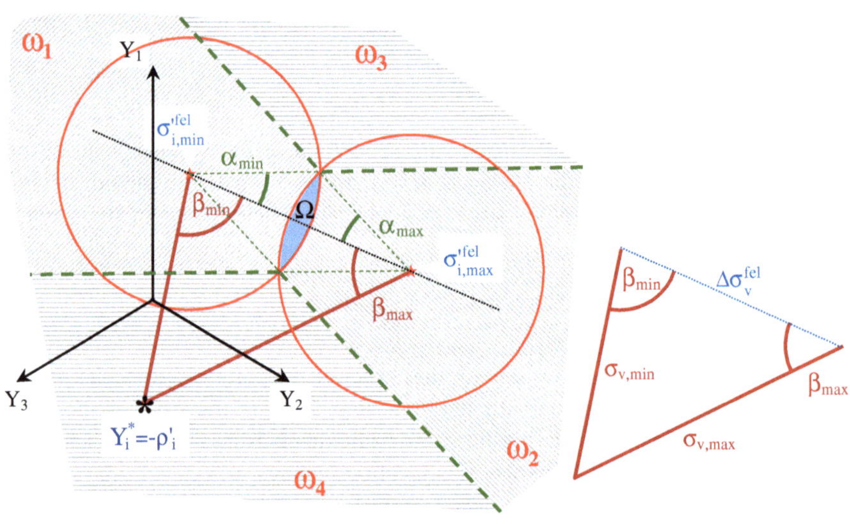

Fig. 4.12 Elastic shakedown; *left* intersection area Ω of the yield surfaces and position of Y^* in the TIV-space; *right* isolated sketch of the triangle to determine the angles $\beta_{\min}$ and $\beta_{\max}$

To demarcate the regions ω_1–ω_4 from each other, the angles $\alpha_{\min}$ and $\alpha_{\max}$ as well as $\beta_{\min}$ and $\beta_{\max}$ are introduced. These angles are measured from the line connecting the fictitious elastic deviatoric stresses of the minimum and the maximum state of loading. Because of elastic shakedown,

$$\Delta\sigma_{\mathrm{v}}^{\mathrm{fel}} = \Delta\sigma_{\mathrm{v}} \, . \tag{4.62}$$

Provided the yield stresses are identical in both load conditions we have

$$\cos(\alpha_{\min}) = \cos(\alpha_{\max}) = \frac{\Delta\sigma_{\mathrm{v}}^{\mathrm{fel}}}{2f_y} \, . \tag{4.63}$$

The angles $\beta_{\min}$ and $\beta_{\max}$ can be determined by the cosine theorem (see Fig. 4.12, right):

$$\cos(\beta_{\min}) = \frac{\Delta\sigma_{\mathrm{v}}^2 + \sigma_{\mathrm{v,min}}^2 - \sigma_{\mathrm{v,max}}^2}{2 \cdot \Delta\sigma_{\mathrm{v}} \cdot \sigma_{\mathrm{v,min}}} \tag{4.64}$$

$$\cos(\beta_{\max}) = \frac{\Delta\sigma_{\mathrm{v}}^2 + \sigma_{\mathrm{v,max}}^2 - \sigma_{\mathrm{v,min}}^2}{2 \cdot \Delta\sigma_{\mathrm{v}} \cdot \sigma_{\mathrm{v,max}}} \, . \tag{4.65}$$

$\sigma_{\mathrm{v,min}}$ and $\sigma_{\mathrm{v,max}}$ are known from the previous iteration step. $\alpha_{\min}$ and $\alpha_{\max}$ vary between 0° and 90°, $\beta_{\min}$ and $\beta_{\max}$ between 0° and 180°.

4.4.2.1 Y^* Is in ω_2

This is true when

$$\beta_{\min} < \alpha_{\min} \quad \text{and} \quad \sigma_{\mathrm{v,min}} > f_y \, . \tag{4.66}$$

Then the estimate of Y is obtained by projection of Y^* on the Mises circle of the minimum load condition (Fig. 4.13):

$$Y_i = \sigma_{i,\min}^{\prime\mathrm{fel}} - \sigma_{i,\min}^{\prime} \left(\frac{f_y}{\sigma_{\mathrm{v,min}}} \right) . \tag{4.67}$$

4.4.2.2 Y^* Is in ω_1

This is the case if

$$\beta_{\max} < \alpha_{\max} \quad \text{and} \quad \sigma_{\mathrm{v,max}} > f_y \, . \tag{4.68}$$

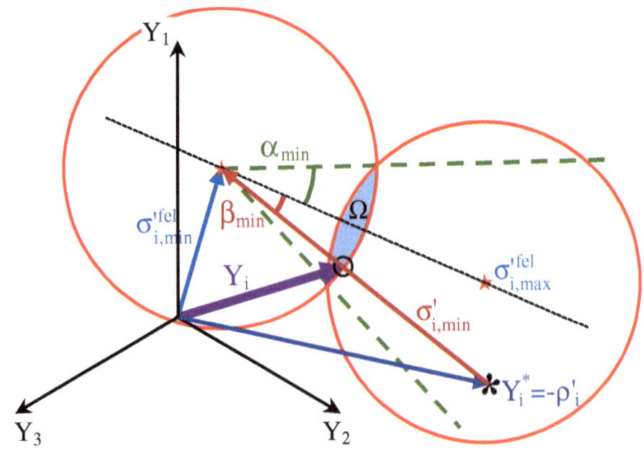

Fig. 4.13 Projection on the Mises circle of the minimum load condition

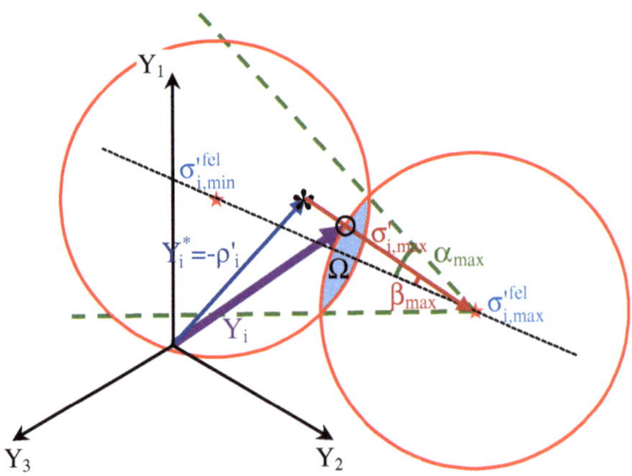

Fig. 4.14 Projection on the Mises circle of the maximum load condition

Then the estimate of Y is gained by projection of Y^* on the Mises circle of the maximum load condition (Fig. 4.14):

$$Y_i = \sigma_{i,\max}^{\prime\text{fel}} - \sigma_{i,\max}^{\prime}\left(\frac{f_y}{\sigma_{\text{v},\max}}\right). \tag{4.69}$$

4.4.2.3 Y^* Is in ω_3 or ω_4

This is the case if

$$\beta_{\min} \geq \alpha_{\min} \quad \text{and} \quad \beta_{\max} \geq \alpha_{\max}. \tag{4.70}$$

Then the estimate of Y is obtained by projection of Y^* on the intersection points of the two Mises circles of the minimum and the maximum loading state, i.e., on the

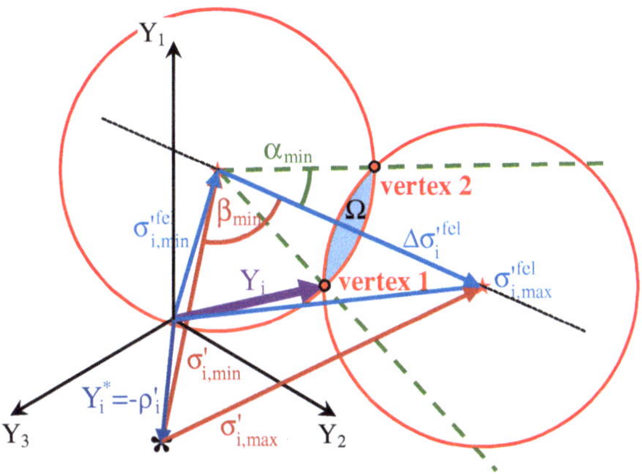

Fig. 4.15 Projection on the vertices of the intersection area Ω

vertices of the area Ω (Fig. 4.15). The location of these vertices can be expressed by linear combinations of known vectors. The following formulation turned out to be convenient (see Fig. 4.16):

$$Y_{i,\text{vertex1}} = \sigma_{i,\min}^{\prime\text{fel}} - a \cdot \sigma_{i,\min}^{\prime} + b \cdot \Delta\sigma_i^{\prime\text{fel}} \tag{4.71}$$

$$Y_{i,\text{vertex2}} = \sigma_{i,\min}^{\prime\text{fel}} + a \cdot \sigma_{i,\min}^{\prime} + c \cdot \Delta\sigma_i^{\prime\text{fel}}\,. \tag{4.72}$$

The scalar parameters a, b, and c can be identified by virtue of the two colored triangles in Fig. 4.17 employing the sinus theorem. Because $\cos(\alpha_{\min})$ and $\cos(\beta_{\min})$ are already known from Eqs. (4.63) and (4.64), the following notation makes sense:

$$a = \frac{f_y}{\sigma_{\text{v},\min}} \cdot \frac{\sqrt{1-\cos^2\alpha_{\min}}}{\sqrt{1-\cos^2\beta_{\min}}} \tag{4.73}$$

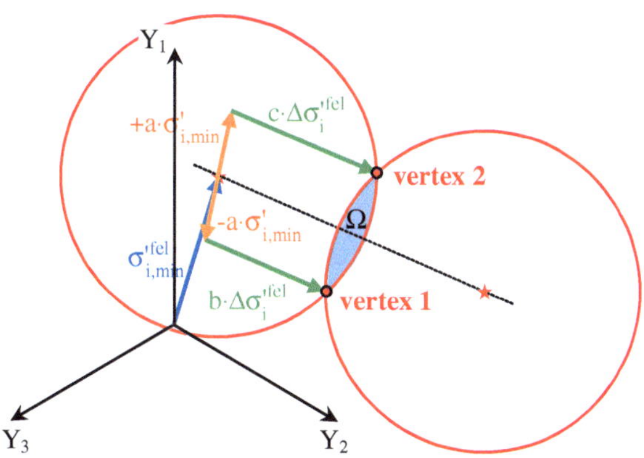

Fig. 4.16 For determining the location of the vertices of the area Ω

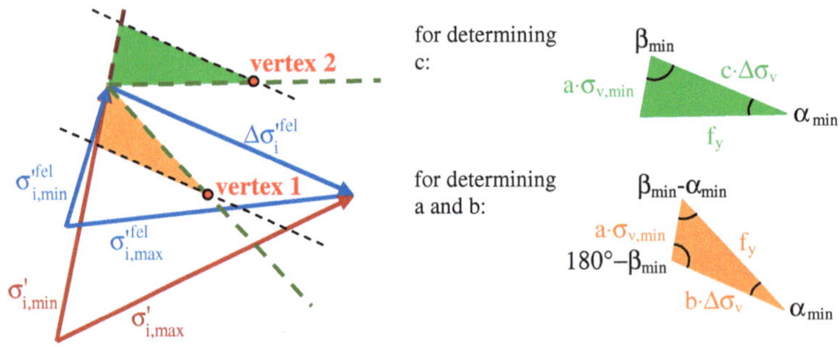

Fig. 4.17 For determining the factors a, b, and c

$$b = \frac{f_y}{\Delta\sigma_v} \cdot \left(\cos\alpha_{\min} - \cos\beta_{\min} \frac{\sqrt{1-\cos^2\alpha_{\min}}}{\sqrt{1-\cos^2\beta_{\min}}} \right) \tag{4.74}$$

$$c = \frac{f_y}{\Delta\sigma_v} \cdot \left(\cos\alpha_{\min} + \cos\beta_{\min} \frac{\sqrt{1-\cos^2\alpha_{\min}}}{\sqrt{1-\cos^2\beta_{\min}}} \right). \tag{4.75}$$

Because of the requirements 180° ≥ β_{min} ≥ α_{min} ≥0°, the parameters a and b are always ≥ 0. The factor c may be negative, however. For this reason, Y^* is always closer to vertex 1 than to vertex 2, so that the projection is always carried out on vertex 1 according to Eqs. (4.71) and (4.72), and therefore the factor c and the region ω_3 are not really needed.

Although the decision as to whether the TIV is in a vertex of the intersection area Ω can change during an iteration process, the location of the vertex itself cannot change and can therefore be determined solely on the basis of a fictitious elastic calculation:

$$Y_{i,\text{vertex}} = (1-a) \cdot \sigma_{i,\min}^{\prime\text{fel}} + b \cdot \Delta\sigma_i^{\prime\text{fel}}, \tag{4.76}$$

where β_{min} and β_{max} according to Eqs. (4.64) and (4.65) as well as a and b according to Eqs. (4.73) and (4.74) need only be determined with the fictitious elastic calculated stresses.

4.5 Examples of Accumulated Strains at Elastic Shakedown

In the following sections the calculation process for estimating the accumulated strain in the state of elastic shakedown is explained by way of some examples, and the quality of the results obtained with the STPZ and the computational effort required is assessed by comparison with incremental analyses.

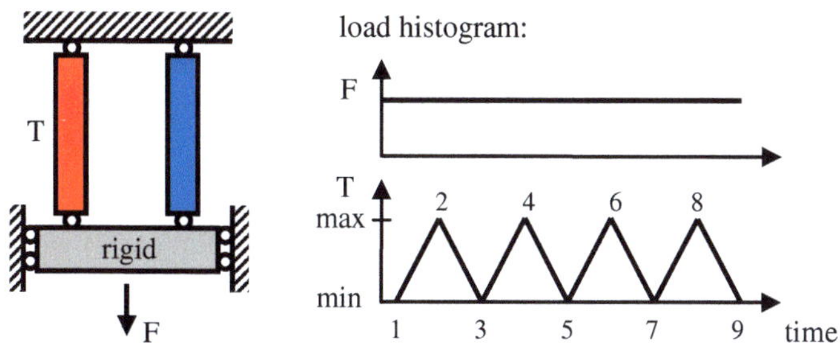

Fig. 4.18 Two-bar model

It should be noted that Khalij et al. [7] have analyzed the elastic shakedown behavior of a hook under cyclic loading using their implementation of Zarka's method in the FE program SYSTUS, whereas Cano and Taheri [8] have studied Zarka's method with respect to elastic shakedown by investigating a tube with variable thickness subjected to internal pressure and a cyclic thermal transient.

4.5.1 Two-Bar Model

The two-bar model (Fig. 4.18) has already been treated in Sects. 2.2–2.4 and examined in Sect. 4.3.1 with respect to the strain range in the state of plastic shakedown.[1]

The external force F causes a so-called primary stress σ_P in both bars. The temperature T applied in the left bar evokes a cyclic fictitious elastic calculated so-called secondary stress with the maximum value σ_t of the same absolute value in the two bars, but opposite signs. At the point in time of the "minimal" state of loading, the fictitious elastic calculated stresses are

$$\sigma_{\text{le,min}}^{\text{fel}} = \sigma_P \tag{4.77}$$

$$\sigma_{\text{ri,min}}^{\text{fel}} = \sigma_P \tag{4.78}$$

and at the point in time of the "maximum" state of loading

$$\sigma_{\text{le,max}}^{\text{fel}} = \sigma_P - \sigma_t \tag{4.79}$$

$$\sigma_{\text{ri,max}}^{\text{fel}} = \sigma_P + \sigma_t \,. \tag{4.80}$$

[1]Jiang and Leckie obtained a solution for different cross-sectional areas and different lengths of the two bars in [9] adopting Zarka's method.

According to Eq. (4.56), elastic shakedown is guaranteed as long as the absolute value of the fictitious elastic stress range does not exceed twice the yield stress:

$$\left|\Delta\sigma_{\mathrm{le/ri}}^{\mathrm{fel}}\right| = \sigma_t \leq 2f_y \,. \tag{4.81}$$

A first estimate of the plastic zone $V_p^{(1)}$ according to Eq. (4.57) yields:

$$\rho_{\mathrm{le}}^{(0)} = \rho_{\mathrm{ri}}^{(0)} \tag{4.82}$$

$$\sigma_{\mathrm{le,min}}^{(0)} = \sigma_{\mathrm{ri,min}}^{(0)} = \sigma_{\mathrm{le,min}}^{\mathrm{fel}} = \sigma_{\mathrm{ri,min}}^{\mathrm{fel}} \tag{4.83}$$

$$\sigma_{\mathrm{le,max}}^{(0)} = \sigma_{\mathrm{le,max}}^{\mathrm{fel}} \quad ; \quad \sigma_{\mathrm{ri,max}}^{(0)} = \sigma_{\mathrm{ri,max}}^{\mathrm{fel}} \tag{4.84}$$

$$\sigma_{\mathrm{v,le,max}}^{(0)} = |\sigma_P - \sigma_t| \geq f_y \quad \rightarrow \quad \text{left bar } V_p^{(1)};\ \text{otherwise } V_e^{(1)} \tag{4.85}$$

$$\sigma_{\mathrm{v,ri,max}}^{(0)} = |\sigma_P + \sigma_t| \geq f_y \quad \rightarrow \quad \text{right bar } V_p^{(1)} \,. \tag{4.86}$$

The primary stress σ_P must not exceed the yield stress, otherwise the plastic limit load of the system would be exceeded. Therefore, it is excluded that one of the two bars can plasticize at the minimum load condition. If σ_P and σ_t are positive, the right bar plasticizes first and thereby undergoes plastic tensile strains. If Eq. (4.85) is satisfied, the left bar undergoes plastic compressive straining. The elastic limit load is given by

$$\sigma_P + \sigma_t = f_y \,. \tag{4.87}$$

According to Eqs. (4.64) and (4.65) we get with respect to $\beta_{\min}$:

$$\cos(\beta_{\min}) = +1 \quad \text{in the left bar} \tag{4.88}$$

$$\cos(\beta_{\min}) = -1 \quad \text{in the right bar} \tag{4.89}$$

and thus

$$\beta_{\min} = 0° \quad \text{in the left bar} \tag{4.90}$$

$$\beta_{\min} = 180° \quad \text{in the right bar,} \tag{4.91}$$

and with respect to $\beta_{\max}$:

if $\sigma_P < \sigma_t$, so that Eqs. (4.77) and (4.79) have different signs:

$$\cos(\beta_{\max}) = -1 \quad \text{in the left bar} \tag{4.92}$$

or, if $\sigma_P > \sigma_t$, because then Eqs. (4.77) and (4.79) have the same sign:

$$\cos(\beta_{\max}) = +1 \quad \text{in the left bar} \tag{4.93}$$

$$\cos(\beta_{\max}) = +1 \quad \text{in the right bar} \tag{4.94}$$

and thus

$$\beta_{\max} = 180^\circ \quad \text{in the left bar if } \sigma_P < \sigma_t \tag{4.95}$$

$$\beta_{\max} = 0^\circ \quad \text{in the left bar if } \sigma_P < \sigma_t \tag{4.96}$$

$$\beta_{\max} = 0^\circ \quad \text{in the right bar,} \tag{4.97}$$

which is evident from Fig. 4.12 because of the uniaxial stress state, cf. Fig. 4.19. For illustrative purposes, $\alpha_{\min}$ and $\alpha_{\max}$ are also depicted there, although the vertices

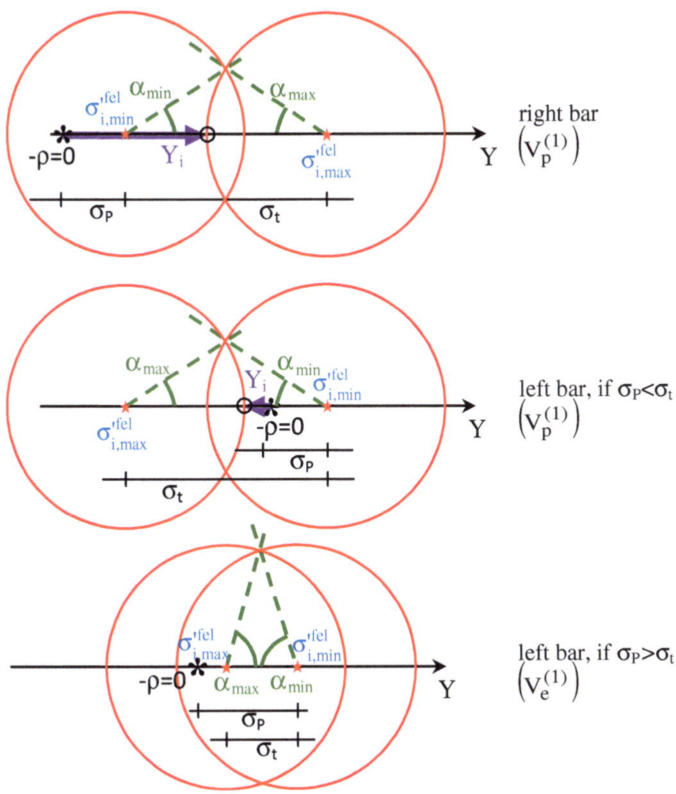

Fig. 4.19 Two-bar model: TIV-space at elastic shakedown, Y_i for the first MEA; *top* right bar; *middle* left bar, if $\sigma_P < \sigma_t$; *bottom* left bar, if $\sigma_P > \sigma_t$

of the area Ω are not required here because of the uniaxial stress state. It is sufficient to register for the formal queries that $\alpha_{\min} > 0$ and $\alpha_{\max} > 0$. The bottom sketch of Fig. 4.19 illustrates that at $\sigma_P > \sigma_t$ the left bar is to be assigned to the elastic subvolume V_e, because the residual stress is within the area Ω (note that Eq. (4.85) is not met).

The TIV is obtained according to the different cases distinguished in Sects. 4.4.2.1–4.4.2.3. The case examined in Sect. 4.4.2.3 fails immediately, because the vertices of Ω, as already mentioned, cannot be addressed here, the first condition of Eq. (4.70) being violated in the left bar and the second condition in the right bar. Section 4.4.2.1 also does not apply because the second condition in Eq. (4.66) is violated in both bars. However, both conditions in Eq. (4.67) are met for the right bar, as well as for the left bar, provided it is assigned to the plastic zone V_p according to Eq. (4.85). Accordingly, Y^* of both bars is in region ω_1, so that the negative residual stress, i.e., the origin of the coordinate system in the first iteration, is to be projected onto the Mises circle centered in the maximum stress; cf. Fig. 4.19.

For the following considerations a distinction is made in regard to the first MEA, according to whether the left bar has been or has not been assigned to the plastic zone after Eq. (4.85) (Sects. 4.5.1.1 and 4.5.1.2).

4.5.1.1 Left Bar Is Assigned to the Elastic Zone During the First MEA

If the left bar is located in V_e, the modified elastic material law applies:

$$\varepsilon_{\text{le}}^{*(1)} = \frac{1}{E}\rho_{\text{le}}^{(1)} \,. \tag{4.98}$$

For the right bar, which is definitely assigned to V_p, the origin of the coordinate system is in region ω_1 of the TIV-space, and the TIV is therefore obtained by projection on the Mises circle that belongs to the maximum load:

$$Y_{\text{ri}}^{(1)} = \sigma_P + \sigma_t - f_y \,. \tag{4.99}$$

With the initial strains according to Eq. (3.40), the modified elastic material law of Eq. (3.39) with E^* according to Eq. (3.15) becomes

$$\varepsilon_{\text{ri}}^{*(1)} = \frac{1}{E_t}\rho_{\text{ri}}^{(1)} + \frac{1}{C}\left(\sigma_P + \sigma_t - f_y\right) . \tag{4.100}$$

With the field equations for the MEA, namely the equilibrium condition

$$\rho_{\text{ri}} = -\rho_{\text{le}} \tag{4.101}$$

and the compatibility condition (the displacement field must be compatible with the boundary conditions)

$$\varepsilon_{\text{le}}^{*} = \varepsilon_{\text{ri}}^{*} \tag{4.102}$$

we obtained as a result of the MEA for the left bar, which should suffice here:

$$\rho_{\text{le}}^{(1)} = \frac{\left(1 - \frac{E_t}{E}\right)}{\left(1 + \frac{E_t}{E}\right)} \left(\sigma_P + \sigma_t - f_y\right) \tag{4.103}$$

$$\varepsilon_{\text{le}}^{*(1)} = \frac{\left(1 - \frac{E_t}{E}\right)}{\left(1 + \frac{E_t}{E}\right)} \frac{\left(\sigma_P + \sigma_t - f_y\right)}{E} \,. \tag{4.104}$$

The stress in the elastic-plastic state is obtained by superposition of residual stresses with the fictitious elastic stresses, both for the minimum as well as for the maximum state of loading:

$$\sigma_{\text{le,min}}^{(1)} = \sigma_P + \rho_{\text{le}}^{(1)} \tag{4.105}$$

$$\sigma_{\text{le,max}}^{(1)} = \sigma_P - \sigma_t + \rho_{\text{le}}^{(1)} \,. \tag{4.106}$$

Now it must be checked whether the assignment of the left bar to V_e according to Eq. (4.85) was correct, i.e., whether

$$\sigma_{\text{v,le,min}}^{(1)} = \left|\sigma_P + \rho_{\text{le}}^{(1)}\right| < f_y \tag{4.107}$$

and

$$\sigma_{\text{v,le,max}}^{(1)} = \left|\sigma_P - \sigma_t + \rho_{\text{le}}^{(1)}\right| < f_y \,. \tag{4.108}$$

If this is the case, we already have the exact results by means of the results of the MEA in Eqs. (4.103) and (4.104). Otherwise, this solution is not correct and a second MEA is to be carried out by assigning the left bar now to V_p instead of V_e ($V_e^{(1)} \rightarrow V_p^{(2)}$). This is required when

$$\frac{\sigma_t}{f_y} > \left(1 - \frac{\sigma_P}{f_y}\right) \frac{2}{1 - \frac{E_t}{E}} \,. \tag{4.109}$$

The negative residual stress according to Eq. (4.103) is then located in region ω_2 of the TIV-space. The TIV in the left bar is then determined through projection on the Mises circle, which belongs to the minimum load condition:

$$Y_{\mathrm{le}}^{(2)} = \sigma_P - f_y, \tag{4.110}$$

resulting in the modified elastic material law for the left bar

$$\varepsilon_{\mathrm{le}}^{*(2)} = \frac{1}{E_t}\rho_{\mathrm{le}}^{(2)} + \frac{1}{C}\left(\sigma_P - f_y\right). \tag{4.111}$$

For the right bar, Eq. (4.100) remains valid and with the field Eqs. (4.101) and (4.102) the result of the second MEA becomes

$$\rho_{\mathrm{le}}^{(2)} = \left(1 - \frac{E_t}{E}\right)\frac{\sigma_t}{2} \tag{4.112}$$

$$\varepsilon_{\mathrm{le}}^{*(2)} = \frac{1 - \frac{E_t}{E}}{E_t}\left(\sigma_P + \frac{\sigma_t}{2} - f_y\right). \tag{4.113}$$

The obligatory check, in which TIV region the bars now reside, reveals that the left bar remains in any case in region ω_2, and the right bar in any case in region ω_1. Thus, there is no further change in the assignment of the two bars to V_p and no change of Y, so that the final and exact solution is found with Eqs. (4.112) and (4.113).

4.5.1.2 Left Bar Is Assigned to the Plastic Zone During the First MEA

Both bars are in V_p, and both in the TIV-region ω_1. Thus Y is obtained by projecting the origin of the coordinate system to the Mises circle associated with the maximum load condition in each bar:

$$Y_{\mathrm{le}}^{(1)} = \sigma_P - \sigma_t + f_y \tag{4.114}$$

$$Y_{\mathrm{ri}}^{(1)} = \sigma_P + \sigma_t - f_y \tag{4.115}$$

and the modified elastic material law reads

$$\varepsilon_{\mathrm{le}}^{*(1)} = \frac{1}{E_t}\rho_{\mathrm{le}}^{(1)} + \frac{1}{C}\left(\sigma_P - \sigma_t + f_y\right) \tag{4.116}$$

$$\varepsilon_{\mathrm{ri}}^{*(1)} = \frac{1}{E_t}\rho_{\mathrm{ri}}^{(1)} + \frac{1}{C}\left(\sigma_P + \sigma_t - f_y\right). \tag{4.117}$$

With Eqs. (4.101) and (4.102), we obtain as a result of the first MEA for the left bar:

$$\rho_{\text{le}}^{(1)} = \left(1 - \frac{E_t}{E}\right)(\sigma_t - f_y) \tag{4.118}$$

$$\varepsilon_{\text{le}}^{*(1)} = \left(1 - \frac{E_t}{E}\right)\frac{\sigma_P}{E_t} \, . \tag{4.119}$$

Now it must be examined whether the assignment of the left bar to V_p according to Eq. (4.85) and the assignment to the TIV-region ω_1 was correct. The left bar is then in V_p, if at least one of the following conditions is met:

$$\sigma_{\text{v,le,min}}^{(1)} = \left|\sigma_P + \rho_{\text{le}}^{(1)}\right| \geq f_y \tag{4.120}$$

$$\sigma_{\text{v,le,max}}^{(1)} = \left|\sigma_P - \sigma_t + \rho_{\text{le}}^{(1)}\right| \geq f_y \, . \tag{4.121}$$

Equation (4.121) is satisfied if

$$\frac{\sigma_t}{f_y} \geq 1 + \frac{\sigma_P}{f_y}\frac{E}{E_t} \, . \tag{4.122}$$

Then the left bar remains in the TIV-region ω_1, so that a further MEA would not change the results, which is, however, also unnecessary, because Eqs. (4.118) and (4.119) provide the correct solution.

Equation (4.120) is satisfied if

$$\frac{\sigma_t}{f_y} \geq \frac{2 - \frac{E_t}{E} - \frac{\sigma_P}{f_y}}{1 - \frac{E_t}{E}} \, . \tag{4.123}$$

Then the left bar is to be assigned to the TIV-region ω_2. Thus the solutions in Eqs. (4.118) and (4.119) that are based on the TIV-region ω_1 still do not represent the final result and another MEA must be executed. The solution for the MEA corresponding to this situation is already specified in Eqs. (4.112) and (4.113).

If, however, neither of the two conditions of Eq. (4.107) is satisfied, then the left bar is to be allocated to V_e ($V_p^{(1)} \rightarrow V_e^{(2)}$), and a new MEA is to be performed. This is the case for parameters outside the range given by Eqs. (4.122) and (4.123). The solution of the second MEA is then given by Eqs. (4.103) and (4.104). Adopting this solution, it is to be re-examined whether neither Eq. (4.120) nor Eq. (4.121) is still complied with. It turns out that in the parameter range

$$\frac{2 - \frac{E_t}{E} - \frac{\sigma_P}{f_y}}{1 - \frac{E_t}{E}} > \frac{\sigma_t}{f_y} > \left(1 - \frac{\sigma_P}{f_y}\right)\frac{2}{1 - \frac{E_t}{E}} \tag{4.124}$$

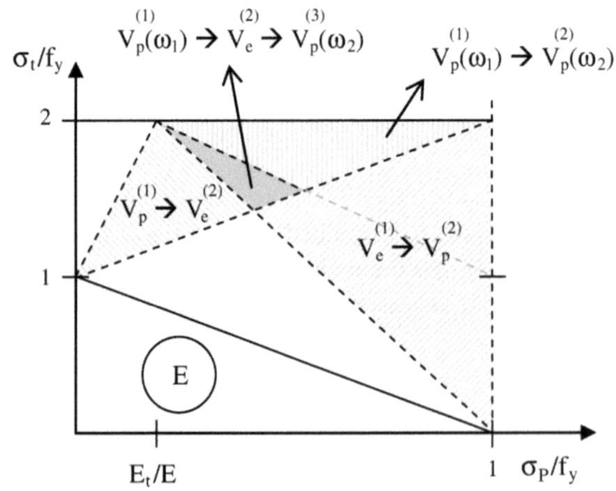

Fig. 4.20 Two-bar model: regions of parameters, where a second MEA (*hatched*) or a third MEA (*gray*) is required to obtain the exact results at elastic shakedown, marking the assignments of the left bar

Eq. (4.120) is met, so a third MEA must be carried out, the result of which is already specified in Eqs. (4.112) and (4.113).

After the third MEA the exact result is always achieved. In a sort of ratcheting interaction diagram (cf. Fig. 2.7), Fig. 4.20 indicates the parameter constellations that require a second or a third MEA.

Figure 4.21 shows the scopes of the three different solutions for the accumulated strains.

As a numerical example, the parameter constellation already examined in Sect. 2.3 is selected ($\sigma_P/f_y = 0.8$; $\sigma_t/f_y = 0.9$; $E_t/E = 0.05$). The calculation procedure is shown in Fig. 4.22.

This result coincides exactly with the incremental solution in Fig. 2.5 (see Fig. 4.23). Two linear analyses were sufficient with the STPZ (the two MEAs), apart from the two mandatory fictitious elastic analyses for the extremes of the two

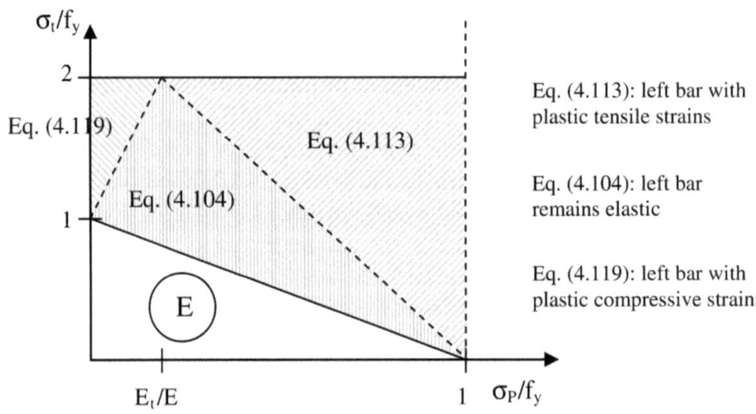

Fig. 4.21 Two-bar model: scopes of different solutions for the strain accumulation at elastic shakedown

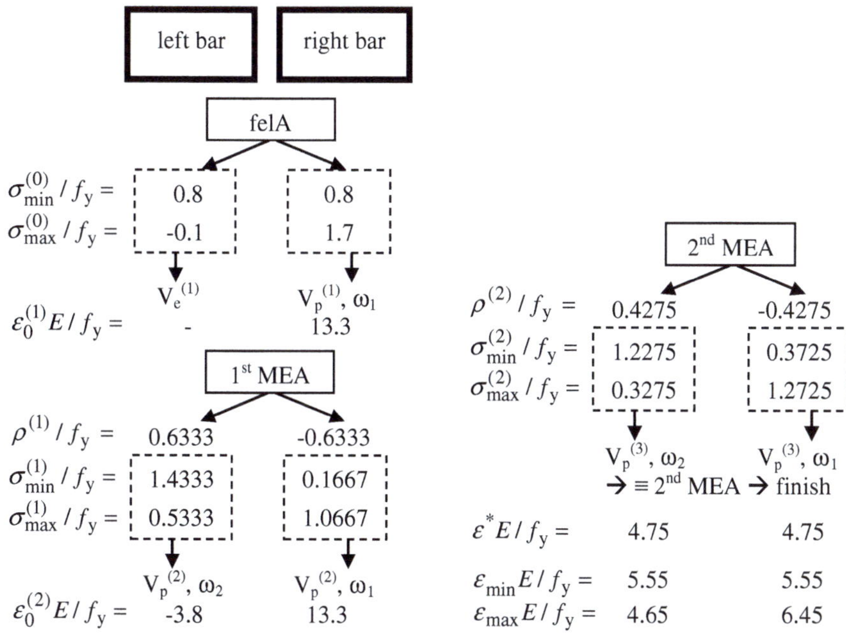

Fig. 4.22 Two-bar model: calculation sequence for numerical example ES

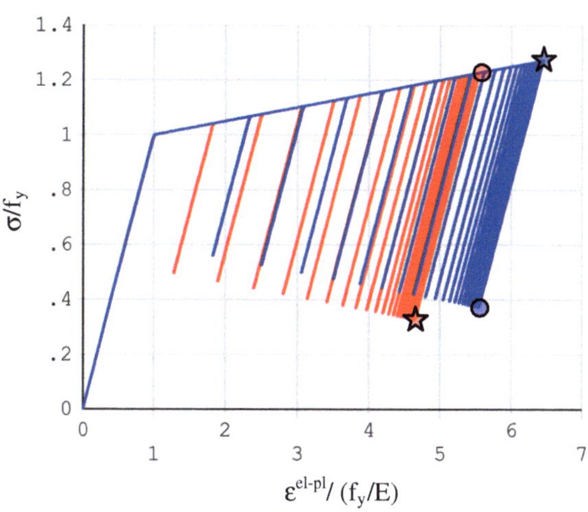

Fig. 4.23 Strain accumulation in the two-bar model with linear kinematic hardening: elastic shakedown ($\sigma_P/f_y = 0.8$; $\sigma_t/f_y = 0.9$; $E_t/E = 0.05$); comparison of the incremental analysis (as in Fig. 2.5) and the STPZ (*circles* for minimum and *stars* for maximum states of loading)

loading conditions. With the incremental analysis, about 30 cycles had to be analyzed for a good approximation to the state of elastic shakedown, which is reached strictly only after an infinite number of cycles. Having chosen two loadsteps per cycle, each with several equilibrium iterations, the computational equivalent of several hundred linear analyses had to be used.

4.5.2 Multiaxial Ratcheting

The example of multiaxial ratcheting, already presented in Sect. 2.5.2 (see Fig. 4.24), was investigated in Sect. 4.3.2 with respect to determining the strain range at plastic shakedown. Herbland et al. [10] analyzed a similar situation with different methods, including Zarka's method and various FEM implementations.

As minimum load condition, the sole action of the external force F is defined, so that at the maximum load condition the maximum value of the displacement u is applied simultaneously in addition to the force F.

To illustrate the calculation procedure, we limit ourselves to the Poisson's ratio $\nu = 0$ because of the simpler notation, so that lateral straining effects are only produced by plasticizing. In direction (1) the stress-controlled primary stress σ_P is acting, and in direction (2) the strain-controlled fictitious elastic calculated secondary stress σ_t.

The fictitious elastic calculated stresses for the minimum state amount to

$$\sigma^{\text{fel}}_{\text{min}} = \sigma_P \begin{pmatrix} 1 \\ 0 \\ 0 \end{pmatrix}; \quad \sigma'^{\text{fel}}_{\text{min}} = \sigma_P \begin{pmatrix} \frac{2}{3} \\ -\frac{1}{3} \\ -\frac{1}{3} \end{pmatrix} \tag{4.125}$$

$$\sigma^{\text{fel}}_{\text{v,min}} = \sigma_P \tag{4.126}$$

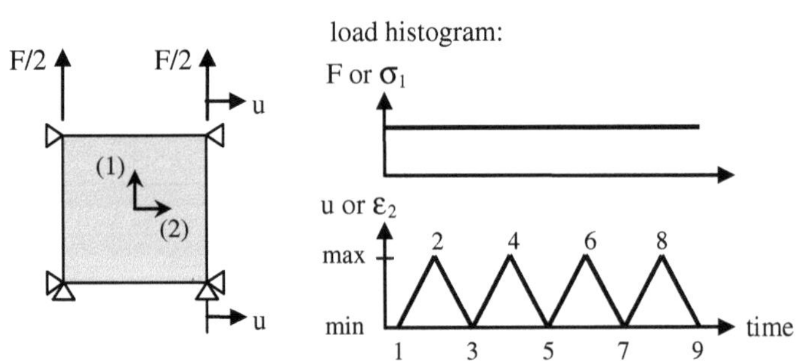

Fig. 4.24 One element under plane stress

and for the maximum state to

$$\sigma_{\max}^{\mathrm{fel}} = \sigma_P \begin{pmatrix} 1 \\ 0 \\ 0 \end{pmatrix} + \sigma_t \begin{pmatrix} 0 \\ 1 \\ 0 \end{pmatrix}; \quad \sigma_{\max}^{\prime\mathrm{fel}} = \sigma_P \begin{pmatrix} \frac{2}{3} \\ -\frac{1}{3} \\ -\frac{1}{3} \end{pmatrix} + \sigma_t \begin{pmatrix} -\frac{1}{3} \\ \frac{2}{3} \\ -\frac{1}{3} \end{pmatrix} \tag{4.127}$$

$$\sigma_{\mathrm{v,max}}^{\mathrm{fel}} = \sqrt{\sigma_P^2 + \sigma_t^2 - \sigma_P \sigma_t}\,. \tag{4.128}$$

The elastic limit load is therefore given by

$$\sigma_t^{\mathrm{el\ limit}} = \frac{1}{2}\sigma_P + \sqrt{f_y^2 - \frac{3}{4}\sigma_P^2}\,. \tag{4.129}$$

Because the plastic limit load must not be exceeded, it is required that

$$|\sigma_P| \le \frac{2}{\sqrt{3}} f_y\,. \tag{4.130}$$

According to Eq. (4.56), elastic shakedown is guaranteed as long as the fictitious elastic equivalent stress range does not exceed twice the yield strength. If restricted to positive secondary stresses:

$$\sigma_t \le 2 f_y\,. \tag{4.131}$$

Because of the homogeneous stress state, the plastic zone V_p is known in advance to consist of the entire element.

Therefore, the following relates exclusively to how the TIV Y_i are estimated, so that appropriate initial strains for the MEA can be defined.

Based on the fictitious elastic analyses, the required angles are calculated from Eqs. (4.63)–(4.65) (see Fig. 4.25):

$$\cos(\alpha_{\min}) = \cos(\alpha_{\max}) = \frac{\sigma_t}{2 f_y} \tag{4.132}$$

and for the first MEA:

$$\cos(\beta_{\min}) = \frac{1}{2}\operatorname{sgn}(\sigma_P)\,, \tag{4.133}$$

so that $\beta_{\min}$ is independent of the load level and only depends on the sign of the primary stress and thus is $\beta_{\min} = 60°$ or $120°$, and

$$\cos(\beta_{\max}) = \frac{2\sigma_t - \sigma_P}{2\sqrt{\sigma_P^2 + \sigma_t^2 - \sigma_P \sigma_t}}\,. \tag{4.134}$$

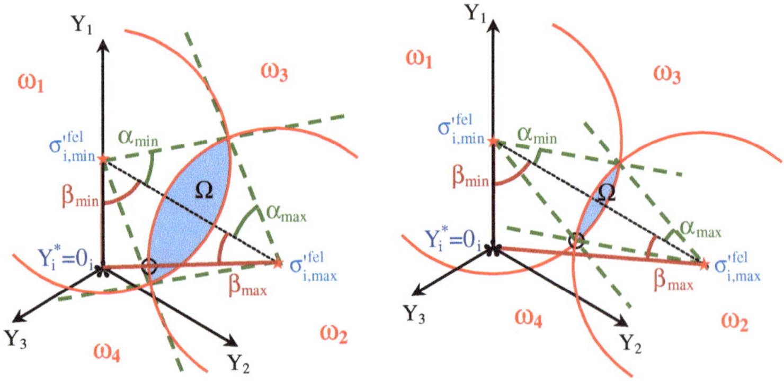

Fig. 4.25 Multiaxial ratcheting: intersection area Ω in the TIV-space; *left* projection onto the edge of the circle; *right* projection onto a vertex of Ω

So, it can now be found in which of the TIV regions ω_1–ω_4 Y^* resides, so that the appropriate projection on the intersection area Ω can be selected. The second condition in Eq. (4.66) cannot be met because of Eq. (4.130), so that the region ω_2 can be ruled out. Whether region ω_1 or ω_4 applies depends on the load constellation. The second condition in Eq. (4.68) is already fulfilled, because otherwise there would be elastic behavior. Y^* is therefore in ω_1, when

$$\beta_{\max} < \alpha_{\max} \,, \tag{4.135}$$

which is the case if

$$\frac{s_t\left(s_t^2-4\right)-s_t^2\sqrt{3\left(4-s_t^2\right)}}{2\left(s_t^2-1\right)} < \frac{\sigma_P}{f_y} < \frac{s_t\left(s_t^2-4\right)+s_t^2\sqrt{3\left(4-s_t^2\right)}}{2\left(s_t^2-1\right)} \tag{4.136}$$

with the abbreviation s_t for the normalized secondary stress

$$s_t = \frac{\sigma_t}{f_y} \,. \tag{4.137}$$

Depending on whether the condition of Eq. (4.135) or Eq. (4.136) is complied with or not, the rest of the procedure is different.

4.5.2.1 Element Is Located in the TIV-Region ω_4

If Eq. (4.135) is not met, the projection of Y^* onto the intersection area Ω in the TIV-space leads to the vertex of Ω because $\beta_{\max} > \alpha_{\max}$.

The position of the vertices of the intersection area Ω, formed in the TIV-space by the minimum and maximum state of loading, can be determined by Eqs. (4.71), (4.73), and (4.74). In the first iteration we get

$$\sigma_{i,\min}^{\prime\text{fel}} = \sigma_{i,\min}^{\prime} \tag{4.138}$$

$$a = \frac{f_y}{|\sigma_P|} \cdot \frac{2}{\sqrt{3}} \sqrt{1 - \left(\frac{\sigma_t}{2f_y}\right)^2} \tag{4.139}$$

$$b = \frac{1}{2}\left(1 - \frac{\text{sgn}(\sigma_P)}{\sqrt{3}} \sqrt{\left(\frac{2f_y}{\sigma_t}\right)^2 - 1}\right) \tag{4.140}$$

$$Y_1^{(1)} = \frac{2}{3}(1-a)\sigma_P - \frac{1}{3} b\sigma_t \tag{4.141}$$

$$Y_2^{(1)} = -\frac{1}{3}(1-a)\sigma_P + \frac{2}{3} b\sigma_t \,. \tag{4.142}$$

Because of its deviatoric nature, $Y_3 = -Y_1 - Y_2$ always applies. As usual, the initial strains are determined and inserted into the modified elastic material law. With the boundary conditions

$$\rho_1 = 0 \quad ; \quad \varepsilon_2^* = 0 \tag{4.143}$$

we get two equations for determination of the residual stress in the direction of the displacement-controlled load and for the residual strain in the direction of the stress-controlled load. For the first MEA we get the solution

$$\frac{\rho_2}{f_y} = \left(1 - \frac{E_t}{E}\right)\left[\frac{1}{2}(1-a)\frac{\sigma_P}{f_y} - b\frac{\sigma_t}{f_y}\right] \tag{4.144}$$

$$\frac{E}{f_y}\varepsilon_1^* = \left(\frac{E}{E_t} - 1\right)\left[-\frac{1}{2}\frac{\rho_2}{f_y} + (1-a)\frac{\sigma_P}{f_y} - \frac{1}{2} b\frac{\sigma_t}{f_y}\right]. \tag{4.145}$$

By superposition of Eq. (4.144) with the fictitious elastic stress in direction (2), i.e., σ_t, the elastic-plastic stress component σ_2 at the maximum state of loading is obtained, whereas Eq. (4.144) represents the elastic-plastic stress component σ_2 at the minimum state of loading. Accordingly, $\sigma_{\text{v,min}}$ and $\sigma_{\text{v,max}}$, and thus the angle $\beta_{\max}$, are changed in Eq. (4.65), although the range $\Delta\sigma_\text{v}$ remains unchanged because of the constant residual stress. $\beta_{\max}$ is in fact increased, so that the condition $\beta_{\max} > \alpha_{\max}$ is certainly fulfilled and the TIV Y_i remains unchanged in another MEA as positioned in the vertex of Ω, resulting in the same residual stresses and residual strains as in Eqs. (4.144) and (4.145). The result of the first MEA cannot therefore be iteratively improved, which is not bad however, because it already represents the exact solution.

The exact solution can be determined with an incremental analysis only with significantly greater effort, and only approximately. The development of elastic-plastic strains in direction (1) is shown in Fig. 4.26 as an example with the parameter combination

$$\frac{\sigma_P}{f_y} = 0.8; \quad \frac{\sigma_t}{f_y} = 1.9; \quad \frac{E_t}{E} = 0.05\,. \tag{4.146}$$

According to Eq. (4.145), the exact result, obtained after superposition with the fictitious elastic solution, is, because of the Poisson's ratio $\nu = 0$, for both the minimum and the maximum load condition $\varepsilon_1 = 6.8008\, f_y/E$. After 200 cycles, the incremental method reaches $\varepsilon_1 = 6.7762\, f_y/E$ with pre-set convergence tolerances and 20 load increments per half cycle, making it accurate only to two digits (after 400 cycles $\varepsilon_1 = 6.8005\, f_y/E$), which, of course, is perfectly sufficient for practical needs in an engineering sense, but makes the attribute "exact" questionable in relation to incremental analyses. The computational effort required for this approximation by an incremental analysis corresponds to several thousands of linear elastic analyses.

Figure 4.27 shows the associated development of the TIV in the TIV-space. One can clearly see how the TIV gradually tapers towards the vertex of the area Ω.

4.5.2.2 Element Is Located in the TIV-Region ω_1

If Eq. (4.135) is complied with, i.e., $\beta_{max} < \alpha_{max}$, the projection of Y^* on the intersection area Ω in the space of the TIV leads to the Mises circle centered in the fictitious elastic stress of the maximum state of loading (Fig. 4.25, left). Equation (4.69) provides for the first iteration

$$Y_1^{(1)} = \left(\frac{2}{3}\sigma_P - \frac{1}{3}\sigma_t\right)\left(1 - \frac{f_y}{\sqrt{\sigma_P^2 + \sigma_t^2 - \sigma_P\sigma_t}}\right) \tag{4.147}$$

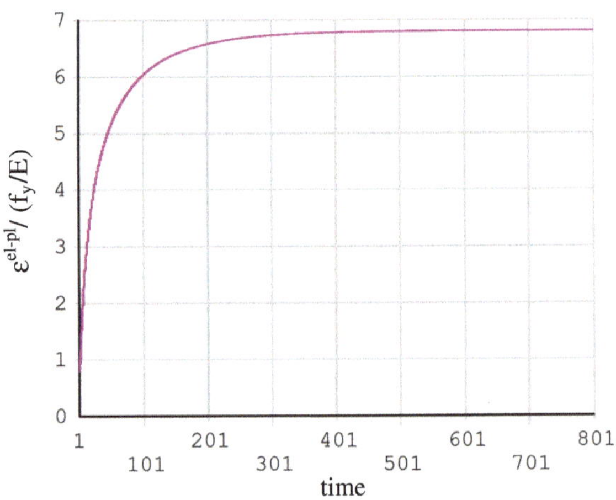

Fig. 4.26 Multiaxial ratcheting: histogram of the normalized elastic-plastic strain in direction (1) for $\sigma_P/f_y = 0.8$, $\sigma_t/f_y = 1.9$, $E_t/E = 0.05$

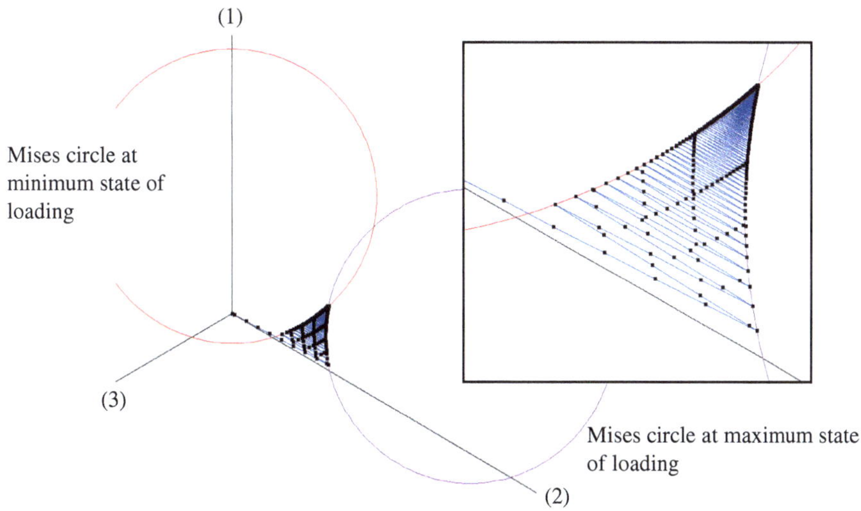

Fig. 4.27 Multiaxial ratcheting: trajectory of the TIV in the TIV-space

$$Y_2^{(1)} = \left(-\frac{1}{3}\sigma_P + \frac{2}{3}\sigma_t\right)\left(1 - \frac{f_y}{\sqrt{\sigma_P^2 + \sigma_t^2 - \sigma_P\sigma_t}}\right). \tag{4.148}$$

Using the boundary conditions Eq. (4.143), the modified elastic material law leads to

$$\frac{\rho_2^{(1)}}{f_y} = \left(1 - \frac{E_t}{E}\right)\left(\frac{1}{2}\frac{\sigma_P}{f_y} - \frac{\sigma_t}{f_y}\right)\left(1 - \frac{f_y}{\sqrt{\sigma_P^2 + \sigma_t^2 - \sigma_P\sigma_t}}\right) \tag{4.149}$$

$$\frac{E}{f_y}\varepsilon_1^{*(1)} = \left(\frac{E}{E_t} - 1\right)\left[-\frac{1}{2}\frac{\rho_2^{(1)}}{f_y} + \left(\frac{\sigma_P}{f_y} - \frac{1}{2}\frac{\sigma_t}{f_y}\right)\left(1 - \frac{f_y}{\sqrt{\sigma_P^2 + \sigma_t^2 - \sigma_P\sigma_t}}\right)\right]. \tag{4.150}$$

After superposition with the fictitious elastic solution for the minimum and maximum load state we obtain the elastic-plastic stress from the first MEA, $\sigma_{2,\min}$ and $\sigma_{2,\max}$, whereby $\sigma_{v,\min}$ and $\sigma_{v,\max}$ differ from the fictitious elastic solution and thus the angle $\beta_{\max}$ in Eq. (4.65) is changed. Depending on the parameter constellation, the vertex of Ω may become decisive in the second iteration, so that Eqs. (4.144) and (4.145) with parameters a and b according to Eqs. (4.139) and (4.140) represent the exact solution. Otherwise a new projection on the Mises circle of the maximum load condition is made on the basis of $\sigma_{2,\max}$ according to Eq. (4.69).

The result of the iterative procedure is shown in Fig. 4.28 as an example with $\sigma_P/f_y = 0.8$ and $E_t/E = 0.05$ and a variation of secondary stresses representing the

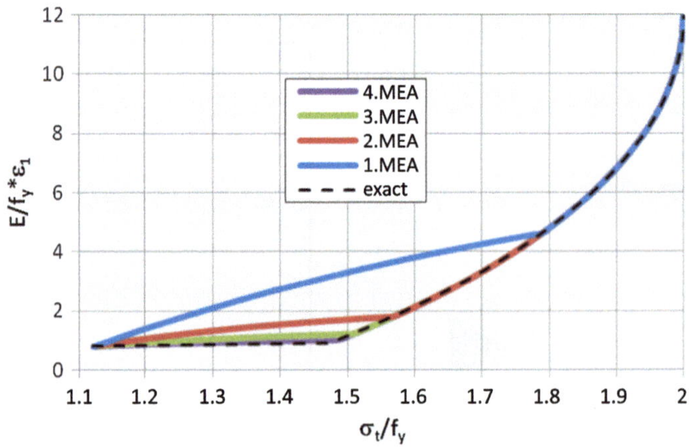

Fig. 4.28 Multiaxial ratcheting: iterative determination of the normalized elastic-plastic strain in direction (1) for $\sigma_P/f_y = 0.8$, $E_t/E = 0.05$

entire region of elastic shakedown bounded by Eqs. (4.129) and (4.131). It is evident that the STPZ achieves either the exact solution or at least a good approximation with few MEAs. The region of almost constant strain with increasing secondary stress is characterized by the fact that the development of plastic strain is already fully completed after the first loading cycle. Here, the STPZ, similar to monotonic loading, cannot offer an appreciable saving in computing time.

4.5.3 Bree Tube

A uniaxial replacement model of the Bree tube has already been dealt with in Sects. 2.5.3 and 4.3.3 (Fig. 2.20). In this section the strains accumulated in the state of elastic shakedown are determined by treating the tube as an axisymmetric model.

The load consists of a constant internal pressure p and a cyclically varying linear temperature distribution over the wall causing a temperature difference ΔT between the inner and outer surfaces (the outside is colder than the inside). If restricted to a part of the tube at a sufficient distance from the edge effects, the stresses are constant along the axial direction, so that the analysis of a narrow portion in the axial direction is sufficient. Because of the thinness, only circumferential and axial stresses occur.

The following results were obtained with FEM analyses (ANSYS [11]).[2] The equivalent stress produced by internal pressure including cap forces was chosen as $\sigma_P/f_y = 0.8$ and the maximum fictitious elastic calculated equivalent stress from the biaxial stress caused by the temperature gradient as $\sigma_t/f_y = 1.9$. Poisson's ratio is $\nu = 0.3$ and the hardening parameter $E_t/E = 0.02$.

[2]Jiang and Leckie obtained an analytical solution in [9] for the axisymmetric model of the Bree tube based on Zarka's method.

The development of the TIV on the inside and outside surfaces, which becomes plastic first, is determined with an incremental analysis, and depicted in Fig. 4.29. This already suggests that the structural behavior is relatively complex because of local and directional stress redistribution. Plasticizing occurs in the first half cycle only on the outer surface because there tensile stresses resulting from the temperature gradient are added to the tensile stresses produced by internal pressure. Local stress redistribution is thus activated on the inner surface, resulting in residual stresses, which causes the TIV to move from the point of origin, despite missing plasticizing. Even at the beginning of the second half cycle (green curve), the TIV changes on the inner surface at first simply because of redistributions, attributed to the progression of the plastic front from the wall interior towards the inner surface, before the inner surface becomes plastic itself, as can be seen by the reversing trajectory in Fig. 4.29. The situation is similar in the following third half cycle (blue curve) on the outer surface.

As is evident from Fig. 4.29, the TIV on the outside and the inside surfaces is located in the state of elastic shakedown in the vertex of the intersection area Ω. This can be detected accurately by the STPZ. In the interior of the tube wall the projection of the negative deviatoric residual stresses on the intersection area Ω does not always come down to its vertex (Y^* is located in region ω_4) but sometimes to the yield surfaces of the minimum or maximum state of loading (Y^* in regions ω_1 or ω_2). In this example, all three possible cases occur simultaneously.

Because the loading path has a significant influence on the plastic behavior of the structure, it is not readily possible to determine an "exact" result, even with the aid

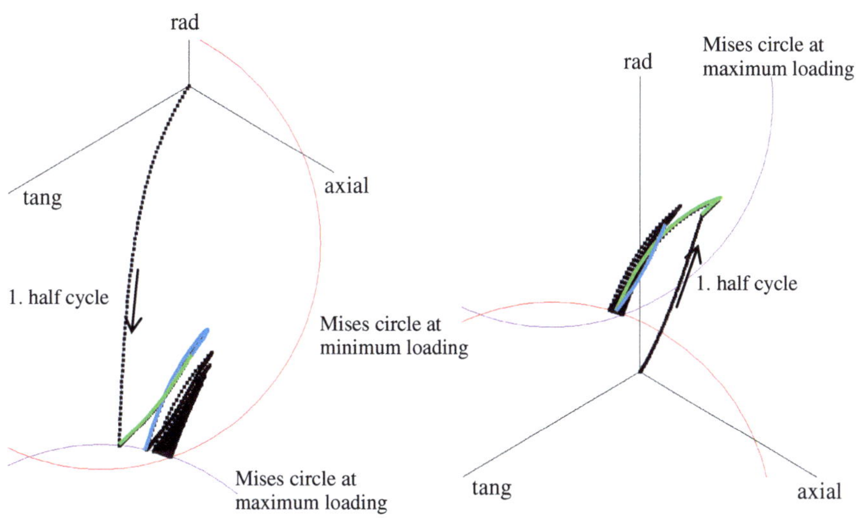

Fig. 4.29 Bree tube (axisymmetric model, $\sigma_P/f_y = 0.8$; $\sigma_t/f_y = 1.9$; $\nu = 0.3$; $E_t/E = 0.02$): trajectories of the TIV in the TIV-space for outer (*left*) and inner surface (*right*); *green* = second half cycle, *blue* = third half cycle

of an incremental analysis. This is because the result depends on the number of the load increments in each half cycle.

This is also evident from the histogram of the plastic strain on the inner and outer surface for the axial and the circumferential direction in Fig. 4.30, where the results of the STPZ are indicated as diamonds for the state of shakedown after five MEAs. From the fifth MEA, the results of the STPZ are always between the incremental solutions with 1 and 20 load increments per half cycle. The incremental analysis already provides a good approximation to the state of shakedown after about 150 half cycles.

Figure 4.31 shows the stress–strain curve normalized to the yield stress f_y or the elastic limit strain f_y/E for the circumferential direction of the inner and outer surfaces. The results of the STPZ for the shakedown state after five MEAs are indicated by diamonds for the minimum and crosses for the maximum load condition.

The largest strain component is the circumferential strain on the outer surface. Table 4.2 shows its development with the MEAs compared to the incremental analyses with 1 and 20 load increments per half cycle. It follows that from the fourth MEA of the STPZ a good compromise between various incremental analyses is achieved.

The computational cost of the STPZ corresponds to four linear elastic analysis, plus the two compulsory fictitious elastic analyses and some local calculations for determining the initial strains to be applied as loading. The two incremental analyses with 150 required half cycles and either 1 load increment with an average of 4 necessary equilibrium iterations, or 20 load increments with an average of 2 required equilibrium iterations require a computational effort which is equivalent to 600–6,000 linear elastic analyses.

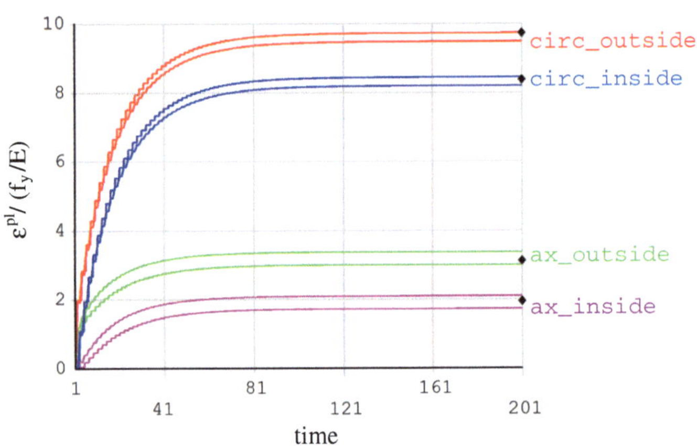

Fig. 4.30 Bree tube (axisymmetric model): histogram of plastic strain after incremental analysis with 1 or 20 load increments per half cycle; and results of the fifth MEA of the STPZ for the state of elastic shakedown (*diamonds*)

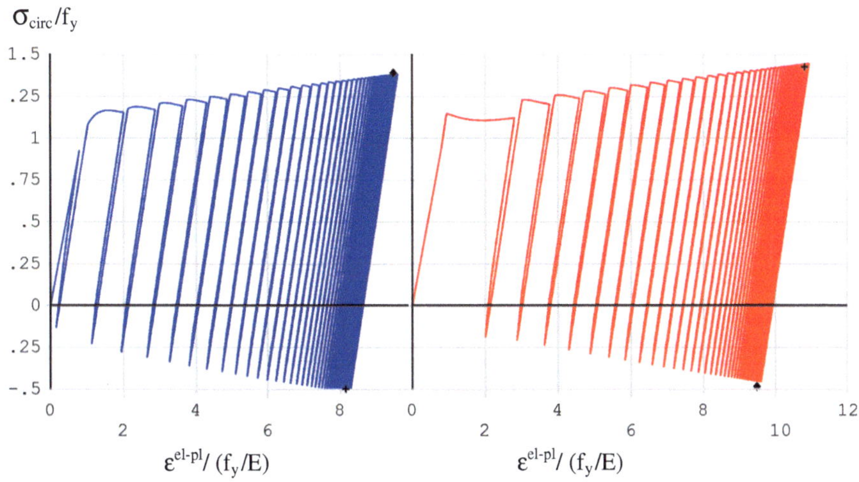

Fig. 4.31 Bree tube (axisymmetric model): stress–strain diagram for circumferential direction after incremental analysis, and stress–strain couples of the fifth MEA of the STPZ (*diamonds* and *crosses*); *left* inside surface; *right* outside surface

Table 4.2 Development of circumferential strain on the outer surface with the MEAs of the STPZ compared with two different incremental analyses in the shakedown state for the maximum load condition

STPZ, MEA	$E/f_y{}^*\varepsilon_z$
1	2.686
2	4.539
3	11.303
4	10.836
5	10.811
6	10.797
7	10.788
8	10.782
Incremental with 1 load increment per half cycle	10.629
Incremental with 20 load increments per half cycle	10.878

4.6 Accumulated Strains at Plastic Shakedown

As mentioned in Sect. 4.1, the criterion for plastic shakedown (PS) is

$$\Delta\sigma_v^{fel} > 2f_y \quad \exists \underline{x} \in V \rightarrow \text{PS}\,. \tag{4.151}$$

The modified elastic analyses (MEAs) for determining the residual stress state in the plastic shakedown condition, and especially the accumulated strain and associated stresses, are performed similar to the approach featured in Sect. 3.1. The following shows, as required, how the geometry of the plastic zone V_p and the initial strains to be applied there as a modified load can be estimated.

4.6.1 Identification of the Elastic and Plastic Subvolumes

With regard to the estimation of the TIV in the plastic subvolume V_p of the structure, it must be distinguished in the plastic shakedown condition as to whether cyclic plasticizing or non-cyclic plasticizing occurs at the point under consideration. Accordingly, the subvolume with cyclic plastic straining is called $V_{p\Delta}$, and the subvolume with plastic strains which no longer change after reaching the shakedown condition, $V_{e\Delta}$. The partial volume in which no plastic strains occur continues to be denoted V_e. $V_{p\Delta}$ is already known from the calculation of the strain range according to Sect. 4.2. The following procedure for determining the accumulated strain therefore assumes that the range of stresses and strains has been determined in advance with the STPZ. Corresponding to Eq. (4.12) we therefore get for the nth MEA:

$$V_{p\Delta}^{(n)} = \left\{ \underline{x} \middle| \Delta\sigma_{\mathrm{v}}^{(n-1)} \geq 2f_y \right\}. \tag{4.152}$$

The yield surfaces belonging to the minimum and maximum states of loading are so disjointed at least at one location of the structure in the space of the TIV. Those points of the structure, which were assigned to the elastic volume V_e when calculating the ranges, but where the yield stress is exceeded at one of the two extreme load conditions, i.e., either at the minimum or the maximum state of loading, are now assigned to the subvolume with non-cyclic plastic strains:

$$V_{e\Delta}^{(n)} = \left\{ \underline{x} \middle| \Delta\sigma_{\mathrm{v}}^{(n-1)} < 2f_y \wedge \left(\sigma_{\mathrm{v,max}}^{(n-1)} > f_y \vee \sigma_{\mathrm{v,min}}^{(n-1)} > f_y \right) \right\}. \tag{4.153}$$

The part of the structure remaining elastic is thus defined by

$$V_{\mathrm{e}}^{(n)} = \left\{ \underline{x} \middle| \sigma_{\mathrm{v,min}}^{(n-1)} < f_{\mathrm{y}} \quad \wedge \quad \sigma_{\mathrm{v,max}}^{(n-1)} < f_{\mathrm{y}} \right\}. \tag{4.154}$$

Thus, the following definition holds instead of Eq. (3.17):

$$V = V_e \cup V_p; \quad V_p = V_{p\Delta} \cup V_{e\Delta}; \quad V_e \cap V_p = \varnothing. \tag{4.155}$$

In contrast to elastic shakedown, the residual stresses and thus the TIV are not constant at any point of the structure in the case of plastic shakedown, not even in V_e. Accordingly, it has no fundamental significance when the two yield surfaces belonging to the extreme loading conditions form an intersection area in the TIV-space at any location of the structure.

4.6.2 Estimation of the Transformed Internal Variable

Because of the different plastic behavior at the two extreme load conditions in the case of plastic shakedown, two different estimates of the TIV are required. However, if use is made of a pre-derived solution for the range values, only one more consideration for any load condition is required to determine the accumulated strain, for example the mean load condition. The stresses and strains of the minimum and maximum load state can then subsequently be obtained by addition or subtraction of half of the stress or strain ranges.

For the estimation of the TIV Y in the mean load condition, three cases are distinguished in the following (Sects. 4.6.2.1–4.6.2.3).

4.6.2.1 Location Under Consideration Is in $V_{p\Delta}$

The TIV is then estimated lying halfway between the two yield surfaces in the TIV-space (see Fig. 4.32):

$$Y_{i,m} = \sigma'^{fel}_{i,\min} + \frac{1}{2} \Delta\sigma'^{fel}_i . \tag{4.156}$$

$Y_{i,m}$ is thus immediately known from the fictitious elastic calculations, so regardless of the elastic-plastic states it is inaccessible to an iterative improvement.

4.6.2.2 Location Under Consideration Is in $V_{e\Delta}$ and It Is $\sigma_{v,\max} > \sigma_{v,\min}$

The mean TIV is then estimated by projecting the negative deviatoric residual stress at the maximum load state

$$Y^*_{i,\max} = Y^*_{i,m} + \frac{1}{2} \Delta Y_i \tag{4.157}$$

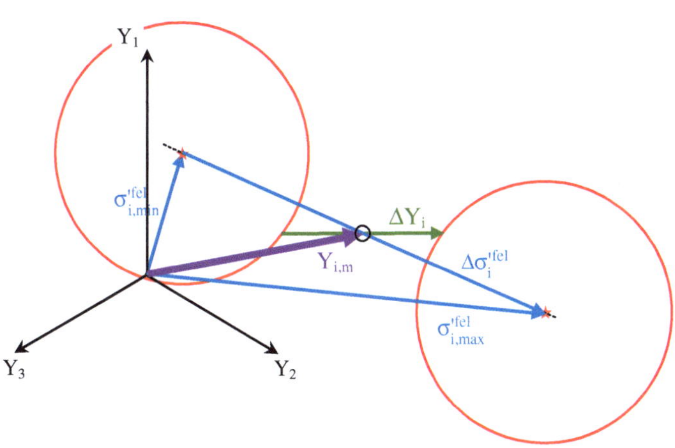

Fig. 4.32 Estimation of the TIV for the mean load condition at PS for a location in $V_{p\Delta}$

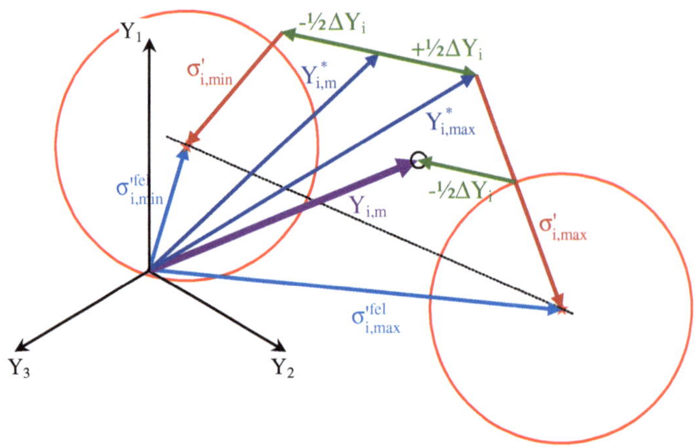

Fig. 4.33 Estimation of the TIV for the mean load condition at PS for a location in $V_{e\Delta}$ by projection to the maximum load condition

with

$$Y^*_{i,m} = -\rho'_{i,m} \tag{4.158}$$

to the Mises circle of the maximum load condition (Fig. 4.33). Because locations in $V_{e\Delta}$ cannot experience plastic strain ranges, i.e., $\Delta\xi_i = 0$, we have

$$\Delta Y_i = -\Delta\rho'_i\,, \tag{4.159}$$

and we get for the TIV in the mean load condition

$$Y_{i,m} = \sigma'^{\text{fel}}_{i,\max} - \sigma'_{i,\max} \cdot \frac{f_y}{\sigma_{\text{v},\max}} + \frac{1}{2}\left(\Delta\sigma'_i - \Delta\sigma'^{\text{fel}}_i\right). \tag{4.160}$$

4.6.2.3 Location Under Consideration Is in $V_{e\Delta}$ and It Is $\sigma_{\text{v,max}} < \sigma_{\text{v,min}}$

The mean TIV is then estimated by projection of the negative deviatoric residual stress at the minimum load condition

$$Y^*_{i,\min} = Y^*_{i,m} - \frac{1}{2}\Delta Y_i \tag{4.161}$$

to the Mises circle of the minimum load condition (Fig. 4.34):

$$Y_{i,m} = \sigma'^{\text{fel}}_{i,\min} - \sigma'_{i,\min} \cdot \frac{f_y}{\sigma_{\text{v},\min}} - \frac{1}{2}\left(\Delta\sigma'_i - \Delta\sigma'^{\text{fel}}_i\right). \tag{4.162}$$

With the thus estimated TIV, an MEA is performed to obtain the residual stress ρ_m and the residual strain ε^*_m for the mean load condition. On this basis, an improved estimate of the plastic zone $V_{e\Delta}$ and an improved estimation of the TIV in $V_{e\Delta}$ can be done as part of an iterative process. $V_{p\Delta}$ is already clear from the previous calculation of the range values.

The elastic-plastic stresses and strains for the mean state arise from the superposition with the fictitious elastic stresses of the mean state

$$\sigma_{i,m} = \rho_{i,m} + \frac{1}{2}\left(\sigma^{\text{fel}}_{i,\max} + \sigma^{\text{fel}}_{i,\min}\right) \tag{4.163}$$

$$\varepsilon_{i,m} = \varepsilon^*_{i,m} + \frac{1}{2}\left(\varepsilon^{\text{fel}}_{i,\max} + \varepsilon^{\text{fel}}_{i,\min}\right), \tag{4.164}$$

and the elastic-plastic stresses and strains of the minimum or maximum load condition by subtraction or addition of the half range values:

$$\sigma_{i,\min/\max} = \sigma_{i,m} \mp \frac{1}{2}\Delta\sigma_i = \sigma_{i,m} \mp \frac{1}{2}\left(\Delta\sigma^{\text{fel}}_i + \Delta\rho_i\right) \tag{4.165}$$

$$\varepsilon_{i,\min/\max} = \varepsilon_{i,m} \mp \frac{1}{2}\Delta\varepsilon_i = \varepsilon_{i,m} \mp \frac{1}{2}\left(\Delta\varepsilon^{\text{fel}}_i + \Delta\varepsilon^*_i\right). \tag{4.166}$$

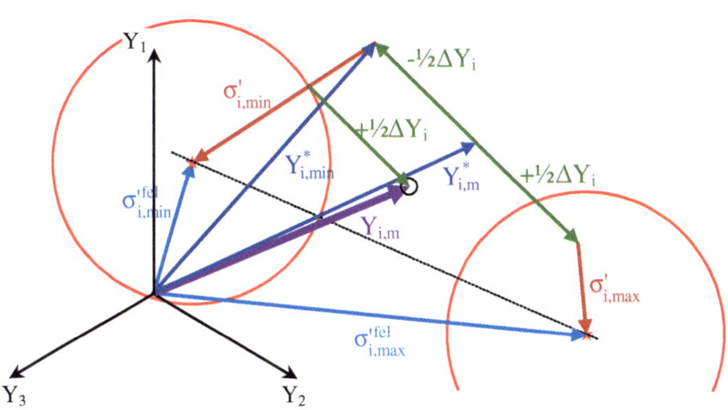

Fig. 4.34 Estimation of the TIV for the mean load condition at PS for a location in $V_{e\Delta}$ by projection to the minimum load condition

4.7 Examples of Accumulated Strains at Plastic Shakedown

In Sect. 4.3, examples for estimating strain ranges in the state of plastic shakedown have already been considered. Because the determination of the accumulated strains relies on the previous determination of the strain ranges, these examples are selected here again.

4.7.1 Two-Bar Model

In Sect. 4.3.1 the range values for the two-bar model (Fig. 4.3) have already been determined in the state of plastic shakedown. It was shown in Sect. 4.5.1 how the strains accumulated in the state of elastic shakedown can be obtained adopting the STPZ.

Because the loading range is strain-controlled (see Eq. (4.30)), the entire system is in $V_{p\Delta}$ in the state of plastic shakedown. Thus the plastic zone V_p is fixed and the TIV for calculating the mean state is to be determined according to Eq. (4.156). After adapting to the uniaxial stress state, we obtain the TIVs in both bars with σ_P as primary stress and σ_t as the maximum fictitious elastic secondary stress ($+\sigma_t$ in the right, $-\sigma_t$ in the left bar):

$$Y_{\text{ri},m} = \sigma_p + \frac{1}{2}\sigma_t \tag{4.167}$$

$$Y_{\text{le},m} = \sigma_p - \frac{1}{2}\sigma_t\,. \tag{4.168}$$

After insertion into the modified elastic material law, we obtain

$$\varepsilon^*_{\text{ri},m} = \frac{1}{E_t}\rho_{\text{ri},m} + \frac{1}{C}\left(\sigma_P + \frac{1}{2}\sigma_t\right) \tag{4.169}$$

$$\varepsilon^*_{\text{le},m} = \frac{1}{E_t}\rho_{\text{le},m} + \frac{1}{C}\left(\sigma_P - \frac{1}{2}\sigma_t\right), \tag{4.170}$$

and with the equilibrium condition

$$\rho_{\text{ri},m} = -\rho_{\text{le},m} \tag{4.171}$$

and the compatibility condition

$$\varepsilon^*_{\text{le},m} = \varepsilon^*_{\text{ri},m} \tag{4.172}$$

we get as the result for the first MEA

$$\rho_{\mathrm{le},m} = \left(1 - \frac{E_t}{E}\right)\frac{\sigma_t}{2} \tag{4.173}$$

$$\varepsilon^*_{\mathrm{le},m} = \frac{1 - \frac{E_t}{E}}{E_t}\sigma_P \tag{4.174}$$

for the mean loading condition. After superposition with the fictitious elastic solution for the mean state and subtraction or addition of half the range values, and employing Eqs. (4.19), (4.28), and (4.30), the solutions obtained for Eqs. (4.165) and (4.166) read

$$\sigma_{\mathrm{le,min}} = \left(1 - \frac{E_t}{E}\right) f_y + \sigma_P \tag{4.175}$$

$$\sigma_{\mathrm{le,max}} = \left(1 - \frac{E_t}{E}\right)(\sigma_t - f_y) + \sigma_P - \sigma_t \tag{4.176}$$

$$\varepsilon_{\mathrm{le,min}} = \frac{\sigma_P}{E_t} \tag{4.177}$$

$$\varepsilon_{\mathrm{le,max}} = \frac{\sigma_P}{E_t} - \frac{\sigma_t}{E}\,. \tag{4.178}$$

As the plastic zone V_p was known in advance, and the TIV was also exactly known a priori because of the uniaxial stress state, because directional redistribution is ruled out then, this is already the exact solution. It is shown in Fig. 4.35 along with the incremental solution of Fig. 2.6 as an example with the parameter configuration indicated there.

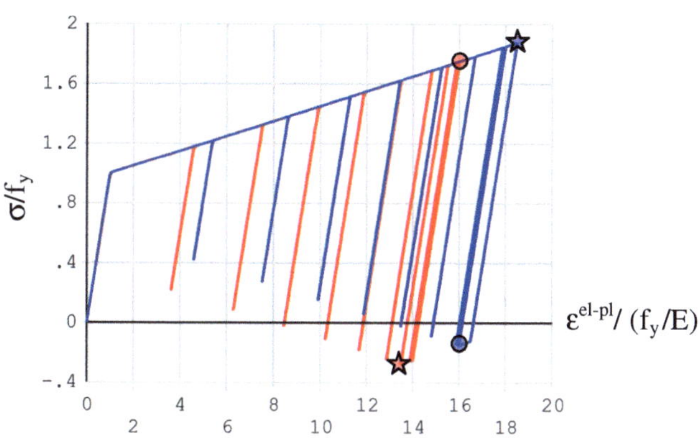

Fig. 4.35 Strain accumulation in the two-bar model with linear kinematic hardening: plastic shakedown ($\sigma_P/f_y = 0.8$; $\sigma_t/f_y = 2.5$; $E_t/E = 0.05$); comparison of the incremental solution (as in Fig. 2.6) and the STPZ (*circles* for minimum and *stars* for maximum load condition)

The STPZ delivers the exact solution for the accumulated stresses and strains in the state of plastic shakedown with a single further MEA beyond the MEA required to determine the stress and strain range. Several half cycles, each with multiple equilibrium iterations, are needed for the incremental solution, as shown in Fig. 4.35.

4.7.2 Multiaxial Ratcheting

For the example of multiaxial ratcheting in a thin-walled tube already treated in Sect. 4.5.2 with respect to elastic shakedown (Fig. 4.24; also see Sect. 2.5.2), the strain range was determined in Sect. 4.3.2 at plastic shakedown. Based on this, the accumulated strains at plastic shakedown are now determined.

V_p is known in advance, because the single element is stressed homogeneously. Thus

$$V_p = V_{p\Delta} \,. \tag{4.179}$$

According to Eq. (4.156), we have for $\nu = 0$,

$$Y_{i,m} = \sigma_P \begin{pmatrix} \frac{2}{3} \\ -\frac{1}{3} \\ -\frac{1}{3} \end{pmatrix} + \frac{\sigma_t}{2} \begin{pmatrix} -\frac{1}{3} \\ \frac{2}{3} \\ -\frac{1}{3} \end{pmatrix}, \tag{4.180}$$

which cannot be improved iteratively, so that the indication of the respective MEA can be omitted here.

With the modified elastic material law (see, e.g., Eq. (3.20) with the initial strains according to Eq. (3.14) we obtain, for the mean state, taking into account the boundary conditions ($\varepsilon_2^* = 0$, $\rho_1 = \rho_3 = 0$):

$$\begin{pmatrix} \varepsilon_{1,m}^* \\ 0 \\ \varepsilon_{3,m}^* \end{pmatrix} = \frac{1}{E_t} \begin{pmatrix} 1 & -\nu^* & -\nu^* \\ -\nu^* & 1 & -\nu^* \\ -\nu^* & -\nu^* & 1 \end{pmatrix} \begin{pmatrix} 0 \\ \rho_{2,m} \\ 0 \end{pmatrix} + \frac{1}{C} \left[\frac{\sigma_P}{2} \begin{pmatrix} +2 \\ -1 \\ -1 \end{pmatrix} + \frac{\sigma_t}{4} \begin{pmatrix} -1 \\ +2 \\ -1 \end{pmatrix} \right]. \tag{4.181}$$

The solution of this system of equations is

$$\frac{\rho_{2,m}}{f_y} = \frac{1}{2} \left(1 - \frac{E_t}{E} \right) \left(\frac{\sigma_P}{f_y} - \frac{\sigma_t}{f_y} \right) \tag{4.182}$$

$$\frac{E}{f_y} \varepsilon_{1,m}^* = \left(\frac{E}{E_t} - 1 \right) \left[\frac{3}{4} \frac{\sigma_P}{f_y} + \frac{1}{2} \frac{E_t}{E} \left(\frac{1}{2} - \nu \right) \left(\frac{\sigma_P}{f_y} - \frac{\sigma_t}{f_y} \right) \right]. \tag{4.183}$$

By superposition with the fictitious elastic solution for the mean state and subtraction or addition of half of the range values given by Eq. (4.42) or Eq. (4.43), we obtain for the two extreme states,

$$\frac{\sigma_{2,\min}}{f_y} = \left(1 - \frac{E_t}{E}\right)\left(\frac{1}{2}\frac{\sigma_P}{f_y} - 1\right) \tag{4.184}$$

$$\frac{\sigma_{2,\max}}{f_y} = \left(1 - \frac{E_t}{E}\right)\left(\frac{1}{2}\frac{\sigma_P}{f_y} + 1\right) + \frac{E_t}{E}\frac{\sigma_t}{f_y} \tag{4.185}$$

$$\frac{E}{f_y}\varepsilon_{1,\min} = \frac{\sigma_P}{f_y}\left(\frac{3}{4}\frac{E}{E_t} + \frac{1}{2} - \frac{1}{4}\frac{E_t}{E}\right) - \frac{1}{2}\left(1 - \frac{E_t}{E}\right) \tag{4.186}$$

$$\frac{E}{f_y}\varepsilon_{1,\max} = \frac{\sigma_P}{f_y}\left(\frac{3}{4}\frac{E}{E_t} + \frac{1}{2} - \frac{1}{4}\frac{E_t}{E}\right) + \frac{1}{2}\left(1 - \frac{E_t}{E}\right)\left(1 - \frac{\sigma_t}{f_y}\right). \tag{4.187}$$

Equation (4.186) is identical to the solution for the limit state of elastic shakedown, resulting from residual strain according to Eq. (4.145) for $\sigma_t = 2f_y$ (so $a = 0$, $b = 0.5$) plus the fictitious elastic strain σ_P/E. It is independent of the maximum secondary stress σ_t. In an extension of Fig. 4.28 into the region of plastic shakedown ($\sigma_t > 2f_y$), the strains thus remain constant.

Equations (4.184)–(4.187) represent the exact solution obtained with a single linear analysis (MEA), in addition to the one required for the range values in Sect. 4.3.2 and the always mandatory fictitious elastic analyses for the two extreme loading conditions.

For comparison purposes, the strain histogram obtained with an incremental analysis for the parameter combination

$$\frac{\sigma_P}{f_y} = 0.6; \quad \frac{\sigma_t}{f_y} = 3.5; \quad \frac{E_t}{E} = 0.05 \tag{4.188}$$

is shown in Fig. 4.36. Accordingly, it takes approximately 100 half cycles until the plastic shakedown condition is approximately achieved. Figure 4.37 shows the associated development of the TIV in the TIV-space. The shakedown state is characterized by the shortest route between the two Mises circles.

4.7.3 Bree Tube

The Bree tube has already been presented in Sect. 2.5.3 and the strain range at the plastic shakedown state was determined with the STPZ for the uniaxial replacement model in Sect. 4.5.3. It turned out that the loading is strain controlled by the temperature gradient. This applies not only to the replacement model with uniaxial stress state, but also to the axial and circumferential strains of the actual Bree tube,

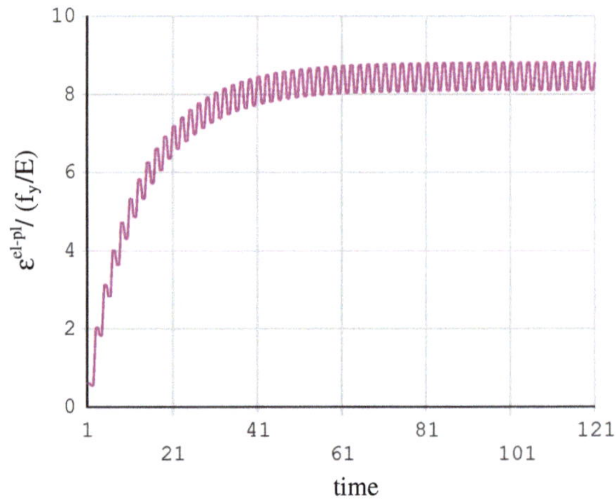

Fig. 4.36 Multiaxial ratcheting: histogram of the normalized elastic-plastic strain in direction (1) for $\sigma_P/f_y = 0.6$; $\sigma_t/f_y = 3.5$; $E_t/E = 0.05$

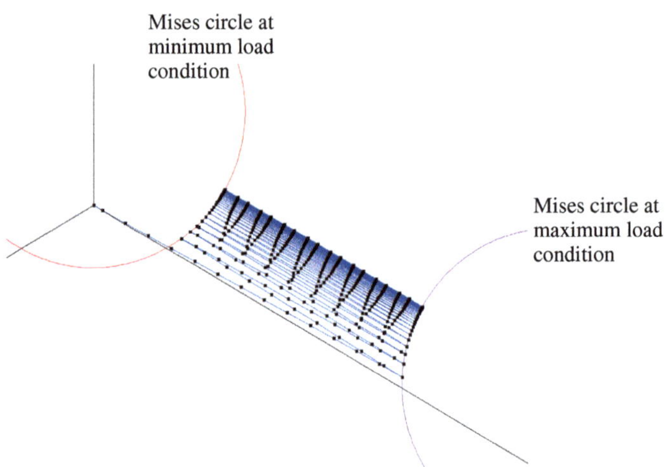

Fig. 4.37 Multiaxial ratcheting: trajectory of the TIV in the TIV-space

even though the plastic strain enhancement factor is not equal to 1 because of the strain component in the radial direction.

As in Sect. 4.5.3 for elastic shakedown, the strain accumulation at plastic shakedown is now determined here with the STPZ for an FE-model (with ANSYS [11]) of the axisymmetric tube. For this purpose the fictitious elastic secondary stress σ_S is merely increased as compared to Sect. 4.5.3 and the loading histogram is modified so that the thermal load is no longer just swelling ($\sigma_S = 0\ldots\sigma_t$) but varies between $\sigma_S = -0.2\,f_y$ and $\sigma_S = 2.6\,f_y$ (Fig. 4.38).

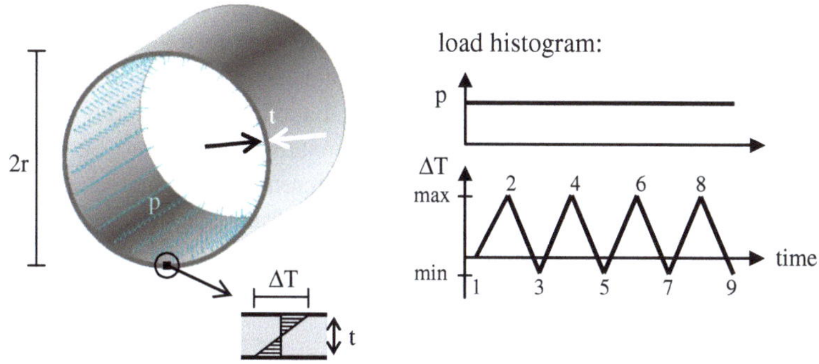

Fig. 4.38 Bree tube: geometry and load histogram

An incremental analysis showed, as for the case of elastic shakedown in Sect. 4.5.3, that the stresses and strains significantly depend on the selected number of load increments per half cycle. The following results were obtained with 80 substeps per half cycle. After about 70 cycles (140 half cycles) a very good approximation to the shakedown state was reached, but not fully achieved even after 100 cycles. Figure 4.39 shows the development of the TIV on the inner and outer surfaces. These locations are attributed to $V_{p\Delta}$. One can see that the trajectories in the shakedown state vary along the shortest connection between the two Mises circles, which can be attributed to the strain-controlled nature of the cyclic

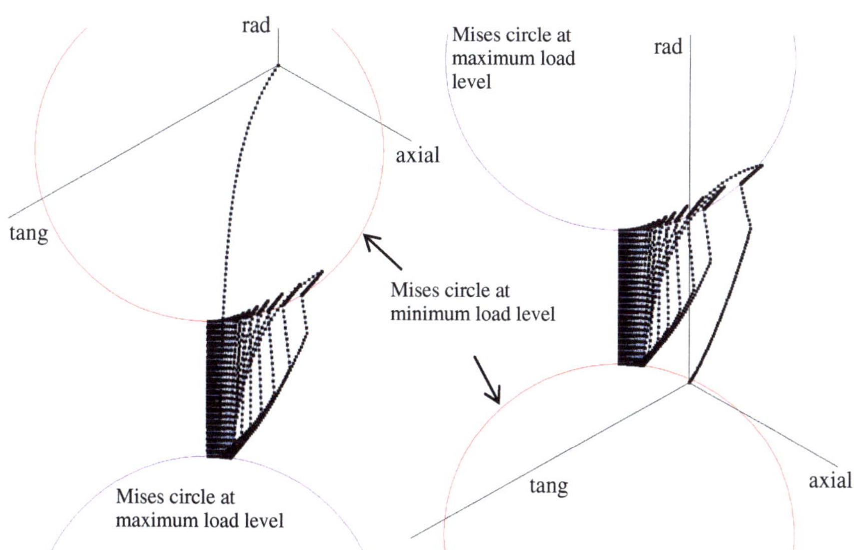

Fig. 4.39 Bree tube (axisymmetric model, $\sigma_P/f_y = 0.8$; $\sigma_S/f_y = -0.2 \ldots 2.6$; $\nu = 0.3$; $E_t/E = 0.02$): trajectories of the TIV in the TIV-space; *left* outside surface; *right* inside surface

load share. At the same time, there are also regions in the interior of the tube wall without cyclic plastic straining which are thus attributed to the subvolume $V_{e\Delta}$. In Fig. 4.40 the corresponding trajectories are shown in the deviatoric stress space.

The stresses and strains at the state of plastic shakedown obtained with the STPZ after five MEAs (on the basis of the TIV estimated according to Sects. 4.6.2.1–4.6.2.3) are shown in Fig. 4.41 along with the incrementally calculated histogram of strains and in Fig. 4.42 together with the stress–strain diagram. It turns out that

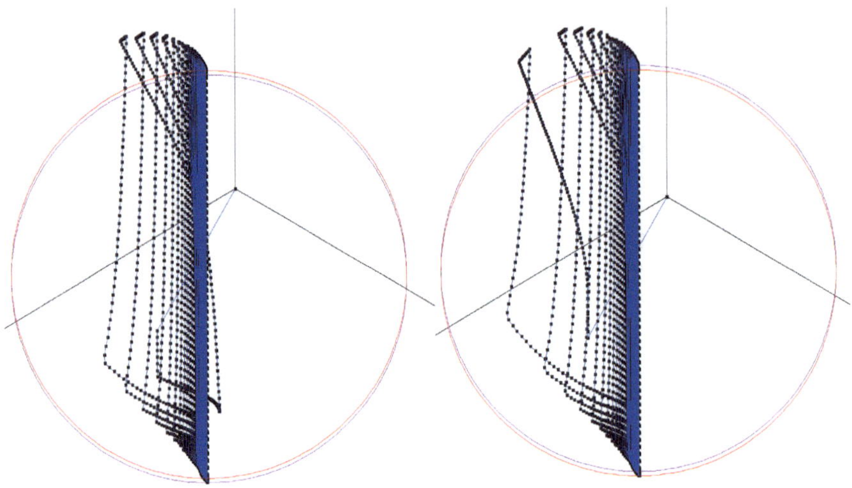

Fig. 4.40 Bree tube (axisymmetric model, $\sigma_P/f_y = 0.8$; $\sigma_S/f_y = -0.2 \ldots 2.6$; $\nu = 0.3$; $E_t/E = 0.02$): trajectories of the stress in the deviatoric stress space; *left* outside surface; *right* inside surface

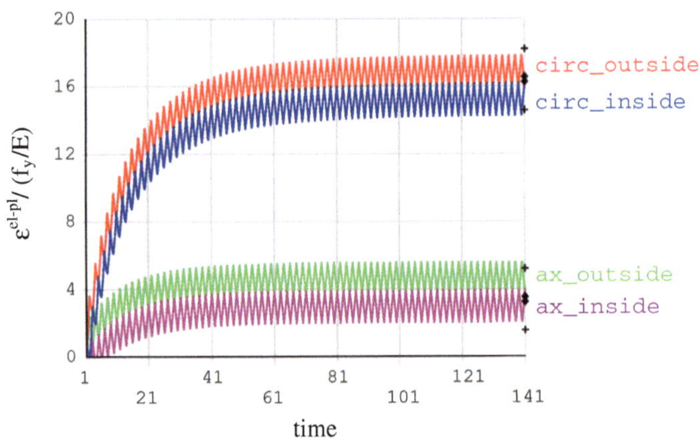

Fig. 4.41 Bree tube (axisymmetric model): histogram of the elastic-plastic strains according to an incremental analysis, and results of the fifth MEA of the STPZ at plastic shakedown (*diamonds* for the minimum, *crosses* for the maximum loading condition)

the approximation of the strain range calculated with the STPZ is very good, but the quality of the accumulated strains obtained with the STPZ is slightly less if one assumes the incremental analysis as an acceptable measure. The circumferential strains are slightly overestimated and the axial strains slightly underestimated by the STPZ.

As an example of the development of strains with the number of MEAs of the STPZ in the state of plastic shakedown, the circumferential strain component on the outside surface is listed in Table 4.3 for the maximum load level. After three MEAs the result changes only slightly.

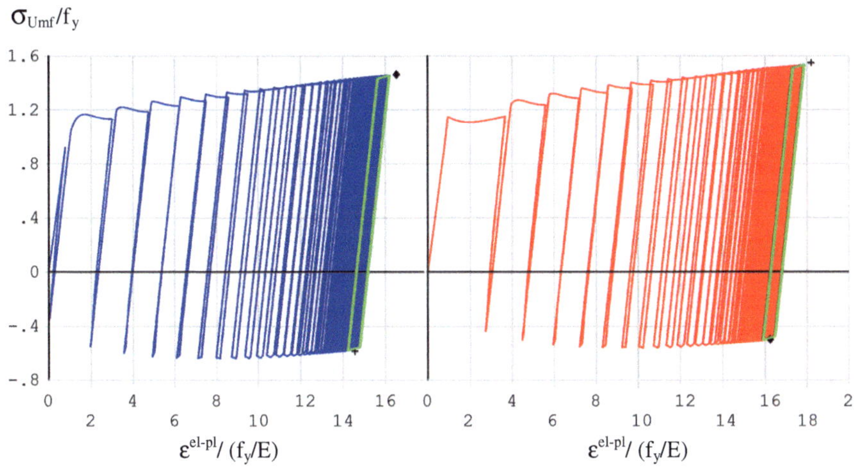

Fig. 4.42 Bree tube (axisymmetric model): stress–strain diagram for the circumferential direction according to an incremental analysis, stress–strain couples of the fifth MEA of the STPZ (*diamonds* and *crosses*); *left* inside surface; *right* outside surface; hysteresis loop in the shakedown state highlighted in *green*

Table 4.3 Development of circumferential strain on the outside surface with the MEAs of the STPZ as compared with an incremental analysis at maximum load level in the state of shakedown

STPZ, MEA	$E/f_y * \varepsilon_z$
1	2.92
2	17.55
3	18.09
4	18.19
5	18.20
6	18.19
7	18.19
8	18.19
Incremental	17.87

4.7.4 Thick-Walled Cylinder Loaded by a Temperature Transient and Internal Pressure

The analysis of a thick-walled cylinder, which is subjected to a cyclic temperature transient, presented in Sect. 4.3.4, is now studied in terms of strain accumulation in the state of plastic shakedown. For this purpose, a constant internal pressure $p = 30$ N/mm^2 is applied in addition to the temperature transient. The magnitude of the pressure corresponds to about 68.2 % of the plastic limit load. The cylinder is considered to be open so that the internal pressure causes only circumferential and radial stresses but no axial stresses. Geometry, load histogram, and material data have already been specified in Fig. 4.6 and Table 4.1.

The maximum stresses are found on the inner surface.

An incremental analysis reveals that plastic shakedown is achieved after approximately 50 cycles (100 half cycles). If the entire temperature transient is gradually passed through, i.e., if a structural analysis is performed for each time step of the thermal analysis, this corresponds to a sequential analysis of about 1,000 different temperature profiles through the wall in each load cycle. Then a total of about 107,000 equilibrium iterations are required for 50 cycles. The calculation effort therefore corresponds roughly to 100,000 linear elastic analyses. The distribution of the elastic-plastic strain across the wall thickness after reaching the state of shakedown for the two extreme load times is shown in Fig. 4.43.

This computational effort can be significantly reduced if the stresses are not calculated for each step of the thermal analysis in each cycle, but if the load is only cycled between the pair of load times, characterized by the two extreme load times (330 and 525 s). However, as already noted at the Bree tube (cf. Fig. 4.30), several intermediate states must also be calculated here in each half cycle. However, the results in the thick-walled cylinder do not depend so much on the chosen number of load increments per half cycle as in the Bree tube, so that for 50 cycles a total of

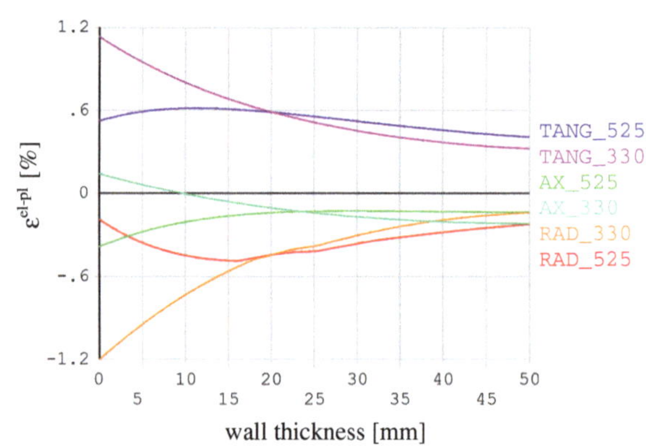

Fig. 4.43 Distribution of elastic-plastic strain components across the wall thickness at the state of plastic shakedown (incremental cyclic analysis for entire temperature transient) for the transient times 330 and 525 s

about 1,000 equilibrium iterations suffice for a reasonable quality of results of the incremental analysis. The distribution of the elastic-plastic strains across the wall thickness after reaching the state of shakedown is shown at the two extreme load times in Fig. 4.44, where 40 substeps per half cycle were chosen. They qualitatively correspond to those for tracing of the entire transient (Fig. 4.43). Quantitatively, however, some differences are noticeable. They are primarily attributable to the fact that the extreme loads do not occur exactly at 330 and 525 s, but shortly afterwards. These points in time are captured by an incremental cyclic analysis of the entire transient, but of course not during a cyclic analysis between the load pair at 330 and 525 s.

Figure 4.45 shows the trajectories of the stresses in the deviatoric stress space and of the TIV in the TIV-space for the inside surface. It can be seen that the cyclic portion of the loading is not strain-controlled, but that directional stress redistribution (Sect. 2.1) takes place in each of the two extreme load conditions.

The application of the STPZ for cyclic loading between the transient times 330 s and 525 s shows that all elements of the FE model are in the plastic zone V_p. Approximately the middle third of the tube wall does not experience cyclic plastic straining in the state of shakedown and is thus associated with the subvolume $V_{e\Delta}$, the inner and the outer third, however, being associated with $V_{p\Delta}$. Figure 4.46 shows the distribution of the elastic-plastic strain across the wall thickness after one or two MEAs for calculating the accumulated state, after the ranges have been determined beforehand in Sect. 4.3.4. From the third MEA, further changes are hardly visible. Comparison with Fig. 4.44 reveals that very good agreement between the incremental calculation and the STPZ is obtained after two MEAs, in addition to those required for calculating the range values.

The histogram of the strains in the circumferential direction on the inside and outside surfaces is shown in Fig. 4.47, the stress–strain curve for the circumferential direction on the inner surface in Fig. 4.48, and the development of strains

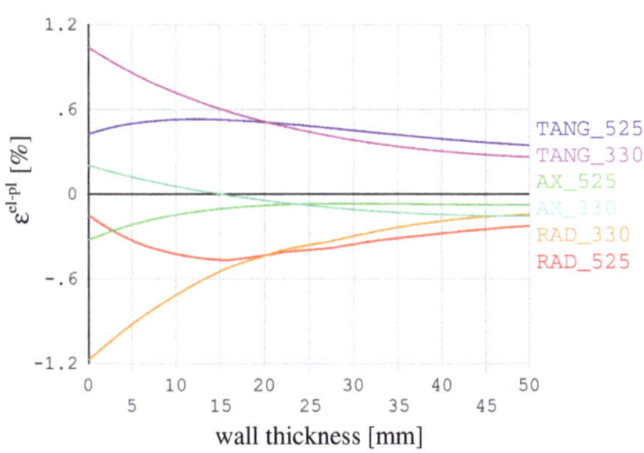

Fig. 4.44 Distribution of elastic-plastic strain components across the wall thickness at the state of plastic shakedown (incremental cyclic analysis only for transient times 330 and 525 s) for comparison with Fig. 4.43

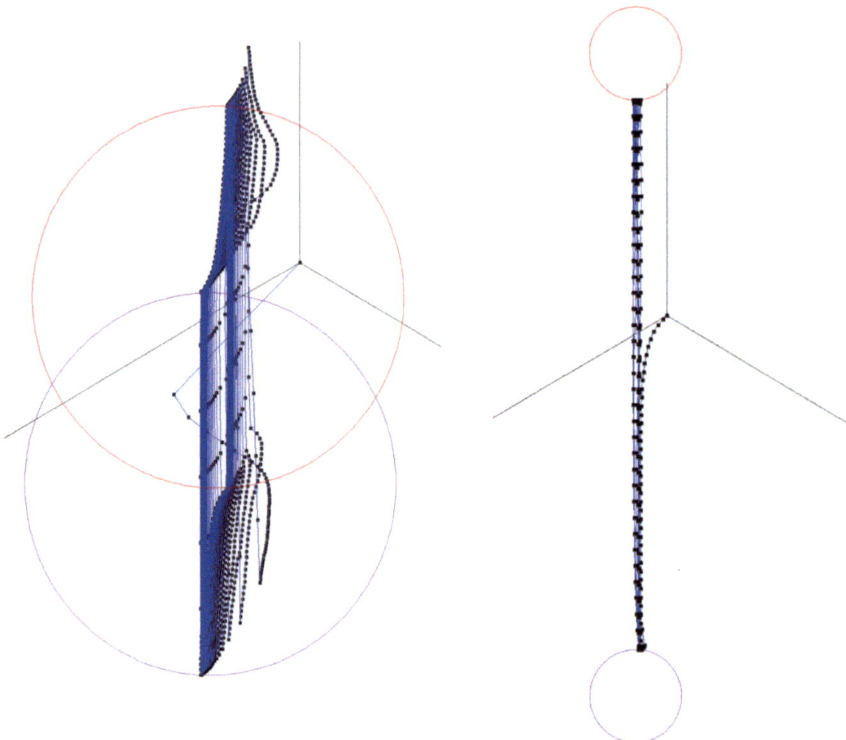

Fig. 4.45 Thick-walled cylinder on the inside surface, incremental cyclic analysis only for the transient times 330 and 525 s; *left* trajectories of the stress in the deviatoric stress space; *right*: trajectories of the TIV in the TIV-space

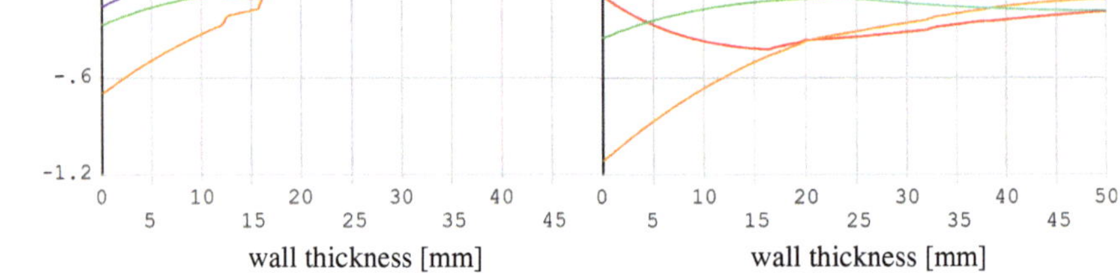

Fig. 4.46 Distribution of elastic-plastic strain components across the wall thickness for the transient times of 330 and 525 s after one MEA (*left*) or two MEAs (*right*) of the STPZ

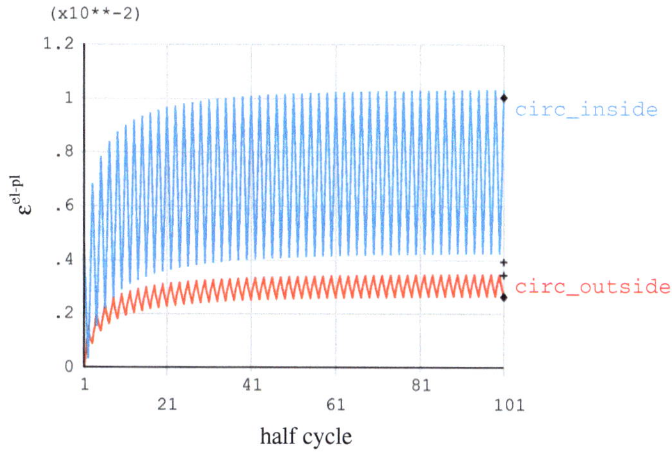

Fig. 4.47 Histogram of the circumferential strain on the inner and outer surfaces from incremental calculation for the pair of loading at the transient times 330 and 525 s; and results after four MEAs of the STPZ at the state of plastic shakedown (*diamonds* for transient time 330 s, *crosses* for 525 s)

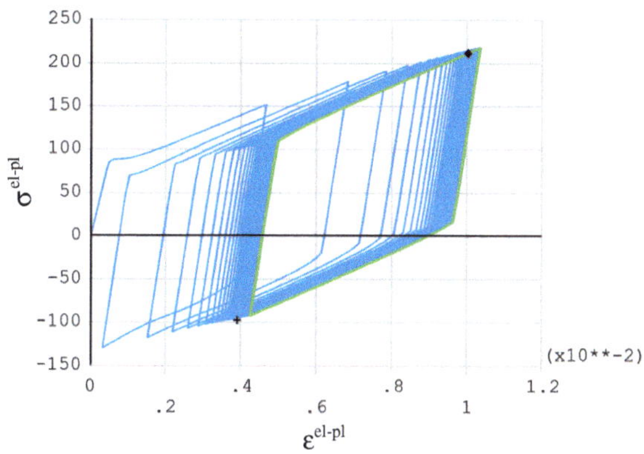

Fig. 4.48 Stress–strain diagram for the circumferential direction on the inner surface from incremental calculation for the pair of loading at the transient times 330 and 525 s (last cycle highlighted in *green*); and results after four MEAs of the STPZ at the state of plastic shakedown (*diamonds* for transient time 330 s, *crosses* for 525 s)

with the MEAs in Table 4.4. Accordingly, the strain ranges determined by the STPZ are in good agreement with the incremental analysis, as already noted in Sect. 4.3.4. The accumulated strains, however, are, at least on the inner surface, somewhat underestimated by the STPZ in relation to the incremental analysis.

The overall conclusion is that a good approximation to the results of an incremental analysis can be obtained by the STPZ, associated with very low computational costs.

Table 4.4 Development of circumferential strain on the inner surface with the MEAs of the STPZ, compared with an incremental analysis in the shakedown condition at the two extreme load states 330 and 525 s

STPZ, MEA	ε_z (%) at 330 s	ε_z (%) at 525 s
1	0.482	−0.158
2	1.007	0.392
3	1.005	0.392
4	1.004	0.393
5	1.004	0.392
6	1.003	0.393
Incremental	1.036	0.427

Apart from the two fictitious elastic analyses for the two extreme load states, just two modified elastic analyses for the range calculation and two other modified elastic analyses to calculate the accumulated strains are necessary.

4.8 STPZ at Multi-Parameter Loading

It was assumed in Chap. 4 that the loading varies between two loading conditions, termed min and max (Sect. 4.1). Because the load condition L at any time t can then be defined by a single time-dependent loading parameter $\lambda(t)$ (see Eq. (4.1)), it is a so-called one-parameter loading.

Such an approach is consistent with codes and standards for service life calculation (e.g., [12]), and also makes sense for fatigue analyses, because the fatigue curves are generally based on cycles between two states (see Sect. 2.8.1). However, it has already been mentioned in Sect. 2.6 (Myth 3) with reference to the three-bar model considered in Sect. 2.5.1 that some ratcheting mechanisms might not be detected with only pairwise consideration of load conditions.

Zarka [2, 13], Tribout [14], Inglebert with different co-authors [15, 16, 17, 18], Yu et al. [19] (see also Cecot with various co-authors [20, 21, 22, 23]), and Yu et al. [24, 25] have, for instance, dealt with the question of how Zarka's method can be used with multi-parameter loading. In some cases, reference has been made to moving loads, such as occur in rolling contact operations.

The following outlines the application of the STPZ where the order of different load conditions and thus the time pattern of fictitious elastic calculated stresses are known. If the order is not known, so that the loads can take any path in the load space, further considerations are necessary. So it must be clarified whether the investigation of the vertices of the m-dimensional load space is sufficient and, if so, whether these vertices must be approached in all possible sequences.

4.8.1 Identifying the Nature of the Shakedown State

The nature of the shakedown state, i.e., whether elastic or plastic shakedown takes place, can be identified in a similar way to that described in Sect. 4.1 (Fig. 4.49).

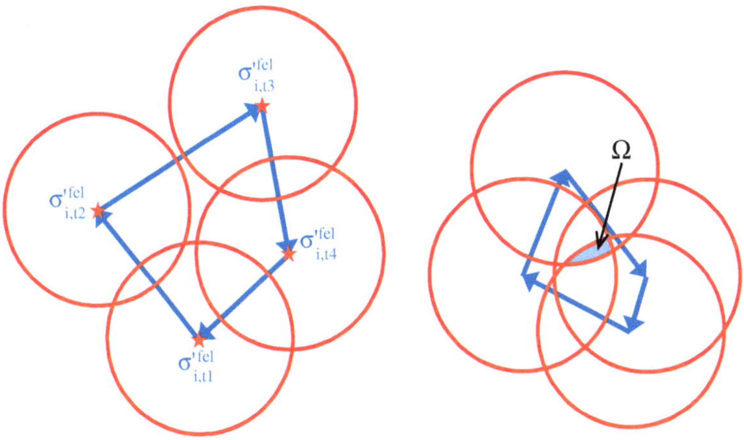

Fig. 4.49 Path of the fictitious elastic deviatoric stresses for loading times *t*1 to *t*4 and associated Mises circles in the space of the TIV at one location of the structure; *left* plastic shakedown; *right* elastic shakedown (provided $\Omega \neq 0$ at each location of the structure)

For each location of the structure, the path of the deviatoric fictitious elastic stress states in the *Y*-space is pursued (arrows in Fig. 4.49). The yield surface is formed around each of these stress states, here the Mises circle. If a common intersection, area Ω, of all these yield surfaces exists at each location of the structure, then it comes to elastic shakedown. However, if there exists at least one point of the structure with no area Ω, it comes to plastic shakedown.

If the yield stress is constant and the stress path is a polygon, a consideration in its vertices is sufficient for this. It is pointed out in [18] that the identification of the six-dimensional (hyper)area, which in fact is only five-dimensional because of the deviatoric character of the stresses, can be very costly if numerous loading conditions need to be considered. A method based on optimization strategies is proposed there for such cases.

4.8.2 Successive Projection at Elastic Shakedown

An estimate of the transformed internal variable *Y* can be obtained at elastic shakedown in V_p, i.e., for locations actively plasticizing, by projection onto the intersection area Ω. This can mean a projection onto one of the Mises circles, or on a vertex of Ω, formed by the intersection point of two Mises circles. This is because *Y* must be within or on the edge of the area Ω and is independent of the current load condition at elastic shakedown, i.e., constant, because the plastic strain and therefore the backstress ξ and the residual stress ρ are constant.

Yu et al. proposed in [19, 24, 25] to obtain *Y* by successive projection onto the intersection area of two consecutive states of loading, namely in the order in which they occur. If the load conditions in the *Y*-space are sufficiently close to each other, which is always obtainable, if need be by increasing the number of load conditions, the otherwise required need for treating the vertices of the area Ω separately is eliminated (see Sect. 4.4.2).

4.8.3 Successive Projection at Plastic Shakedown

Such a successive projection is not readily useful with regard to plastic shakedown, but requires some extensions. The problem here is that Y is not constant in the course of the loading cycle and the residual stress ρ as one of the two constituents of Y can change because of stress redistribution, even if the other constituent of Y, i.e., the backstress ξ or equivalently the plastic strain $\varepsilon^{\mathrm{pl}}$, does not change. Yu et al. provide no further description for a successive projection in such cases, but rather the message that this is tedious.

Following a proposal by Tribout in [14], the two loading conditions would have to be identified first, leading to the largest fictitious elastic stress range and starting the successive projection on one of these two states. According to [14], however, this proposal applies only in the case where all locations of the structure experience their extreme stress states simultaneously at these two loading conditions, which means a considerable restriction.

4.8.4 Example of the Three-Bar Model

Estimating the TIV in the case of multiple loading conditions by successive projection is illustrated at least by way of example for a simple structure, namely for the three-bar model looked at in Sect. 2.5.1. Geometry, load histogram, and the histogram of the fictitious elastic calculated stresses are shown in Fig. 4.50.

The simplicity of this structure is on the one hand because of the uniaxial stress state in all three bars so that directional stress redistribution does not occur, and on the other hand that the cyclic portion of the loading is strain-controlled ($K_e = 1$). Moreover, the extreme conditions in all bars appear at the same times (times 2, 4, 6), so that all bars actively plasticize in all load reversal points and the TIVs are on the edge of the corresponding yield surfaces. The plastic zone is then known a priori: all bars are in V_p.

The fictitious elastic calculated stress in the directly loaded bar is σ_t, negative when heated and positive when cooled, and $-0.5\sigma_t$ in the other two bars. The fictitious elastic stress range is $\Delta\sigma_t = 1.5\sigma_t$. Thus, the limit of elastic shakedown ($|\Delta\sigma_t| = 2f_y$) can easily be identified (ES if $|\sigma_t|/f_y < 4/3$).

At time 1, only the primary stress σ_P is acting at equal magnitude in all three bars because of the force F.

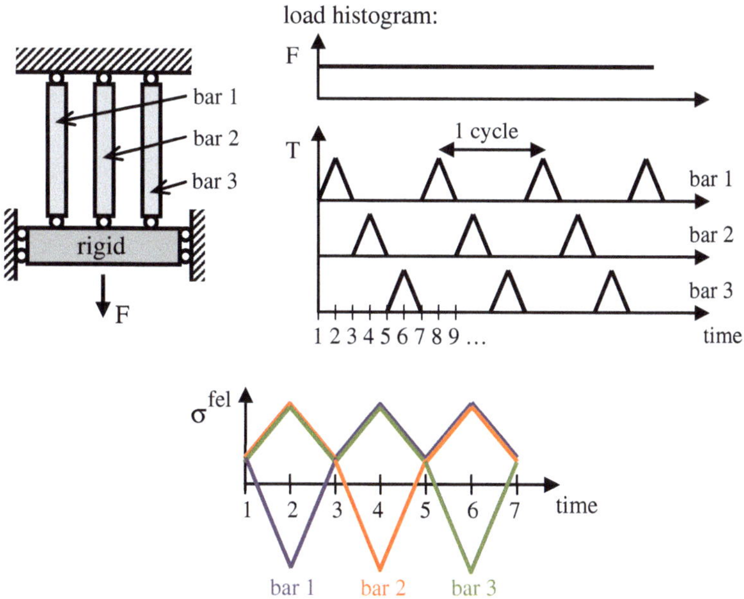

Fig. 4.50 Three-bar model; *left* geometry; *right*: load histogram; *bottom*: histogram (schematic) of the fictitious elastic calculated stresses

4.8.4.1 Elastic Shakedown

For a parameter combination that leads to elastic shakedown, the developments of the TIV are shown as a result of the successive projection in Fig. 4.51 and in Table 4.5. At time $t = 2$ in the second cycle, i.e., at $t = 8$, the structure is in a different state than at time $t = 2$ in the first cycle. After that, i.e., from the second cycle ($n = 2$), the successively projected TIVs are constant in all three bars (the index does not refer here to the direction, but to the number of the bar):

$$\frac{Y_1}{f_y} = \frac{Y_2}{f_y} = \frac{Y_3}{f_y} = 0.2\,. \tag{4.189}$$

All three bars belong to the subvolume V_p. From the modified elastic material laws

$$\begin{aligned} \varepsilon_1^* &= \frac{\rho_1}{E_t} + \frac{Y_1}{C} \\ \varepsilon_2^* &= \frac{\rho_2}{E_t} + \frac{Y_2}{C} \\ \varepsilon_3^* &= \frac{\rho_3}{E_t} + \frac{Y_3}{C} \end{aligned} \tag{4.190}$$

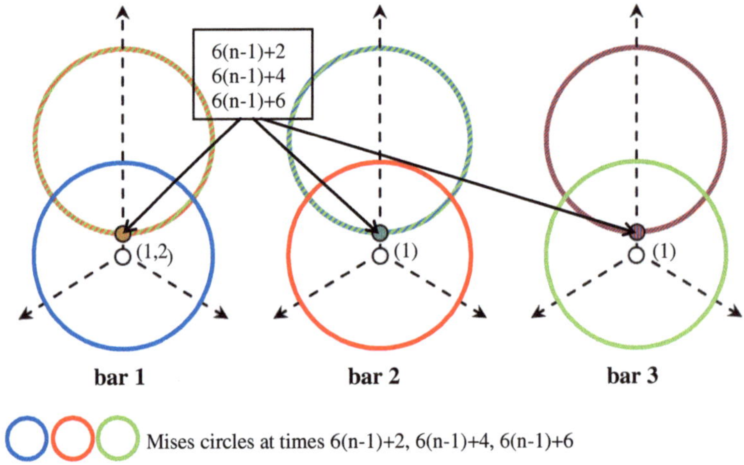

Fig. 4.51 Three-bar model at elastic shakedown: Mises circles and successive projection of Y in the TIV-space for $\sigma_p/f_y = 0.8$ and $\sigma_t/f_y = -0.8$; *color filled circles*: TIV at the indicated points in time for the nth cycle; *white filled circles*: different TIV at the indicated points in time in the first cycle

Table 4.5 Development of Y/f_y from successive projection for $\sigma_p/f_y = 0.8$ and $\sigma_t/f_y = -0.8$

	First cycle (n = 1)				Second cycle (n = 2)		
	$t = 1$	$t = 2$	$t = 4$	$t = 6$	$t = 8$	$t = 10$	etc.
Bar 1	0	0	0.2	0.2	0.2	0.2	…
Bar 2	0	0.2	0.2	0.2	0.2	0.2	…
Bar 3	0	0.2	0.2	0.2	0.2	0.2	…

and the compatibility condition, which requires that the residual strains ε^* are the same in all bars, and the equilibrium condition that the sum of the residual stresses of all bars must be zero, we get

$$\varepsilon^* = \frac{1}{3}\frac{Y_1 + Y_2 + Y_3}{C} \,. \tag{4.191}$$

The last terms in each of the equations at Eq. (4.190) represent the initial strains. Because of Eq. (4.189) they are the same in all bars, and the MEA results in

$$\rho_1 = \rho_2 = \rho_3 = 0 \,. \tag{4.192}$$

In the state of elastic shakedown, the accumulated strain is therefore at the end of the cycle (where $\sigma^{\text{fel}} = \sigma_p$), and if a hardening modulus $E_t = 0.1E$ is chosen,

$$\varepsilon_1 = \varepsilon_2 = \varepsilon_3 = \frac{\sigma_p}{E} + \varepsilon^* = 2.6\frac{f_y}{E} \,. \tag{4.193}$$

This result is exact and was found with a single MEA for the Y values projected successively over two load cycles.

4.8.4.2 Plastic Shakedown—Mechanism (a)

There are different mechanisms for ratcheting in the case of plastic shakedown. Mechanism (a) is characterized by the fact that all three bars respond elastically at the removal of the temperature in any bar.

For an appropriate combination of parameters, the developments of the TIV are shown as a result of the successive projection in Fig. 4.52 and in Table 4.6 for the points in time with active plastic straining.

From the second cycle ($n = 2$) the successively projected TIV are periodical in all three bars. If an MEA is performed making use of the TIV identified successively from the second cycle, and choosing $E_t/E = 0.1$, the same residual strain is obtained at shakedown in all three bars, for example at the point in time $6(n-1)+2$, i.e., when the temperature is applied to bar 1,

$$\varepsilon^* = -3\frac{f_y}{E}\,, \tag{4.194}$$

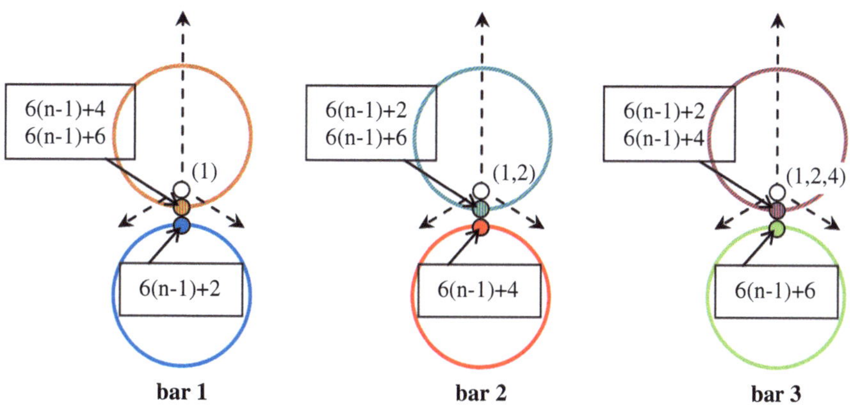

Fig. 4.52 Three-bar model at plastic shakedown: Mises circles and successive projection of Y in the TIV-space for $\sigma_p/f_y = 0$ and $\sigma_t/f_y = -1.5$; *color filled circles*: TIV at the indicated points in time for nth cycle; *white filled circles*: different TIV at the indicated points in time in the first cycle

Table 4.6 Development of Y/f_y from successive projection for $\sigma_p/f_y = 0$ and $\sigma_t/f_y = -1.5$

	First cycle ($n = 1$)				Second cycle ($n = 2$) etc.		
	$t = 1$	$t = 2$	$t = 4$	$t = 6$	$t = 8$	$t = 10$	$t = 12$
Bar 1	0	-0.5	−0.25	−0.25	−0.5	−0.25	−0.25
Bar 2	0	0	−0.5	−0.25	−0.25	−0.5	−0.25
Bar 3	0	0	0	−0.5	−0.25	−0.25	−0.5

which is now, however, not associated with vanishing residual stresses, and the accumulated strains become

$$\begin{aligned} \varepsilon_1 &= \frac{\sigma_p}{E} + \frac{\sigma_t}{E} + \varepsilon^* = -4.5\frac{f_y}{E} \\ \varepsilon_2 = \varepsilon_3 &= \frac{\sigma_p}{E} - \frac{1}{2}\frac{\sigma_t}{E} + \varepsilon^* = -2.25\frac{f_y}{E} \end{aligned} . \tag{4.195}$$

This result is exact and was found with a single MEA for the Y values projected successively over two load cycles.

4.8.4.3 Plastic Shakedown—Mechanism (b)

Mechanism (b) is characterized by the fact that the two other bars respond elastically at the removal of the temperature in any bar.

For an appropriate combination of parameters, the developments of the TIV are shown as a result of the successive projection in Fig. 4.53 and in Table 4.7. Already from the first cycle (n = 1) the successively projected TIV are periodic in all three bars.

Performing an MEA based on these successively projected TIV, the same residual strain is obtained at shakedown in all three bars, for example at the point in time $6(n-1)+2$, i.e., when bar 1 is at temperature, and choosing $E_t/E = 0.1$,

$$\varepsilon^* = -3\frac{f_y}{E} , \tag{4.196}$$

and the accumulated strains

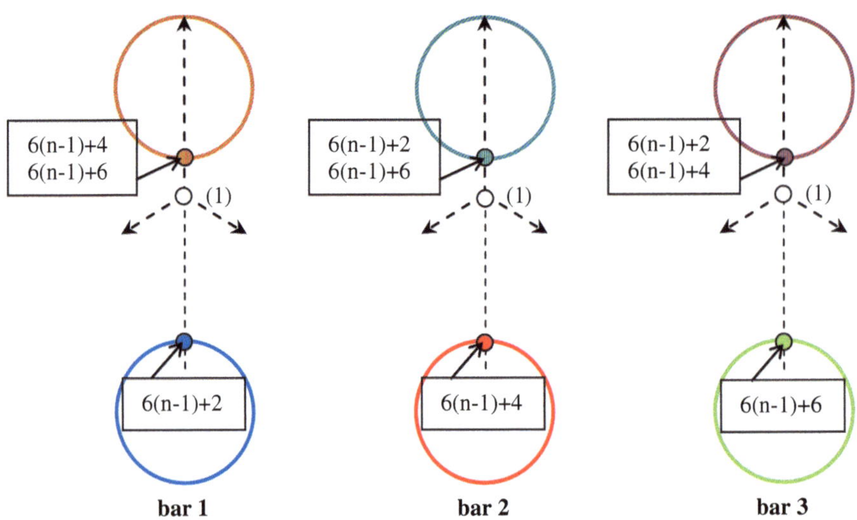

Fig. 4.53 Three-bar model at plastic shakedown: Mises circles and successive projection of Y in the TIV-space for $\sigma_p/f_y = 0$ and $\sigma_t/f_y = -3$

Table 4.7 Development of Y/f_y from successive projection for $\sigma_p/f_y = 0$ and $\sigma_t/f_y = -3$

	First cycle (n = 1)			
	t = 1	t = 2	t = 4	t = 6
Bar 1	0	−2	0.5	0.5
Bar 2	0	0.5	−2	0.5
Bar 3	0	0.5	0.5	−2

$$\begin{aligned} \varepsilon_1 &= \frac{\sigma_p}{E} + \frac{\sigma_t}{E} + \varepsilon^* = -6\frac{f_y}{E} \\ \varepsilon_2 = \varepsilon_3 &= \frac{\sigma_p}{E} - \frac{1}{2}\frac{\sigma_t}{E} + \varepsilon^* = -1.5\frac{f_y}{E} \end{aligned}. \tag{4.197}$$

This result is exact and was found with a single MEA for the Y values projected successively within the first load cycle.

4.8.4.4 Plastic Shakedown—Mechanism (C)

Mechanism (c) is characterized by the fact that all bars plasticize at removal of the temperature in any bar.

For releasing this mechanism, a high load level is required ($\sigma_t/f_y < -^4/_3(2 + E/E_t)$; e.g., $\sigma_p/f_y = 0$, $\sigma_t/f_y = -18$, $E_t/E = 0.1$). As with mechanism (b), so also here the successively projected TIV are periodical from the very first cycle (n = 1) in all three bars. Therefore, further details are omitted here.

Again, the exact elastic-plastic result is found with a single MEA.

References

1. Maier, G., Comi, C., Corigliani, A., Perego, U., Hübel, H.: Bounds and estimates on inelastic deformations, Commission of the European Communities, contract RA1-0162-I and RA1-0168-D, Report EUR 16555 EN. European Commission, Brussels (1992)
2. Zarka, J., Frelat, J., Inglebert, G., Kasmai-Navidi, P.: A new Approach to Inelastic Analyses of Structures. Martinus Nijhoff Publishers, Dordrecht (1988). (greatly expanded, 1990)
3. Zarka, J., Inglebert, G., Engel, J.J.: On a simplified inelastic analysis of structures. Nucl. Eng. Des. **57**, 333–368 (1980)
4. Zarka, J., Casier, J.: Elastic-plastic response of a structure to cyclic loading: practical rules. In: Nemet-Nasser, S. (ed.) Mechanics Today, vol. 6, pp 93–198. Pergamon Press, New York (1981)
5. Inglebert, G., Frelat, J.: Quick analysis of inelastic structures using a simplified method. Nucl. Eng. Des. **116**, 281–291 (1989)
6. Hübel, H., et al.: Performance study of the simplified theory of plastic zones and the Twice-Yield method for the fatigue check. Int. J. Press. Vessels Pip. **116**, 10–19 (2014). doi:10.1016/j.ijpvp.2014.01.003
7. Khalij, L., Hariri, S., Vaucher, R.: Shakedown of three-dimensional structures under cyclic loading with the simplified analysis. Comput. Mater. Sci. **24**, 393–400 (2002)
8. Cano, V., Taheri, S.: Elastic shakedown domain in an axisymmetrical structure subjected to a cyclic thermal and mechanical loading—comparison between an incremental model and a simplified method ECCOMAS 2000, Barcelona, September (2000)

9. Jiang, W., Leckie, F.A.: A direct method for the shakedown analysis of structures under sustained and cyclic loads. J. Appl. Mech. **59**, 251–260 (1992)
10. Herbland, T., et al.: An Evaluation of Simplified Methods to Compute the Mechanical Steady State, CETIM. http://www.cetim.fr/fr/Recherche?cx=004804313408259911386%3Ak-rpkzta8vw&cof=FORID%3A11&ie=UTF-8&q=herbland&x=0&y=0 (2012)
11. ANSYS Release 14.5, ANSYS Inc. Canonsburg, USA (2012)
12. Sicherheitstechnische Regel des KTA, KTA 3201.2, Komponenten des Primärkreises von Leichtwasserreaktoren, Teil 2: Auslegung, Konstruktion und Berechnung. Fassung 6/96 (including correction from BAnz. Nr. 129 vom 13.07.2000). Office of the KTA c/o Bundesamt für Strahlenschutz, Salzgitter (2000)
13. Zarka, J., Karaouni, H.: New Rules to be Included in Codes and Standards to Represent Multiaxial Variable Amplitudes Loadings on Structures Proceedings of ICONE 10 (ICONE10-22029). Arlington (2002)
14. Tribout, J.: A Simplified Method for the Inelastic Analysis of Structures Under Non radial Cyclic Loadings, SMiRT 7, vol. L, paper L14/4. Chicago (1983)
15. Inglebert, G., Frelat, J., Proix, J.M.: Structures under cyclic loading. Arch. Mech. **37**(4–5), 365–382 (1985). (Warszawa)
16. Legrand, E., Hassine, T., Inglebert, G.: A New Algorithm to Perform Shakedown Analysis on a Structure Under Non radial Loading 10th International. Conference on Mathematical and Computer Modelling and Scientific Computing, Boston, July (1995)
17. Hassine, T., Legrand, E., Inglebert, G.: Direct Estimation of the Stabilised State of a Structure Under Non radial Loading for Elastic Shakedown Case 10th International. Conference on Mathematical and Computer Modelling and Scientific Computing, Boston, July (1995)
18. Hassine, T., Inglebert, G., Pons, M.: Shakedown and damage analysis applied to rocket engines. In: Weichert, D., Maier, G. (eds.) Inelastic Analysis of Structures under Variable Load, pp 255–267. Kluwer Academic Publishers, Dordrecht (2000)
19. Yu, M., Moran, B., Keer, L.M.: A direct analysis of two-dimensional elastic-plastic rolling contact. Trans. ASME J. Tribol. **115**, 227–236 (1993)
20. Cecot, W.: On application of the Zarka approach to estimation of residual strains and stresses railroad rails. In: Proceedings of COMPLAS V, pp. 1873–1878 Barcelona (1997)
21. Cecot, W., Orkicz, J., Midura, G.: Estimation of Railroad Rail Residual Deformation after Roller Straightening Process ECCM-2001, Cracow, June. (2001)
22. Krok, J., Cecot, W., Pazdanowski, M.: Shakedown Analysis of Residual Stresses in Railroad Rails with Kinematic Hardening taken into Account ECCM-2001, Cracow, June. (2001)
23. Cecot, W.: Application of h-adaptive FEM and Zarka's approach to analysis of shakedown problems. Int. J. Numer. Meth. Eng. **6**, 2139–2158 (2004)
24. Yu, C.-C., Moran, B., Keer, L.M.: A simplified direct method for cyclic strain calculation: repeated rolling/sliding contact on a case-hardened half plane. Trans. ASME J. Tribol. **118**, 329–334 (1996)
25. Yu, C.-C., Keer, L.M.: Edge effect on elastic-plastic rolling/sliding contacts. Comput. Mech. **18**, 259–268 (1996)

STPZ with Temperature-Dependent Material Data

5

Unsteady thermal stresses are often responsible for the life of structures under cyclic loads beyond the elastic regime, such as in power plants. Temperature changes not only play the role of a load, such as in a number of examples in previous sections but also always bring about changes in the elastic and plastic material data, which was not considered in the previous chapters.

This chapter examines the role played by the temperature dependence of material data in terms of the elastic-plastic strain range and the accumulation of strain, what effects can be taken into account, and in which way by the STPZ.

"Temperature dependence" of material data means that material data not only depend on the local temperature distribution during anisothermal processes but also that they are variable in time at a fixed location. Material data varying only locally are in any case always permitted by the STPZ and do not require special treatment.

5.1 Linear Kinematic Hardening

Apart from the fact that the temperature or the free thermal expansion acts as a load, whereby the temperature dependence of the thermal expansion coefficient can easily be taken into consideration, the temperature may affect the strain range and the strain accumulation produced by cyclic loading in a variety of ways. Apart from the Young's modulus and Poisson's ratio, where the latter is often considered as temperature-independent, the tangent modulus E_t (or C) and the yield stress f_y are also dependent on temperature.

Under certain circumstances, even ratcheting can be caused by a cyclic temperature change, even if the temperature does not constitute a structural load, i.e., if the thermal strain can develop freely and is not constrained. A reason for this may be that a macroscopically isotropic polycrystalline material consists of single crystals with highly anisotropic thermal expansion coefficients [1]. At a temperature change, stresses are caused by mutual interference between the crystals, which at

H. Hübel, *Simplified Theory of Plastic Zones*,
DOI 10.1007/978-3-319-29875-7_5

most may lead to alternating plasticity in the absence of other loads. If, however, additional loads are applied, such as a constant external stress, then directional stress redistribution (see Sect. 2.1) occurs in every single crystal, so that the strain increments at each load reversal of a cyclic temperature change are not exactly opposite to each other. Instead, a net strain increment is created in each cycle, designated as multiaxial ratcheting in Sect. 2.5.2 and described there in more detail. In addition, it was pointed out in [1] that similar processes occur because of density changes as a result of thermally induced phase transformation of the crystal lattice. These two phenomena are not considered further in the following, but only the role of temperature-dependent hardening and temperature-dependent yield stress.

5.1.1 Temperature-Dependent Hardening Modulus

The effects of a temperature-dependent tangent modulus and a temperature-dependent yield stress at linear kinematic hardening have already been examined in [2] with respect to ratcheting. If a temperature-dependent hardening modulus $C(T)$ is introduced in the differential hardening law Eq. (1.24),

$$\mathrm{d}\xi_i = \frac{2}{3}\, C(T)\, \mathrm{d}\varepsilon_i^{\mathrm{pl}} \tag{5.1}$$

with

$$C(T) = \frac{E(T)\, E_t(T)}{E(T) - E_t(T)}, \tag{5.2}$$

then it can in general not be integrated in closed form to a finite hardening law. The yield surface then moves only in the deviatoric stress space if plastic strain increments also occur, so only simultaneously with load changes and not when the temperature is changing at constant load. During cyclic temperature changes, this would lead to infinite ratcheting [2], but this is hardly confirmed by experimental evidence. Another common approach is therefore to supplement the hardening law Eq. (5.1) by an additive term, which is referred to as Rice hardening in ANSYS:

$$\mathrm{d}\xi_i = \frac{2}{3}\, C(T)\, \mathrm{d}\varepsilon_i^{\mathrm{pl}} + \frac{1}{C(T)} \frac{\partial C(T)}{\partial T}\, \xi_i \mathrm{d}T, \tag{5.3}$$

which is very convenient, because Eq. (5.3) can be integrated in closed form to the finite hardening law

$$\xi_i = \frac{2}{3}\, C(T)\, \varepsilon_i^{\mathrm{pl}}, \tag{5.4}$$

and Eq. (5.3) can also be written as

$$d\xi_i = \frac{2}{3}\, C(T)\, d\varepsilon_i^{pl} + \frac{2}{3}\frac{\partial C(T)}{\partial T}\varepsilon_i^{pl} dT. \tag{5.5}$$

Accordingly, in anisothermal processes, kinematic hardening or softening can take place even in purely elastic state changes ($d\varepsilon_i^{pl} = 0_i$), without establishing a ratcheting mechanism thereby.

As an example, a tension-compression bar is considered under stress-controlled alternating loading and, for reasons of greater clarity, phase-shifted temperature cycles (Fig. 5.1). The temperature itself causes no stresses. The yield stress is constant, i.e., temperature-independent.

The stress–strain hysteresis loops shown in Fig. 5.2 are closed, so that at times 5, 9, etc., the stresses and strains are the same as at time 1, and thus there is no ratcheting. Between times 2 and 3, the yield point is only seemingly reduced. In fact, the yield surface has already moved between times 1 and 2 from $C_{hot}.\varepsilon^{pl}$ to $C_{cold}.\varepsilon^{pl}$, without thereby changing the plastic strains. The stress $\sigma_1 = \sigma_2$ is no longer located on the edge of the yield surface, but now lies in its interior, and therefore the full diameter of the Mises circle is no longer available as an elastic range at a stress reversal (Fig. 5.3). During the temperature rise from time 3 to time 4 the Mises yield surface remains at the same position, but this is associated with a change of the plastic state produced by the thermally-induced softening ($E_{t,hot} < E_{t,cold}$).

A notable consequence of the finite hardening law is that cyclic plastic straining may occur at a constant state of stress solely produced by a cyclic temperature change, which itself does not cause stresses. This case is shown as a stress–strain curve in Fig. 5.4 and also represented in Fig. 5.5, both in the space of deviatoric stresses and in the space of the TIV.

The differential approach Eq. (5.3), leading to the finite hardening law Eq. (5.4) in the case of linear kinematic hardening, is widely used and is also implemented in the anisothermal variant of the Chaboche model (Sect. 1.2.8) [3]:

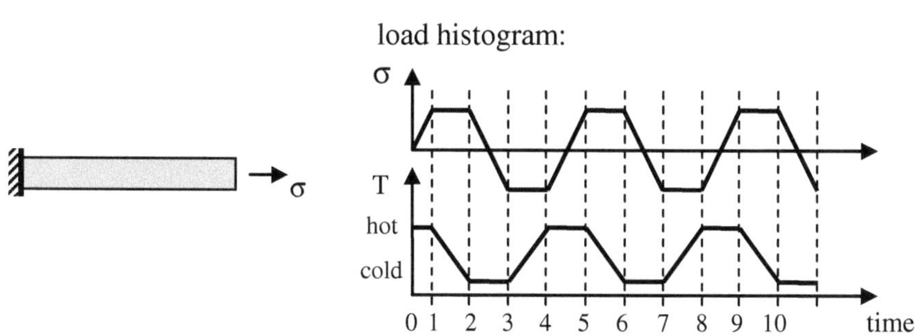

Fig. 5.1 Anisothermal process: tension-compression bar under alternating load

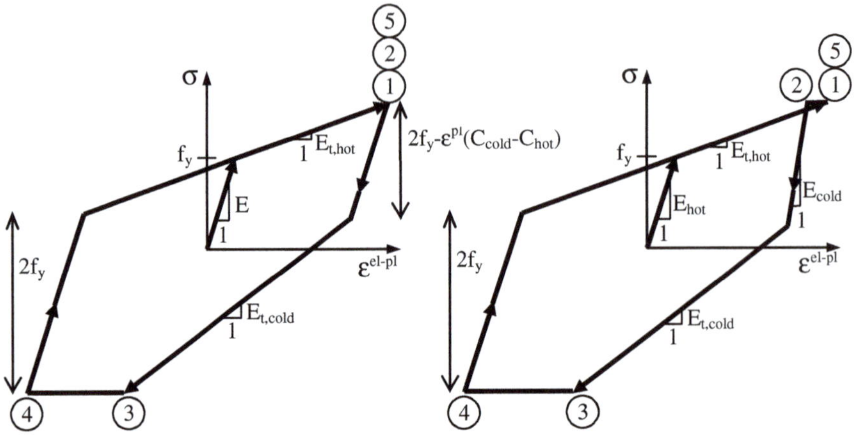

Fig. 5.2 Stress–strain hysteresis of the tension-compression bar under alternating stress with temperature-dependent hardening modulus E_t ($E_{t,\mathrm{hot}} < E_{t,\mathrm{cold}}$) according to a finite hardening law; *left* constant E-modulus; *right* temperature-dependent E-modulus ($E_{\mathrm{hot}} < E_{\mathrm{cold}}$)

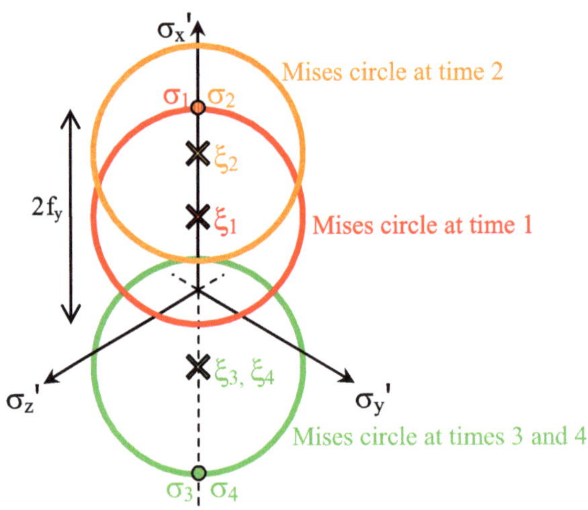

Fig. 5.3 Tension-compression bar under alternating stress condition with temperature-dependent hardening modulus E_t ($E_{t,\mathrm{hot}} < E_{t,\mathrm{cold}}$) according to a finite hardening law with constant E-modulus: yield surface in the deviatoric stress space

$$\mathrm{d}\xi_i = \frac{2}{3} C \mathrm{d}\varepsilon_i^{\mathrm{pl}} - \gamma \xi_i \mathrm{d}p + \frac{1}{C(T)} \frac{\partial C(T)}{\partial T} \xi_i \mathrm{d}T\,. \tag{5.6}$$

Because of the recovery term, this differential hardening law cannot be integrated in closed form.

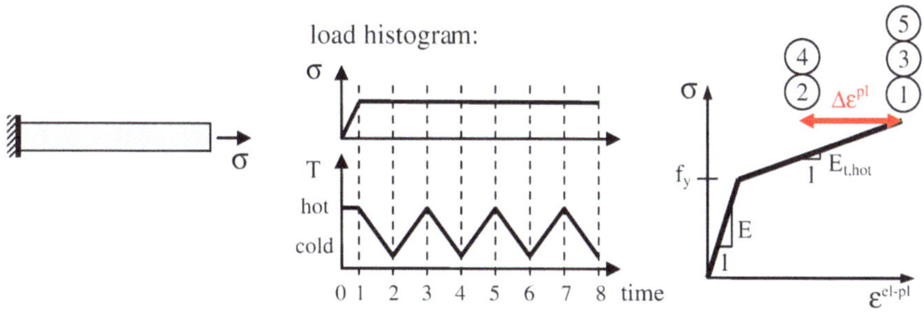

Fig. 5.4 Plastic strain range at constant tension stress produced by temperature-dependent hardening modulus E_t ($E_{t,\mathrm{hot}} < E_{t,\mathrm{cold}}$)

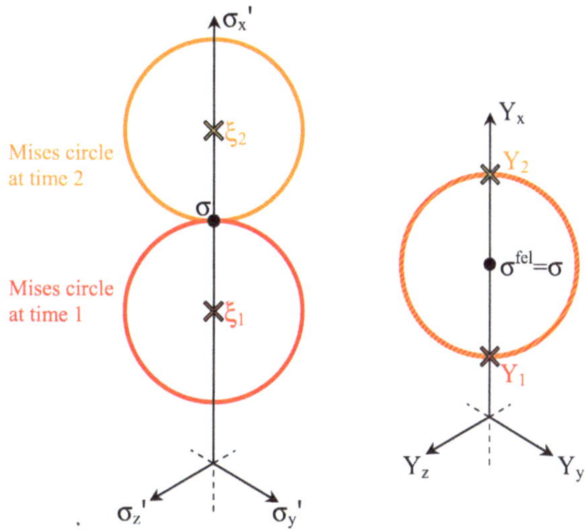

Fig. 5.5 Plastic shakedown at constant tension stress produced by temperature-dependent hardening modulus E_t ($E_{t,\mathrm{hot}} < E_{t,\mathrm{cold}}$); *left* deviatoric stress space; *right* TIV-space

5.1.2 Temperature-Dependent Yield Stress

For a temperature-dependent yield stress f_y ($f_{y,\mathrm{hot}} < f_{y,\mathrm{cold}}$), the stress–strain hysteresis of the tension-compression bar (Fig. 5.1) is represented in Fig. 5.6. Again, no ratcheting occurs. Although the hysteresis at first glance is very similar to the case of temperature-dependent hardening in Fig. 5.2, the underlying mechanism is quite different, as is clear from the motion of the yield surface shown in Fig. 5.7.

Noteworthy is also another effect of temperature-dependent yield stress, which comes to light if, instead of the previously considered out-of-phase temperature change, a change in temperature is considered that is synchronous with the stress change. The corresponding stress–strain hysteresis is shown in Fig. 5.8 as a result of alternating stress-controlled load if the yield stress depends linearly on time and

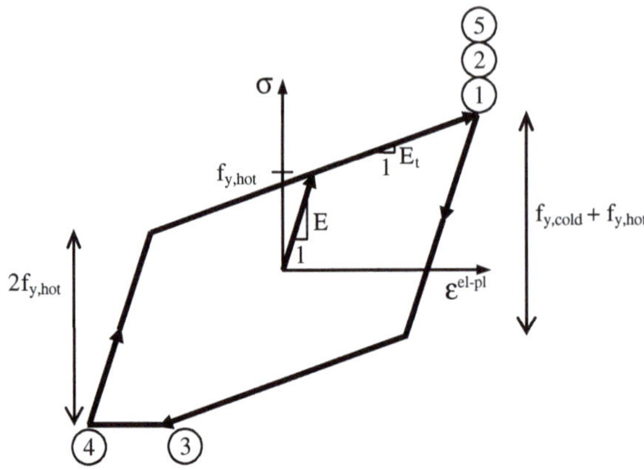

Fig. 5.6 Stress–strain hysteresis of the tension-compression bar under alternating stress with temperature-dependent yield stress ($f_{y,\text{hot}} < f_{y,\text{cold}}$)

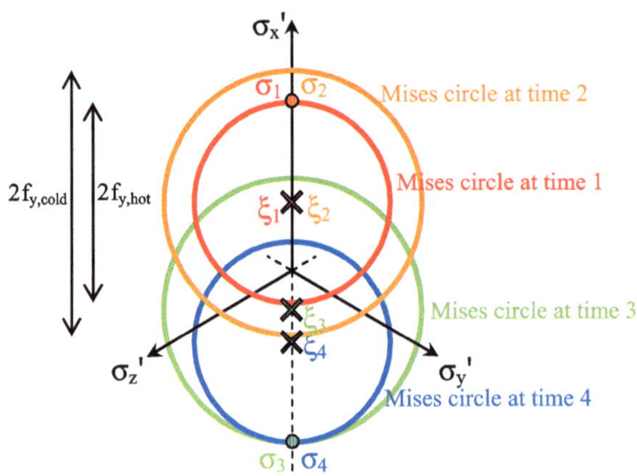

Fig. 5.7 Tension-compression bar under alternating stress with temperature-dependent yield stress ($f_{y,\text{hot}} < f_{y,\text{cold}}$): yield surface in the deviatoric stress space

therefore on the stress level, which gives the impression of different hardening modules during tension and compression.

If the loading is strain-controlled, other amazing behavior may be seen (Fig. 5.9). Depending on the ratio of the "hot" to the "cold" yield stress, anomalous behavior can occur where, with increasing strain, the stress decreases, accompanied by a decrease in the elastic strain share with a simultaneous increase of the plastic strain share. In the stress–strain curve this appears similar to a negative hardening modulus E_t. This has the remarkable consequence that the extreme stresses do not occur at the same times as the extreme strains, and thus the extreme stress or strain ranges are not necessarily linked to the load reversal points.

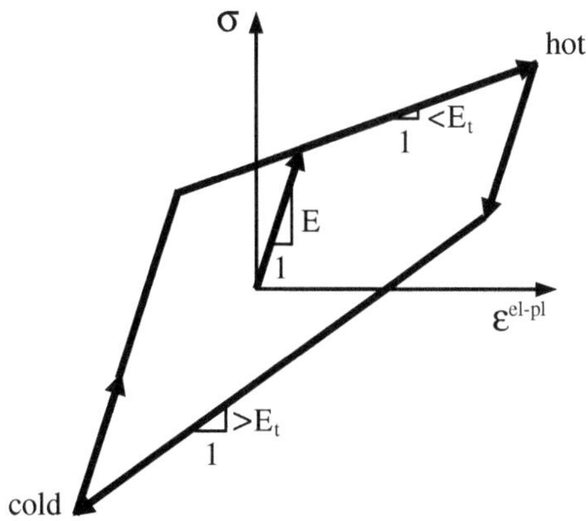

Fig. 5.8 Stress–strain hysteresis of the tension-compression bar under alternating stress with temperature-dependent yield stress and synchronized stress and temperature change

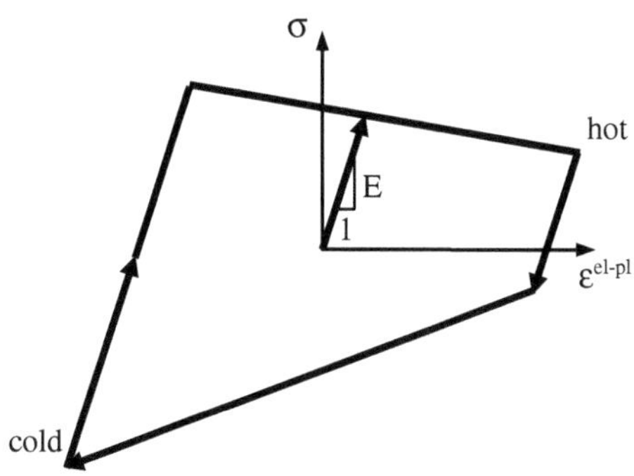

Fig. 5.9 Anomaly of the stress–strain hysteresis of the tension-compression bar under alternating strain with temperature-dependent yield stress and synchronized strain and temperature change

As an explanation of this fact, Fig. 5.10 illustrates in the deviatoric stress space that the stress decreases from time 1 to time 2, even though the internal variable ξ and hence the plastic strain is increased, because the radius of the yield surface f_y has become much smaller during this loading step.

As can be deduced from Fig. 5.10, such behavior apparently occurs in uniaxial stress states during time intervals $\Delta t = t_2 - t_1$ associated with active plastic straining, when

$$C\Delta\varepsilon^{\mathrm{pl}}\mathrm{sgn}(\Delta\sigma) < -\Delta f_y \mathrm{sgn}(\Delta\sigma). \tag{5.7}$$

With linear kinematic hardening and temperature-dependent yield stress, further behavior may occur which may be unfamiliar to some readers, namely re-elastic behavior after previous plastic straining at monotonous load increase. The situation

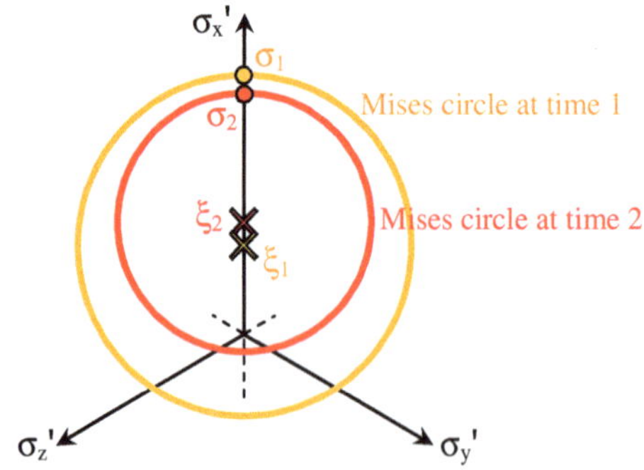

Fig. 5.10 Deviatoric stress space to explain the stress–strain anomaly with highly temperature-dependent yield stress

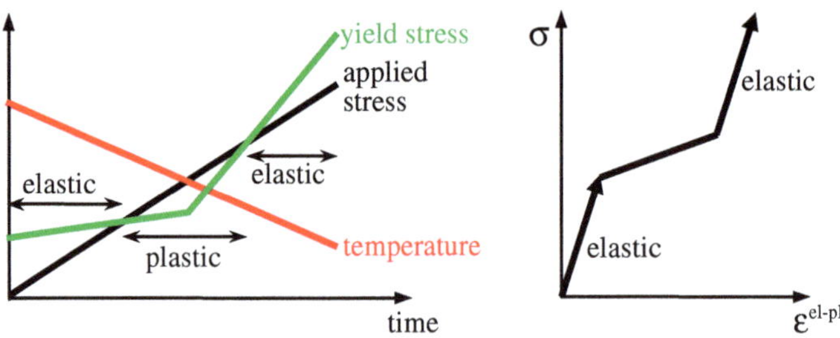

Fig. 5.11 Elastic behavior after previous plastic straining at monotonic increase in load because of temperature-dependent yield stress

shown in Fig. 5.11 for a tension bar may serve as an example, in which the stress increases linearly and the temperature decreases linearly with time. The temperature does not cause stress, but only controls the change in the yield stress during the loading process. If the yield stress depends monotonously, but not linearly, on the temperature, it may happen that, after initial elastic behavior plastic straining begins and that after further load increase, the plastic process comes to a halt. Then only elastic behavior occurs because the yield stress increases, because of the temperature, faster than the applied stress.

5.1.3 Basic Equations for Tension Bar

For a tension bar under monotonic loading, the stress–strain relationship (see Fig. 1.7) for linear kinematic hardening and isothermal behavior is given at loading time t by

$$\varepsilon(t) = \frac{\sigma(t) - f_y\left(1 - \frac{E_t}{E}\right)}{E_t}. \tag{5.8}$$

For anisothermal behavior, the stresses and strains at a specific loading time only depend on the material properties belonging to the temperatures present at this time in the case of the finite hardening law Eq. (5.4). The temperature history thus has no influence on the current material data (no temperature history effects). For monotonous plastic straining under tensile stress we therefore have

$$\varepsilon(t) = \frac{\sigma(t) - f_y(t)\left(1 - \frac{E_t(t)}{E(t)}\right)}{E_t(t)} . \tag{5.9}$$

The strain range at cyclic plastic straining between the states t_1 and t_2 is then

$$\Delta\varepsilon = \varepsilon(t_2) - \varepsilon(t_1) = \frac{\sigma(t_2)}{E_t(t_2)} - \frac{\sigma(t_1)}{E_t(t_1)} - \frac{f_y(t_2)}{C(t_2)}\,\mathrm{sgn}(\sigma(t_2)) - \frac{f_y(t_1)}{C(t_1)}\,\mathrm{sgn}(\sigma(t_1)), \tag{5.10}$$

which reduces to

$$\Delta\varepsilon = \varepsilon(t_2) - \varepsilon(t_1) = \frac{\Delta\sigma}{E_t} - \frac{2f_y}{E_t}\left(1 - \frac{E_t}{E}\right) \tag{5.11}$$

in the case of isothermal processes. Also interesting is the special case that only the yield stress is dependent on temperature:

$$\Delta\varepsilon = \varepsilon(t_2) - \varepsilon(t_1) = \frac{\Delta\sigma}{E_t} - \frac{f_y(t_1) + f_y(t_2)}{E_t}\left(1 - \frac{E_t}{E}\right)\mathrm{sgn}(\Delta\sigma) . \tag{5.12}$$

It is evident from Eq. (5.10) that the elastic-plastic strain range $\Delta\varepsilon$ in principle cannot be determined from the stress range

$$\Delta\sigma = \sigma_{\max} - \sigma_{\min} \tag{5.13}$$

in anisothermal processes, not even in uniaxial stress states. However, this is not all that surprising, because not even with linear elastic behavior can the strain range be calculated from the stress range in the case of a temperature-dependent Young's modulus, but is

$$\Delta\varepsilon = \frac{\sigma_{\max}}{E_{\max}} - \frac{\sigma_{\min}}{E_{\min}} , \tag{5.14}$$

so that the strain range for a pulsating load between 0 and $\sigma_{\max}$ differs from that for an alternating load between $-\frac{1}{2}\sigma_{\max}$ and $+\frac{1}{2}\sigma_{\max}$.

With respect to the STPZ, it is discussed in the following sections how a temperature dependence of the yield stress and the hardening modulus can be taken into account in anisothermal cyclic processes. This has implications for the identification of the plastic zone V_p, the modified elastic material parameters E^* and ν^*, and the determination of the initial strains or initial stresses for the modified elastic analysis.

5.2 STPZ at Anisothermal Monotonic Loading

The assignment of a location $\underline{x}$ either to the elastic or plastic subvolume V_e or V_p at the nth MEA takes place solely on the basis of the currently effective yield stress, e.g., for a load condition defined as "maximum":

$$V_p^{(n)} = \left\{ \underline{x} \;\middle|\; \sigma_{\mathrm{v}(\underline{x})}^{(n-1)} \geq \left(f_{\mathrm{y,max}}\right)_{(\underline{x})} \right\} \tag{5.15}$$

$$V_e^{(n)} = \left\{ \underline{x} \;\middle|\; \sigma_{\mathrm{v}(\underline{x})}^{(n-1)} < \left(f_{\mathrm{y,max}}\right)_{(\underline{x})} \right\}. \tag{5.16}$$

The elastic-plastic stress components are as usual obtained by superposition with the fictitious elastic calculated stress components, which are then used to form the equivalent stress σ_{v}.

The modified elastic material parameters E^* and ν^* of the modified elastic material law for monotonic behavior, Eq. (3.20), are defined in Eqs. (3.15) and (3.16) for temperature-independent behavior. The STPZ requires in principle a finite hardening law such as Eq. (5.4), so that a differentially formulated hardening law as in Eq. (5.1) cannot be accounted for. Then the modified elastic material law for anisothermal behavior is formally the same as for isothermal behavior:

$$\varepsilon_i^* = \begin{cases} \left(E_{ij}^*\right)^{-1} \rho_j + \varepsilon_{i,0} & \forall \underline{x} \in V_p \\ E_{ij}^{-1} \rho_j & \forall \underline{x} \in V_e \end{cases}, \tag{5.17}$$

whereby now those modified elastic material parameters are to be used, which result from the temperature at the considered time of loading:

$$E^* = E_{t,\max} \tag{5.18}$$

$$\nu^* = \frac{1}{2} - \frac{E_{t,\max}}{E_{\max}} \left(\frac{1}{2} - \nu_{\max} \right). \tag{5.19}$$

The initial strains in the nth iteration step accordingly arise from the yield stress effective at this loading time and the plastic hardening modulus C:

$$\varepsilon_{i,0} = \frac{3}{2 C_{\max}} Y_i^{(n)} \tag{5.20}$$

with

$$C_{\max} = \frac{E_{\max}\, E_{t,\max}}{E_{\max} - E_{t,\max}} \tag{5.21}$$

$$Y_i^{(n)} = \sigma_i^{'\mathrm{fel}} - \sigma_i^{'(n-1)} \left(\frac{f_{y,\max}}{\sigma_{\mathrm{v}}^{(n-1)}} \right) \quad \forall \underline{x} \in V_p^{(n)} . \tag{5.22}$$

It is worth noting that the modified elastic material parameters E^* and ν^* are location-dependent when a load is applied through a temperature field, whereas the same modified elastic stiffness exists throughout the entire plastic zone in the case of temperature-independent behavior. Apart from that, there are no particular differences from isothermal behavior. This is unfortunately different at cyclic loading.

5.3 STPZ at Anisothermal Cyclic Loading

The determination of the nature of the shakedown state, i.e., whether under cyclic loading elastic or plastic shakedown occurs, is very complex with temperature-dependent material parameters. For example, it was shown for a simple configuration in Fig. 5.4, that plastic shakedown occurs without any change in stress, just because the hardening modulus is temperature-dependent.

The following description and application of the STPZ is limited therefore, essentially to the consideration of a temperature-dependent yield stress; see also [4].

If thermal expansion is constrained during temperature changes, stresses and strains evolve automatically in phase with the temperature change, unlike in Fig. 5.1. The size of the yield surfaces at the two extreme states of a one-parameter cyclic loading is then different. Under certain circumstances, the nature of the shakedown state is only determined by the ratio of the fictitious elastic calculated equivalent stress range between these two extreme load conditions to the average diameter of the two yield surfaces. If the yield stress $f_{y,\min}$ is assigned to the "minimum" load condition, and $f_{y,\max}$ to the "maximum" load condition, then their arithmetic mean is inserted in the conditions Eqs. (4.3) and (4.4):

$$\Delta\sigma_{\mathrm{v}(\underline{x})}^{\mathrm{fel}} \leq \left(f_{y,\min} + f_{y,\max}\right)_{(\underline{x})} \quad \forall \underline{x} \in V \quad \rightarrow \quad \mathrm{ES} \tag{5.23}$$

$$\Delta\sigma_{\mathrm{v}(\underline{x})}^{\mathrm{fel}} > \left(f_{y,\min} + f_{y,\max}\right)_{(\underline{x})} \quad \exists \underline{x} \in V \quad \rightarrow \quad \mathrm{PS}. \tag{5.24}$$

That this approach, which is to consider only the two extreme loading conditions, in general only allows an approximate solution, has already been discussed in previous chapters. As shown in Sect. 5.1.2, phenomena can occur as a result of temperature dependence of the yield stress, which cannot be detected by sole consideration of the extreme load conditions. The reasons for this may be that the extreme values of stress and strain do not occur at the same time as the extreme states of loading (Fig. 5.9), or because, despite load increase, again elastic behavior is achieved (Fig. 5.11). If such phenomena occur, the STPZ includes additional inaccuracies. With incremental analyses, the demands on the numerical stability increases, resulting in increased computational effort. In many applications, however, the temperature dependence of the yield stress is "good-natured", so that after appropriate adjustment of the STPZ as described in the following sections, no special loss of accuracy is expected.

No special requirements are imposed on the course of the temperature development between the two extreme load conditions, i.e., whether or not the temperature changes linearly or nonlinearly with time. The temperature change, as load changes generally, should only occur monotonously. Then, usually, the yield stress also changes monotonically during the loading process. The dependence of the yield stress on the temperature may then be either linear or nonlinear.

5.4 Strain Range at Plastic Shakedown

The modified elastic material law, taking into account temperature-dependent material data, can be expressed for the finite hardening law, Eq. (5.4), by

$$\Delta\varepsilon_i^* = \begin{cases} \left[\left(E_{ij}^*\right)_{\max}^{-1} \rho_{j,\max} - \left(E_{ij}^*\right)_{\min}^{-1} \rho_{j,\min}\right] + \left[\varepsilon_{i,0,\max} - \varepsilon_{i,0,\min}\right] & \forall \underline{x} \in V_p \\ \left[E_{ij,\max}^{-1}\, \rho_{j,\max} - E_{ij,\min}^{-1}\, \rho_{j,\min}\right] & \forall \underline{x} \in V_e \end{cases} \tag{5.25}$$

with

$$E_{\max}^* = E_{t,\max}; \quad E_{\min}^* = E_{t,\min} \tag{5.26}$$

$$\nu_{\max}^* = \frac{1}{2} - \frac{E_{t,\max}}{E_{\max}}\left(\frac{1}{2} - \nu_{\max}\right); \quad \nu_{\min}^* = \frac{1}{2} - \frac{E_{t,\min}}{E_{\min}}\left(\frac{1}{2} - \nu_{\min}\right) \tag{5.27}$$

$$\varepsilon_{i,0,\max} = \frac{3}{2C_{\max}} Y_{i,\max}^{(n)}; \quad \varepsilon_{i,0,\min} = \frac{3}{2C_{\min}} Y_{i,\min}^{(n)} \tag{5.28}$$

$$C_{\max} = \frac{E_{\max}\, E_{t,\max}}{E_{\max} - E_{t,\max}}; \quad C_{\min} = \frac{E_{\min}\, E_{t,\min}}{E_{\min} - E_{t,\min}} \tag{5.29}$$

$$Y_{i,\max}^{(n)} = \sigma_{i,\max}^{\prime\text{fel}} - \sigma_{i,\max}^{\prime(n-1)}\left(\frac{f_{y,\max}}{\sigma_{\text{v},\max}^{(n-1)}}\right); \quad Y_{i,\min}^{(n)} = \sigma_{i,\min}^{\prime\text{fel}} - \sigma_{i,\min}^{\prime(n-1)}\left(\frac{f_{y,\min}}{\sigma_{\text{v},\min}^{(n-1)}}\right). \tag{5.30}$$

Unfortunately, none of the three terms in square brackets in Eq. (5.25) can be rephrased using the corresponding range values $\Delta\rho$, $\Delta\varepsilon_0$ or ΔY. Accordingly, neither a unique modified load nor unique modified elastic material data can be assigned to a specific location of a structure, so that the strain range cannot be determined directly by a modified elastic analysis or an iterative series of MEAs. However, this is not a peculiarity or limitation of the STPZ, as the discussion in Sect. 5.1.3 has already shown.

Because the position of the stress–strain loop moves towards higher stresses with increasing accumulated strain in the case of ratcheting, a separate analysis of strain range and accumulated strain does not make sense.

However, when taking a closer look, one can see that Eq. (5.25) can be reformulated solely by range expressions, when the temperature dependence of the material data is limited to the yield stress. In this case, the modified elastic material law is the same as for isothermal behavior in Eq. (4.5):

$$\Delta\varepsilon_i^* = \begin{cases} \left(E_{ij}^*\right)^{-1} \Delta\rho_j + \Delta\varepsilon_{i,0} & \forall \underline{x} \in V_p \\ E_{ij}^{-1}\,\Delta\rho_j & \forall \underline{x} \in V_e \end{cases} \tag{5.31}$$

with

$$\Delta\varepsilon_{i,0} = \frac{3}{2C}\Delta Y_i^{(n)}, \tag{5.32}$$

whereby the TIV ΔY_i corresponds to Eq. (4.17) and can be determined analogously to temperature-independent behavior in Fig. 4.2 from consideration of the TIV-space in Fig. 5.12:

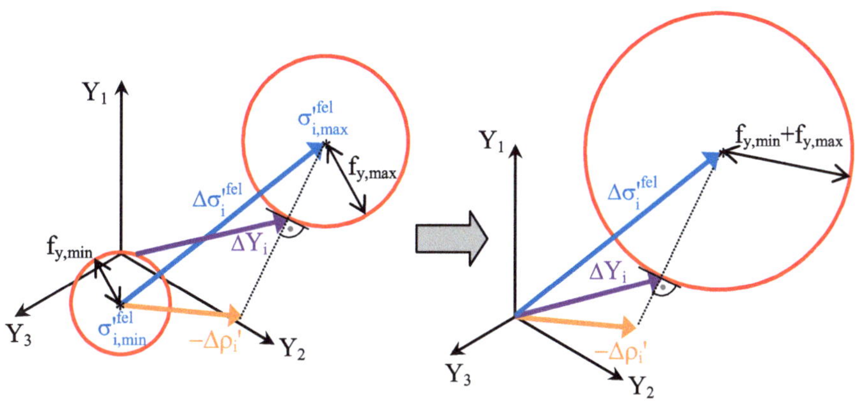

Fig. 5.12 Estimation of the TIV-range at cyclic loading in the TIV-space in the case of temperature-dependent yield stress

$$\Delta Y_i^{(n)} = \Delta\sigma_i^{'\mathrm{fel}} - \Delta\sigma_i^{'(n-1)} \left(\frac{f_{y,\max} + f_{y,\min}}{\Delta\sigma_{\mathrm{v}}^{(n-1)}} \right) \quad \forall \underline{x} \in V_p^{(n)} . \tag{5.33}$$

This corresponds exactly to the formulation of temperature-independent behavior when twice the yield stress is replaced by the sum of the two yield stresses effective at the extremes of the load cycle, which is equivalent to an arithmetic averaging of the two yield stresses. It also follows directly how a location $\underline{x}$ of the structure is assigned to the elastic or the plastic subvolumes V_e or V_p in the nth MEA, so that, instead of Eqs. (4.12) and (4.13), the following equations now apply:

$$V_p^{(n)} = \left\{ \underline{x} \,\middle|\, \Delta\sigma_{\mathrm{v}(\underline{x})}^{(n-1)} \geq \left(f_{y,\min} + f_{y,\max}\right)_{(\underline{x})} \right\} \tag{5.34}$$

$$V_e^{(n)} = \left\{ \underline{x} \,\middle|\, \Delta\sigma_{\mathrm{v}(\underline{x})}^{(n-1)} < \left(f_{y,\min} + f_{y,\max}\right)_{(\underline{x})} \right\} . \tag{5.35}$$

In this special case, or in the approximation that the yield stress, but not the hardening behavior, is considered as temperature-dependent, the relationships shown simplify to a great extent. This is because all calculations can be performed practically in the same way as in isothermal behavior discussed in Sect. 4.2. In the entire region V_p, the structure then has the same modified elastic stiffness, because E^* and ν^* are independent of the yield stress. Thus only the geometry of the plastic zone V_p and the TIV-range ΔY_i, and thus also the modified loading, vary from location to location. A separate analysis of strain range and accumulated strain again makes sense, just as the determination of the nature of the state of shakedown as described in Sect. 5.3. In this case, no specific law for the dependence of the yield stress on temperature is required.

In the special case of a uniaxial stress state, Eqs. (5.31)–(5.33) degenerate to

$$\Delta\varepsilon^* = \begin{cases} \dfrac{\Delta\rho}{E_t} + \Delta\varepsilon_0 & \forall \underline{x} \in V_p \\[2ex] \dfrac{\Delta\rho}{E} & \forall \underline{x} \in V_e \end{cases} \tag{5.36}$$

$$\Delta\varepsilon_0 = \frac{\Delta Y}{C} \quad \forall \underline{x} \in V_p^{(n)} \tag{5.37}$$

$$\Delta Y = \Delta\sigma^{\mathrm{fel}} - \left(f_{y,\min} + f_{y,\max}\right) \operatorname{sgn}\left(\Delta\sigma^{\mathrm{fel}}\right) \quad \forall \underline{x} \in V_p^{(n)} . \tag{5.38}$$

Because the fatigue curves determined by isothermal experiments are based on strain ranges (see Sect. 2.8.1), these are also required even if they cannot be calculated that easily because of temperature-dependent hardening behavior. It is therefore sometimes recommended in design codes to perform the calculation of anisothermal processes with isothermal material data, determined for a weighted average $\bar{T}$ of the temperatures at the two extreme load conditions. In the KTA safety standards [5, Sects. 7.13.2 and 8.4.3.2], for example, it is recommended to choose

$$\bar{T} = 0.25\check{T} + 0.75\hat{T}, \tag{5.39}$$

where $\check{T}$ represents the lower and $\hat{T}$ the higher temperature. The elastic as well as the modified elastic material parameters E, ν, E^*, and ν^* determined for the temperature $\bar{T}$ are then used in Eqs. (5.31)–(5.33).

5.5 Examples for Strain Ranges at Plastic Shakedown

Some of the previously discussed examples are chosen and the strain ranges are studied from the point of view of variable yield stresses.

5.5.1 Two-Bar Model

Some of the findings in Sect. 4.3.1 no longer apply with temperature-dependent yield stress. The fictitious elastic solution of Eqs. (4.19) and (4.20) is, of course, still true, stating that the absolute values of the stress ranges are the same in both bars. However, because of Eqs. (5.34) and (5.35) it is no longer necessarily true that either both bars are elastic or both bars are plastic, but now it is possible that one bar is plastic and the other remains elastic.

At minimum load condition both bars are at "cold" temperature, whereas at maximum load condition the left bar is "hot" and the right bar maintains it "cold" starting temperature.

Prior to the first MEA, the residual stresses are still zero, so that according to Eq. (5.34) not only the left but also the right bar is assigned to the plastic zone V_p, if

$$\Delta\sigma_{\text{right}}^{\text{fel}} = -\Delta\sigma_{\text{left}}^{\text{fel}} = \sigma_t \geq 2f_{y,\text{cold}} \,. \tag{5.40}$$

Then the modified elastic material law for the uniaxial stress state, Eqs. (5.36)–(5.38), is, taking into account the fact that the right bar persists at the temperature of the "minimal" load condition,

$$\Delta\varepsilon_{\text{left}}^* = \frac{1}{E_t}\Delta\rho_{\text{left}} + \frac{1}{C}\left(\Delta\sigma_{\text{left}}^{\text{fel}} + f_{y,\text{hot}} + f_{y,\text{cold}}\right) \tag{5.41}$$

$$\Delta\varepsilon_{\text{right}}^* = \frac{1}{E_t}\Delta\rho_{\text{right}} + \frac{1}{C}\left(\Delta\sigma_{\text{right}}^{\text{fel}} - 2f_{y,\text{cold}}\right) \,. \tag{5.42}$$

With the equilibrium condition

$$\Delta\rho_{\text{right}} = -\Delta\rho_{\text{left}} \tag{5.43}$$

and the kinematic condition

$$\Delta\varepsilon^*_{\text{left}} = \Delta\varepsilon^*_{\text{right}} \tag{5.44}$$

we obtained as a result of the first MEA:

$$\Delta\rho_{\text{left}} = -\left(1 - \frac{E_t}{E}\right)\left(\Delta\sigma^{\text{fel}}_{\text{left}} + \frac{3}{2}f_{y,\text{cold}} + \frac{1}{2}f_{y,\text{hot}}\right) \tag{5.45}$$

$$\Delta\varepsilon^*_{\text{left}} = \frac{\left(1 - \frac{E_t}{E}\right)}{E_t}\frac{1}{2}\left(f_{y,\text{hot}} - f_{y,\text{cold}}\right), \tag{5.46}$$

so that the loading cannot be considered as strain-controlled ($\Delta\varepsilon^* = 0$) any longer. Now it is necessary to check whether the assignment made on the basis of the criterion of Eq. (5.40) applies, i.e., that both bars belong to the subvolume V_p. For this, the residual stress from the first MEA is to be superimposed with the fictitious elastic solution and in the right bar, which remains at cold temperature even at the maximum load condition, to compare with $2f_{y,\text{cold}}$ according to Eq. (5.34). The right bar remains elastic if

$$\Delta\sigma_{\text{right}} = \Delta\sigma^{\text{fel}}_{\text{right}} + \Delta\rho_{\text{right}} < 2f_{y,\text{cold}}\,. \tag{5.47}$$

This is the case when

$$2f_{y,\text{cold}} \le \Delta\sigma^{\text{fel}}_{\text{right}} < \frac{1}{2}\frac{E}{E_t}\left(f_{y,\text{cold}} - f_{y,\text{hot}}\right) + \frac{3}{2}f_{y,\text{cold}} + \frac{1}{2}f_{y,\text{hot}}\,. \tag{5.48}$$

The second MEA is then performed with the modified elastic material law

$$\Delta\varepsilon^*_{\text{left}} = \frac{1}{E_t}\Delta\rho_{\text{left}} + \frac{1}{C}\left(\Delta\sigma^{\text{fel}}_{\text{left}} + f_{y,\text{hot}} + f_{y,\text{cold}}\right) \tag{5.49}$$

$$\Delta\varepsilon^*_{\text{right}} = \frac{1}{E}\Delta\rho_{\text{right}}\,. \tag{5.50}$$

The solution of Eqs. (5.43) and (5.44) is then

$$\Delta\rho_{\text{left}} = -\frac{1 - \frac{E_t}{E}}{1 + \frac{E_t}{E}}\left(\Delta\sigma^{\text{fel}}_{\text{left}} + f_{y,\text{cold}} + f_{y,\text{hot}}\right) \tag{5.51}$$

$$\Delta\varepsilon^*_{\text{left}} = \frac{1 - \frac{E_t}{E}}{1 + \frac{E_t}{E}} \frac{\left(\Delta\sigma^{\text{fel}}_{\text{left}} + f_{y,\text{cold}} + f_{y,\text{hot}}\right)}{E} \,. \tag{5.52}$$

The elastic-plastic strain range is then obtained by superposition of Eq. (5.46) or Eq. (5.52) with the fictitious elastic strain range.

At the latest after the second MEA the exact result is achieved. Incremental analyses, however, often need more stringent convergence tolerances and many load steps at temperature-dependent yield stress, even with uniaxial stress states, i.e., without directional redistribution.

5.5.2 Multiaxial Ratcheting

A renewed examination of the range quantities of the single element, which served as an example of multiaxial ratcheting, examined in Sect. 4.3.2, brings no new insights if a cyclic temperature change is superimposed. This causes no thermal stress but controls only the change of material parameters. Because the plastic zone is immutable and the cyclic portion of the load is uniaxial strain-controlled, the stress range in the direction of the load (see Eq. (4.42)) and the two transverse strains (see Eq. (4.43)) remain valid if the constant yield stress is replaced there by the arithmetic mean of the two yield stresses effective at the extremes of the load.

5.5.3 Bree Tube

In Sect. 4.3.3, the strain and stress range was determined for the beam-type replacement model of the Bree-tube, in which only uniaxial stresses occur. Temperature-independent material behavior was assumed there.

Now it is investigated what changes result from temperature-dependent yield stress. For simplicity, it is assumed that the yield stress depends linearly on the temperature:

$$f_y(T) = f_y(T_{\text{cold}}) + \frac{T - T_{\text{cold}}}{T_{\text{hot}} - T_{\text{cold}}} \left(f_y(T_{\text{hot}}) - f_y(T_{\text{cold}})\right) , \tag{5.53}$$

where the yield stress of course decreases with increasing temperature:

$$f_y(T_{\text{cold}}) = f_{y,\text{cold}} > f_y(T_{\text{hot}}) = f_{y,\text{hot}} \,. \tag{5.54}$$

Thus the behavior of real materials is not particularly well covered, but of course still better than to assume no temperature dependence of the yield stress. For the purpose of making the reader more familiar with the application of the STPZ, it is good enough anyway.

The minimum load condition is characterized by a uniform temperature distribution across the beam thickness, causing no stresses if acting alone. At the maximum load condition, however, there is a linear temperature profile, which is characterized by heating on one side of the central axis (in the negative coordinate direction y) and by a corresponding equal cooling in the positive y-direction. Thus, the fictitious elastic stresses are antimetric to the center line of the beam and are positive at the top surface $y = +t/2$. The fictitious elastic stress range is already given by Eq. (4.44):

$$\Delta\sigma_{(y)}^{\text{fel}} = \frac{y}{t/2}\sigma_t \,. \tag{5.55}$$

The distribution of the relevant yield stresses for the minimum and maximum load is given by

$$f_{y,\min(y)} = \frac{1}{2}\left(f_{y,\text{cold}} + f_{y,\text{hot}}\right) \tag{5.56}$$

$$f_{y,\max(y)} = \frac{1}{2}\left(f_{y,\text{cold}} + f_{y,\text{hot}}\right) + \frac{y}{t/2}\frac{1}{2}\left(f_{y,\text{cold}} - f_{y,\text{hot}}\right) . \tag{5.57}$$

For the first MEA, the boundaries between V_e and V_p arise solely from the fictitious elastic calculation. Provided the temperature gradient is sufficiently large, we get two separate plastic zones, which are, in contrast to temperature-independent yield stresses (see Eq. (4.47)), now no longer arranged antimetrically:

$$\frac{y_{\text{pl,above}}}{t/2} = + \frac{f_{y,\text{cold}} + f_{y,\text{hot}}}{\sigma_t - \frac{1}{2}\left(f_{y,\text{cold}} - f_{y,\text{hot}}\right)} \tag{5.58}$$

$$\frac{y_{\text{pl,below}}}{t/2} = - \frac{f_{y,\text{cold}} + f_{y,\text{hot}}}{\sigma_t + \frac{1}{2}\left(f_{y,\text{cold}} - f_{y,\text{hot}}\right)} \,. \tag{5.59}$$

The ranges of the TIV in the plastic zones are

$$\Delta Y_{(y > y_{\text{pl,above}})} = \frac{y}{t/2}\left[\sigma_t - \frac{1}{2}\left(f_{y,\text{cold}} - f_{y,\text{hot}}\right)\right] - \left(f_{y,\text{cold}} + f_{y,\text{hot}}\right) \tag{5.60}$$

$$\Delta Y_{(y < y_{\text{pl,below}})} = \frac{y}{t/2}\left[\sigma_t + \frac{1}{2}\left(f_{y,\text{cold}} - f_{y,\text{hot}}\right)\right] + \left(f_{y,\text{cold}} + f_{y,\text{hot}}\right) . \tag{5.61}$$

The modified elastic material law for the two V_p-regions and the intermediate elastic subvolume V_e is therefore

$$\Delta\varepsilon^*_{(y>y_{\text{pl,above}})} = \frac{\Delta\rho_{(y)}}{E_t} + \frac{1}{C}\left\{\frac{y}{t/2}\left[\sigma_t - \frac{1}{2}\left(f_{y,\text{cold}} - f_{y,\text{hot}}\right)\right] - \left(f_{y,\text{cold}} + f_{y,\text{hot}}\right)\right\} \tag{5.62}$$

$$\Delta\varepsilon^*_{(y_{\text{pl,below}}\leq y\leq y_{\text{pl,above}})} = \frac{\Delta\rho_{(y)}}{E} \tag{5.63}$$

$$\Delta\varepsilon^*_{(y<y_{\text{pl,below}})} = \frac{\Delta\rho_{(y)}}{E_t} + \frac{1}{C}\left\{\frac{y}{t/2}\left[\sigma_t + \frac{1}{2}\left(f_{y,\text{cold}} - f_{y,\text{hot}}\right)\right] + \left(f_{y,\text{cold}} + f_{y,\text{hot}}\right)\right\}. \tag{5.64}$$

Because of the Bernoulli hypothesis that cross-sections must remain plane, the residual strain range $\Delta\varepsilon^*$ is a linear function of the section coordinate y. Because the same conditions must be present in all sections, no rotation may occur because of the boundary conditions. Therefore, $\Delta\varepsilon^*$ must be independent of y. According to Eq. (5.63), the residual stress range must then also be distributed constantly in the elastic subvolume and is denoted $\Delta\rho_m$. The residual stress ranges in the two V_p-regions vary linearly and can be expressed depending on $\Delta\rho_m$. In total we have for the three regions:

$$t/2 \geq y > y_{\text{pl,above}}:$$
$$\Delta\rho_{(y)} = \frac{E_t}{E}\Delta\rho_m + \frac{E_t}{C}\left(f_{y,\text{cold}} + f_{y,\text{hot}}\right) - \frac{y}{t/2}\frac{E_t}{C}\left[\sigma_t - \frac{1}{2}\left(f_{y,\text{cold}} - f_{y,\text{hot}}\right)\right] \tag{5.65}$$

$$y_{\text{pl,above}} \geq y \geq y_{\text{pl,below}}:$$
$$\Delta\rho_{(y)} = \Delta\rho_m \tag{5.66}$$

$$y_{\text{pl,below}} > y \geq t/2:$$
$$\Delta\rho_{(y)} = \frac{E_t}{E}\Delta\rho_m - \frac{E_t}{C}\left(f_{y,\text{cold}} + f_{y,\text{hot}}\right) - \frac{y}{t/2}\frac{E_t}{C}\left[\sigma_t + \frac{1}{2}\left(f_{y,\text{cold}} - f_{y,\text{hot}}\right)\right]. \tag{5.67}$$

$\Delta\rho_m$ can now be determined from the requirement that for equilibrium reasons, the range of the axial force must vanish:

$$\Delta N = \int_{-t/2}^{+t/2} \Delta\rho_{(y)}\mathrm{d}y \stackrel{!}{=} 0. \tag{5.68}$$

After execution of the integral, some algebra, and inserting $\Delta\rho_m$ in Eq. (5.63) we get

$$\Delta\varepsilon^* = \frac{1}{2E}\frac{\left(\frac{y_{\text{pl,above}}}{t/2}+\frac{y_{\text{pl,below}}}{t/2}\right)\left(f_{y,\text{cold}}+f_{y,\text{hot}}\right)-f_{y,\text{cold}}+f_{y,\text{hot}}}{2\frac{C}{E}+\frac{y_{\text{pl,above}}}{t/2}-\frac{y_{\text{pl,below}}}{t/2}}. \tag{5.69}$$

Because of the lack of antimetry of the TIV-range in the two plastic zones, and thus of the initial strains in the modified elastic material law, the residual strain range does not now disappear in contrast to temperature-independent yield stresses (see Eq. (4.55)), but there occurs a cyclic membrane strain. The thermal load is thus no longer strain-controlled.

It is now necessary to examine whether the plastic zone as a result of this first MEA has changed. For this purpose, $\Delta\rho_m$ is to be inserted in Eqs. (5.65)–(5.67) to superimpose the result with the fictitious elastic solution and to identify V_e and V_p afresh with Eqs. (5.34) and (5.35). If necessary, more MEAs are to be carried out but this is not dealt with here.

Whereas at temperature-independent yield stress, one MEA was still enough to determine the exact strain range (Sect. 4.3.3), so one or the other additional MEA is now required. Although the range of the TIV Y is still precisely known from the outset because of the uniaxial stress state, the geometry of the plastic zone V_p must be determined iteratively.

5.5.4 Thick-Walled Cylinder Loaded by a Temperature Transient

In Sect. 4.3.4, the strain range has already been calculated for a thick-walled cylinder with temperature-independent material data exposed to a temperature transient of the fluid moving through. If the yield stress is considered to be temperature-dependent to a "reasonable" degree, for example, instead of the constant value of 100 N/mm^2, now in the considered temperature range (50–350 °C) from 130 to 70 N/mm^2, this does have a big impact on the stress ranges but very little effect on the strain ranges. Virtually no differences can be identified with respect to the ranges of the initial strains and the elastic-plastic strains shown in Figs. 4.9–4.11 when applying the STPZ or an incremental analysis. A closer examination is thus not made here.

5.6 Accumulated Strains at Elastic Shakedown

The following describes how temperature-dependent yield stresses can be taken into account in the STPZ to determine the strain accumulated in the state of elastic shakedown. The description is based on the procedure for temperature-independent yield stress presented in Sect. 4.4. Whether it is useful in light of the discussion in

Sect. 5.1 regarding also taking into account the temperature dependence of the elastic material parameters E and ν, as well as the hardening modulus E_t, and how this can possibly be done, is currently still unclear and needs further research.

If the nature of the shakedown state at temperature-dependent yield stress is elastic shakedown according to Eq. (5.34), the plastic zone V_p and the subvolume V_e remaining elastic can be identified for the nth MEA by the conditions

$$V_p^{(n)} = \left\{ \underline{x} \;\middle|\; \sigma_{\text{v,min}}^{(n-1)} \geq f_{y,\text{min}} \quad \vee \quad \sigma_{\text{v,max}}^{(n-1)} \geq f_{y,\text{max}} \right\} \tag{5.70}$$

$$V_e^{(n)} = \left\{ \underline{x} \;\middle|\; \sigma_{\text{v,min}}^{(n-1)} < f_{y,\text{min}} \quad \wedge \quad \sigma_{\text{v,max}}^{(n-1)} < f_{y,\text{max}} \right\}, \tag{5.71}$$

see Fig. 5.13.

At any location $\underline{x}$ in V_p, estimating the TIV is analogous to the procedure for estimating temperature-independent yield stress in Sect. 4.4.2 by projecting

$$Y_i^{*(n)} = -\rho_i^{\prime(n-1)} \tag{5.72}$$

on the intersection area Ω of the yield surfaces of the two extreme loading states in the space of the TIV (Fig. 5.14).

What projection rule is applied depends again, as in Sect. 4.4.2, on the position of Y^* in the regions ω_1–ω_4 of the TIV-space. For their determination, the angles α_{min}, α_{max}, β_{min}, and β_{max} are required. β_{min} and β_{max} are independent of the radius of the yield surfaces, so they can still be determined by Eqs. (4.64) and (4.65):

$$\cos(\beta_{\text{min}}) = \frac{\Delta\sigma_{\text{v}}^2 + \sigma_{\text{v,min}}^2 - \sigma_{\text{v,max}}^2}{2 \cdot \Delta\sigma_{\text{v}} \cdot \sigma_{\text{v,min}}} \tag{5.73}$$

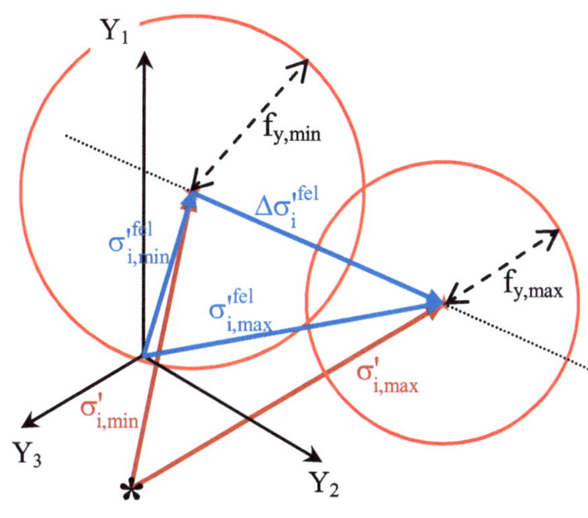

Fig. 5.13 Elastic shakedown at temperature-dependent yield stress: yield surfaces, fictitious elastic, and elastic-plastic stresses in the TIV-space

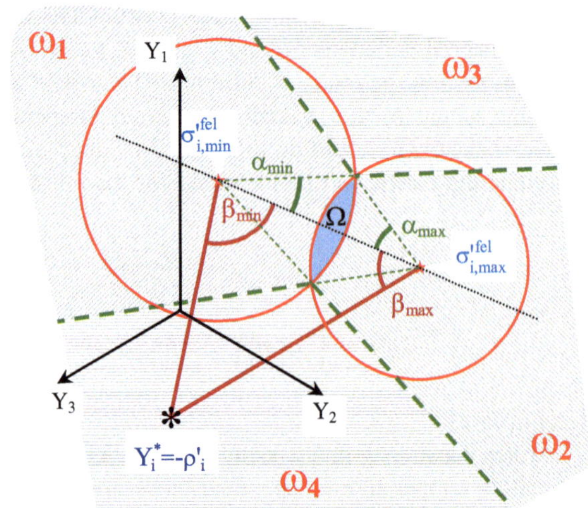

Fig. 5.14 Elastic shakedown at temperature-dependent yield stress: intersection area Ω of the yield surfaces and position of Y^* in the TIV-space

$$\cos(\beta_{\max}) = \frac{\Delta\sigma_{\mathrm{v}}^2 + \sigma_{\mathrm{v,max}}^2 - \sigma_{\mathrm{v,min}}^2}{2 \cdot \Delta\sigma_{\mathrm{v}} \cdot \sigma_{\mathrm{v,max}}} \,. \tag{5.74}$$

In contrast, $\alpha_{\min}$ and $\alpha_{\max}$ are now to be determined from

$$\cos(\alpha_{\min}) = \frac{\Delta\sigma_{\mathrm{v}}^2 + f_{y,\min}^2 - f_{y,\max}^2}{2 \cdot \Delta\sigma_{\mathrm{v}} \cdot f_{y,\min}} \tag{5.75}$$

$$\cos(\alpha_{\max}) = \frac{\Delta\sigma_{\mathrm{v}}^2 + f_{y,\max}^2 - f_{y,\min}^2}{2 \cdot \Delta\sigma_{\mathrm{v}} \cdot f_{y,\max}} \,, \tag{5.76}$$

and are, therefore, unlike temperature-independent yield stress, now different, and can even exceed 90°. It should be remembered that the equivalent stress range in the case of elastic shakedown is identical to the fictitious elastic equivalent stress range, so that $\cos(\alpha_{\min})$ and $\cos(\alpha_{\max})$ are known solely on the basis of a fictitious elastic calculation.

Another difference of Eqs. (5.75) and (5.76) compared to Eq. (4.63) is that the right side can now be less than −1, namely, when one Mises circle is entirely contained within the other Mises circle, no intersection points of the two circular edges exist. If

$$\Delta\sigma_{\mathrm{v}} + f_{y,\min} < f_{y,\max} \,, \tag{5.77}$$

then only the region ω_2 exists and the Mises circle of minimum loading state is completely in that of the maximum state (Fig. 5.15, left). If, however,

$$\Delta\sigma_{\mathrm{v}} + f_{y,\max} < f_{y,\min} \,, \tag{5.78}$$

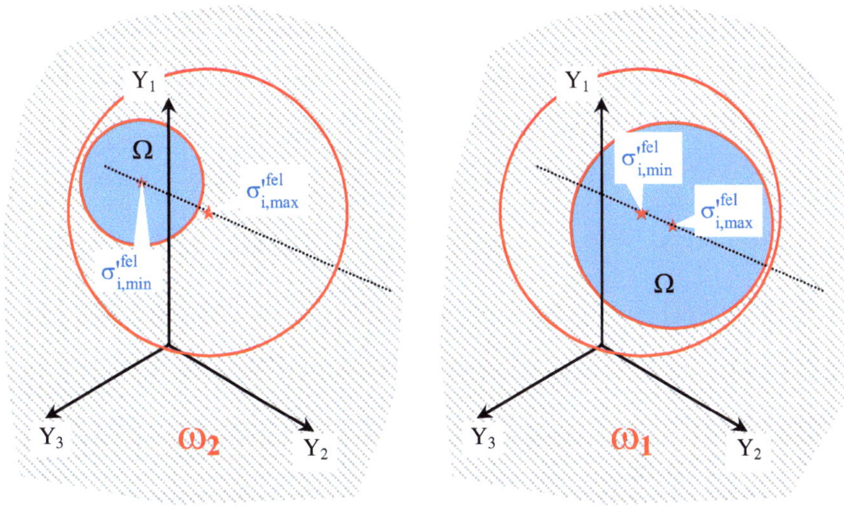

Fig. 5.15 Elastic shakedown at temperature-dependent yield stress: *left* Mises circle of minimum load state is completely in the maximum one; *right* Mises circle of maximum load state is completely in the minimum one

then only the region ω_1 exists and the Mises circle associated with the maximum load condition lies entirely within that for the minimum load condition (Fig. 5.15, right).

Analogous to Sect. 4.4.2, the following three projection rules are distinguished.

5.6.1 Y^* Is in ω_2

This is true if either of the following conditions is met:

$$\beta_{\text{min}} < \alpha_{\text{min}} \quad \text{and} \quad \sigma_{\text{v,min}} > f_{y,\text{min}} \tag{5.79}$$

$$\Delta\sigma_{\text{v}} + f_{y,\text{min}} < f_{y,\text{max}}. \tag{5.80}$$

Then the estimate of Y is carried out by projection of Y^* on the Mises circle of the minimum load condition (Fig. 5.16):

$$Y_i = \sigma'^{\text{fel}}_{i,\text{min}} - \sigma'_{i,\text{min}}\left(\frac{f_{y,\text{min}}}{\sigma_{\text{v,min}}}\right). \tag{5.81}$$

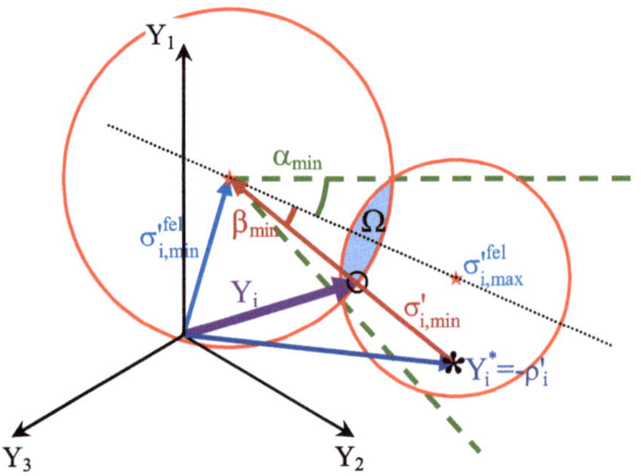

Fig. 5.16 Projection on the Mises circle of minimum loading state

5.6.2 Y^* Is in ω_1

This is the case when one of the two following conditions is true:

$$\beta_{\max} < \alpha_{\max} \quad \text{and} \quad \sigma_{\text{v,max}} > f_{y,\max} \tag{5.82}$$

$$\Delta\sigma_{\text{v}} + f_{y,\max} < f_{y,\min}. \tag{5.83}$$

Then Y is estimated by projection of Y^* on the Mises circle of the maximum loading condition (Fig. 5.17):

$$Y_i = \sigma'^{\text{fel}}_{i,\max} - \sigma'_{i,\max}\left(\frac{f_{y,\max}}{\sigma_{\text{v,max}}}\right). \tag{5.84}$$

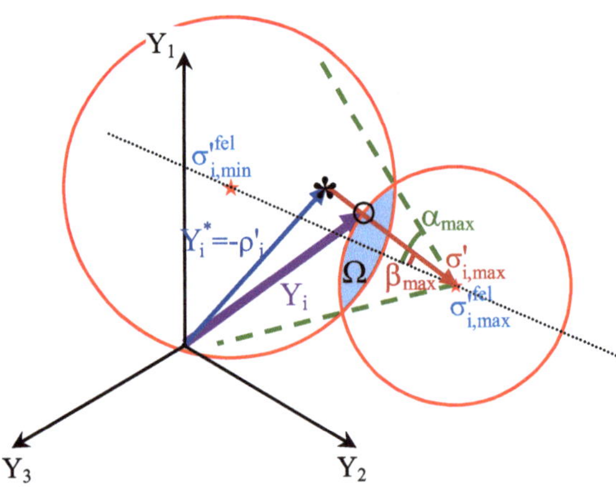

Fig. 5.17 Projection on the Mises circle of maximum loading state

5.6.3 Y^* Is in ω_4

This is the case when all the following three conditions are satisfied:

$$\beta_{\min} \geq \alpha_{\min} \tag{5.85}$$

$$\beta_{\max} \geq \alpha_{\max} \tag{5.86}$$

$$\Delta\sigma_{\mathrm{v}} \geq \left| f_{y,\min} - f_{y,\max} \right| . \tag{5.87}$$

The last of these three conditions precludes that one Mises circle is completely located inside the other. Then Y is estimated by projection of Y^* on the intersection point of the two Mises circles of the minimum and the maximum state, i.e., on the vertex of the intersection area Ω, Fig. 5.18.

How the position of this vertex can be determined is already described in Sect. 4.4.2.3 (see Fig. 4.16). As is already known from there, Y^* cannot be located in region ω_3. With temperature-dependent yield stresses we get:

$$Y_{i,\text{vertex1}} = \sigma'^{\text{fel}}_{i,\min} - a \cdot \sigma'_{i,\min} + b \cdot \Delta\sigma'^{\text{fel}}_i \tag{5.88}$$

$$a = \frac{f_{y,\min}}{\sigma_{\mathrm{v},\min}} \cdot \frac{\sqrt{1 - \cos^2 \alpha_{\min}}}{\sqrt{1 - \cos^2 \beta_{\min}}} \tag{5.89}$$

$$b = \frac{f_{y,\min}}{\Delta\sigma_{\mathrm{v}}} \cdot \left(\cos\alpha_{\min} - \cos\beta_{\min} \frac{\sqrt{1 - \cos^2\alpha_{\min}}}{\sqrt{1 - \cos^2\beta_{\min}}} \right) . \tag{5.90}$$

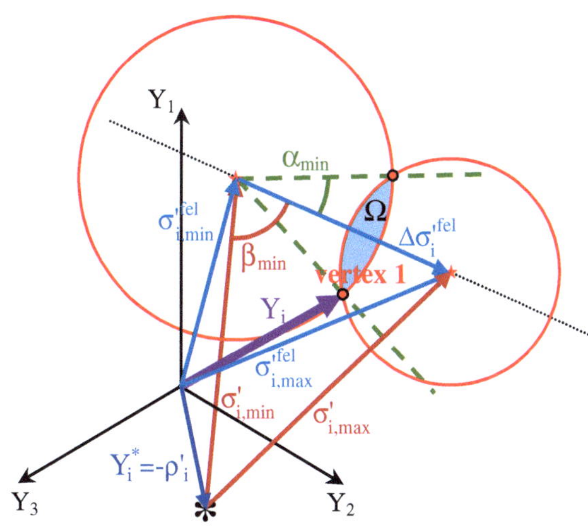

Fig. 5.18 Projection on the vertex of the intersection area Ω

To calculate the vertex coordinates of Ω it is sufficient, again, to insert the fictitious elastic stresses in Eqs. (5.88)–(5.90) and in Eqs. (5.73) and (5.74) to get the required values of $\cos(\beta_{\min})$ and $\cos(\beta_{\max})$. However, this is not sufficient, or only within the first MEA, for the calculation of $\cos(\beta_{\min})$ and $\cos(\beta_{\max})$ to examine the conditions in Eqs. (5.79), (5.82), (5.85), and (5.86).

5.7 Examples of Accumulated Strains at Elastic Shakedown

5.7.1 Two-Bar Model

In Sect. 4.5.1, the STPZ has been applied to the two-bar model (Fig. 4.18) for determining the strains accumulated in the state of elastic shakedown at temperature-independent yield stress. Now, a linear dependence of the yield stress of the temperature is taken into account. Because of the linear dependence, the phenomenon of re-elastic behavior after previous plasticizing in the course of a monotonous load increase, as described in Sect. 5.1.2, is excluded.

The fictitious elastic calculated stresses are given by Eqs. (4.77)–(4.80). For simplicity, we want to restrict ourselves to positive values of σ_P and σ_t in the following. In addition, we assume $f_{y,\mathrm{hot}} \leq f_{y,\mathrm{cold}}$.

At minimum load condition both bars are at "cold" temperature, whereas at maximum load the left bar is "hot" and the right bar retains its initial temperature. Therefore elastic shakedown occurs according to Eq. (5.23) if the following conditions are met for the two bars:

$$\Delta\sigma^{\mathrm{fel}}_{\mathrm{left}} = -\sigma_t \geq -\left(f_{y,\mathrm{cold}} + f_{y,\mathrm{hot}}\right) \tag{5.91}$$

$$\Delta\sigma^{\mathrm{fel}}_{\mathrm{right}} = \sigma_t \leq 2f_{y,\mathrm{cold}}\,. \tag{5.92}$$

Accordingly, only Eq. (5.91) prevails.

Whereas in Sect. 4.5.1 the idea was pursued that a certain load level is given, and the corresponding residual stress state and the elastic-plastic strains are searched in the state of shakedown, an inverse approach is now followed. A residual stress state is given thereby, and the STPZ is used to determine which load level belongs to this.

A total of nine possible combinations of residual stress states should be examined, depending on which bar is non-plasticized (Y^* in Ω) or whether Y^* is to be projected on the yield surface of the minimum or of the maximum load condition, i.e., if the bar is in region ω_1 or ω_2 (Table 5.1, Fig. 5.19).

Table 5.1 Two-bar model: possible combinations of states of the bars in ES

Combination	Left bar in	Right bar in
1	V_p with ω_2	V_p with ω_1
2	V_p with ω_2	V_p with ω_2
3	V_p with ω_2	V_e
4	V_p with ω_1	V_p with ω_1
5	V_p with ω_1	V_p with ω_2
6	V_p with ω_1	V_e
7	V_e	V_p with ω_1
8	V_e	V_p with ω_2
9	V_e	V_e

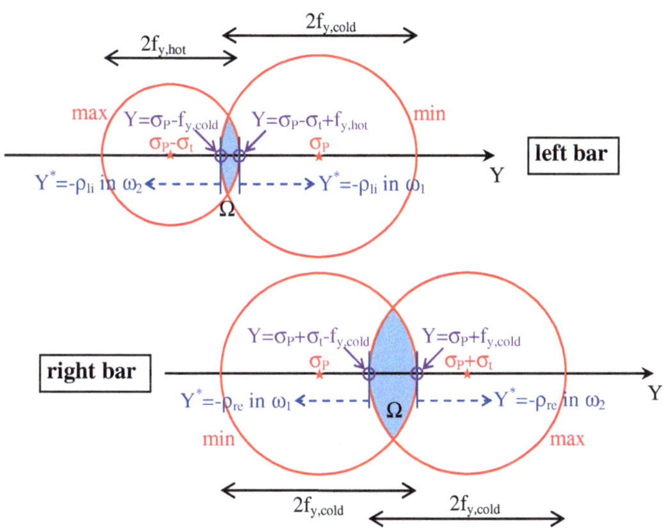

Fig. 5.19 Two-bar model at ES: states of the bars in the TIV-space

5.7.1.1 Combination 1

The left bar is located in V_p, namely in the region ω_2. According to Fig. 5.19 this is the case when

$$Y^*_{\text{left}} = -\rho_{\text{left}} \leq \sigma_P - f_{y,\text{cold}}\,. \tag{5.93}$$

The projection of Y^* on the yield surface of the minimum load condition then returns

$$Y_{\text{left}} = \sigma_P - f_{y,\text{cold}}\,. \tag{5.94}$$

The corresponding modified elastic material law reads

$$\varepsilon^*_{\text{left}} = \frac{\rho_{\text{left}}}{E_t} + \frac{\sigma_P - f_{y,\text{cold}}}{C} \,. \tag{5.95}$$

The right bar is also located in V_p, but in region ω_1. According to Fig. 5.19 this is the case when

$$Y^*_{\text{right}} = -\rho_{\text{right}} \leq \sigma_P + \sigma_t - f_{y,\text{cold}} \,. \tag{5.96}$$

The projection of Y^* on the yield surface of the maximum load condition then returns

$$Y_{\text{right}} = \sigma_P + \sigma_t - f_{y,\text{cold}} \tag{5.97}$$

and the modified elastic material law reads

$$\varepsilon^*_{\text{right}} = \frac{\rho_{\text{right}}}{E_t} + \frac{\sigma_P + \sigma_t - f_{y,\text{cold}}}{C} \,. \tag{5.98}$$

The modified elastic analysis requires the solution of the equilibrium condition

$$\rho_{\text{right}} = -\rho_{\text{left}} \tag{5.99}$$

and the compatibility condition

$$\varepsilon^*_{\text{left}} = \varepsilon^*_{\text{right}} \,, \tag{5.100}$$

and supplies after inserting the modified elastic material laws of both bars

$$\rho_{\text{left}} = \left(1 - \frac{E_t}{E}\right)\frac{\sigma_t}{2} \tag{5.101}$$

$$\varepsilon^*_{\text{left}} = \frac{1 - \frac{E_t}{E}}{E_t}\left(\sigma_P + \frac{\sigma_t}{2} - f_{y,\text{cold}}\right) . \tag{5.102}$$

The range of validity of this solution is given by Eqs. (5.93) and (5.96). Inserting Eq. (5.101) in Eq. (5.93) yields

$$\sigma_t \geq \left(f_{y,\text{cold}} - \sigma_P\right)\frac{2}{1 - \frac{E_t}{E}} \,, \tag{5.103}$$

and inserting Eqs. (5.101) and (5.99) in Eq. (5.97) provides

$$\sigma_t \geq \left(f_{y,\text{cold}} - \sigma_P\right) \frac{2}{1 + \frac{E_t}{E}}\,, \tag{5.104}$$

which, however, cannot be decisive if compared with Eq. (5.103).

5.7.1.2 Combination 2

The conditions of Eqs. (5.93)–(5.95) remain valid for the left bar, although the right bar is now in region ω_2. According to Fig. 5.19 this is the case if

$$Y^*_{\text{right}} = -\rho_{\text{right}} \geq \sigma_P + f_{y,\text{cold}}\,, \tag{5.105}$$

so that the projection of Y^* on the yield surface of the minimum load condition results in

$$Y_{\text{right}} = \sigma_P + f_{y,\text{cold}} \tag{5.106}$$

and the modified elastic material law reads

$$\varepsilon^*_{\text{right}} = \frac{\rho_{\text{right}}}{E_t} + \frac{\sigma_P + f_{y,\text{cold}}}{C}\,. \tag{5.107}$$

The result of the MEA,

$$\rho_{\text{left}} = \left(1 - \frac{E_t}{E}\right) f_{y,\text{cold}}\,, \tag{5.108}$$

however, cannot simultaneously satisfy Eqs. (5.93) and (5.105), so that combination 2 turns out to be impossible.

5.7.1.3 Combination 3

Combination 3 requires the right bar to remain elastic, so that

$$\sigma_P + f_{y,\text{cold}} > Y^*_{\text{right}} = -\rho_{\text{right}} > \sigma_P + \sigma_t - f_{y,\text{cold}}\,, \tag{5.109}$$

and its modified elastic material law reads:

$$\varepsilon^*_{\text{right}} = \frac{\rho_{\text{right}}}{E}\,. \tag{5.110}$$

A simultaneous solution of Eqs. (5.99) and (5.100) along with the modified elastic material law of the left bar, Eq. (5.95), is only possible, if the primary stress σ_P is greater than the “cold” yield stress, whereby, however, the plastic limit load would be exceeded. This combination is therefore regarded as irrelevant, even if the plastic limit load is defined only for non-hardening material, so that in the underlying hardening here the plastic limit load could in theory be exceeded.

5.7.1.4 Combination 4

For the left bar the following applies:

$$Y^*_{\text{left}} = -\rho_{\text{left}} \geq \sigma_P - \sigma_t + f_{y,\text{hot}} \,. \tag{5.111}$$

Then the projection of Y^* on the yield surface of the maximum load condition yields

$$Y_{\text{left}} = \sigma_P - \sigma_t + f_{y,\text{hot}} \,. \tag{5.112}$$

The corresponding modified elastic material law reads

$$\varepsilon^*_{\text{left}} = \frac{\rho_{\text{left}}}{E_t} + \frac{\sigma_P - \sigma_t + f_{y,\text{hot}}}{C} \,. \tag{5.113}$$

For the right bar, Eqs. (5.96)–(5.98) apply. The MEA for the field Eqs. (5.99) and (5.100) results in

$$\rho_{\text{left}} = \left(1 - \frac{E_t}{E}\right)\left[\sigma_t - \frac{1}{2}\left(f_{y,\text{cold}} + f_{y,\text{hot}}\right)\right] \tag{5.114}$$

$$\varepsilon^*_{\text{left}} = \frac{\left(1 - \frac{E_t}{E}\right)}{E_t}\left[\sigma_P + \frac{1}{2}\left(f_{y,\text{hot}} - f_{y,\text{cold}}\right)\right] \,. \tag{5.115}$$

Introducing Eq. (5.114) in Eq. (5.111), we get the requirement

$$\sigma_t \geq \frac{1}{2}\left[\frac{E}{E_t}\left(f_{y,\text{hot}} - f_{y,\text{cold}}\right) + \left(f_{y,\text{hot}} + f_{y,\text{cold}}\right)\right] + \frac{E}{E_t}\sigma_P \,, \tag{5.116}$$

and inserting Eq. (5.114) in the condition of Eq. (5.96), the range of validity of combination 4 is further reduced to

$$\sigma_t \geq \frac{1}{2}\left[\frac{E}{E_t}\left(f_{y,\text{cold}} - f_{y,\text{hot}}\right) + \left(f_{y,\text{hot}} + f_{y,\text{cold}}\right)\right] - \frac{E}{E_t}\sigma_P \,. \tag{5.117}$$

5.7.1.5 Combination 5

For the combination of the modified elastic material laws Eqs. (5.113) and (5.107) we get

$$\rho_{\text{left}} = \frac{1}{2}\left(1 - \frac{E_t}{E}\right)\left[\sigma_t + f_{y,\text{cold}} - f_{y,\text{hot}}\right] \,, \tag{5.118}$$

which, after insertion into the corresponding Eq. (5.105), turns out to be impossible for the state of elastic shakedown, i.e., in the range of validity of Eq. (5.91).

5.7.1.6 Combination 6

By combining the modified elastic material laws Eqs. (5.113) and (5.110), the solution of the MEA is

$$\rho_{\text{left}} = \frac{1-\frac{E_t}{E}}{1+\frac{E_t}{E}}\left(-\sigma_P + \sigma_t - f_{y,\text{hot}}\right) \tag{5.119}$$

$$\varepsilon^*_{\text{left}} = \frac{1-\frac{E_t}{E}}{1+\frac{E_t}{E}}\frac{\left(\sigma_P - \sigma_t + f_{y,\text{hot}}\right)}{E} \; . \tag{5.120}$$

The range of validity is obtained by insertion into Eq. (5.111):

$$\sigma_t \geq \sigma_P + f_{y,\text{hot}} \; , \tag{5.121}$$

and by the right condition in Eq. (5.109):

$$\sigma_t < \frac{1}{2}\left[\frac{E}{E_t}\left(f_{y,\text{cold}} - f_{y,\text{hot}}\right) + \left(f_{y,\text{hot}} + f_{y,\text{cold}}\right)\right] - \frac{E}{E_t}\sigma_P \; . \tag{5.122}$$

5.7.1.7 Combination 7

If the left bar should remain elastic, then the following must apply, as shown in Fig. 5.19:

$$\sigma_P - f_{y,\text{cold}} < Y^*_{\text{left}} = -\rho_{\text{left}} < \sigma_P - \sigma_t + f_{y,\text{hot}} \; , \tag{5.123}$$

and the modified elastic material law reads

$$\varepsilon^*_{\text{left}} = \frac{\rho_{\text{left}}}{E} \; . \tag{5.124}$$

Combined with Eq. (5.98) for the right bar, the solution of the equilibrium and compatibility condition is

$$\rho_{\text{left}} = \frac{1-\frac{E_t}{E}}{1+\frac{E_t}{E}}\left(\sigma_P + \sigma_t - f_{y,\text{cold}}\right) \tag{5.125}$$

$$\varepsilon^*_{\text{left}} = \frac{1-\frac{E_t}{E}}{1+\frac{E_t}{E}}\frac{\left(\sigma_P + \sigma_t - f_{y,\text{cold}}\right)}{E} \; . \tag{5.126}$$

The validity of this solution is limited by Eq. (5.123) to

$$\sigma_t < \left(f_{y,\text{cold}} - \sigma_P\right) \frac{2}{1 - \frac{E_t}{E}} \tag{5.127}$$

$$\sigma_t < \frac{1}{2}\left[\frac{E}{E_t}\left(f_{y,\text{hot}} - f_{y,\text{cold}}\right) + \left(f_{y,\text{hot}} + f_{y,\text{cold}}\right)\right] + \frac{E}{E_t}\sigma_P \tag{5.128}$$

and by Eq. (5.96) to

$$\sigma_t > f_{y,\text{cold}} - \sigma_P\,. \tag{5.129}$$

5.7.1.8 Combination 8

The combination of the modified elastic material laws Eqs. (5.124) and (5.107) leads to the solution

$$\rho_{\text{left}} = \frac{1 - \frac{E_t}{E}}{1 + \frac{E_t}{E}}\left(\sigma_P + f_{y,\text{cold}}\right), \tag{5.130}$$

turning out to be impossible, however, when checking the associated ranges of validity.

5.7.1.9 Combination 9

The limits of purely elastic behavior in the entire structure are given by the requirement that the area Ω in Fig. 5.19 must include the coordinate origin in both bars. This leads to the conditions

$$\sigma_t \leq f_{y,\text{cold}} - \sigma_P \tag{5.131}$$

$$\sigma_t \leq f_{y,\text{hot}} + \sigma_P. \tag{5.132}$$

5.7.1.10 Ratcheting Interaction Diagram

The results obtained in Sects. 5.7.1.1–5.7.1.9 are exact if anomalous behavior, as described in Sect. 5.1.2 (see Fig. 5.9), can be ruled out. They can be represented as a ratcheting interaction diagram (Fig. 5.20). For $f_{y,\text{hot}} = f_{y,\text{cold}}$ the RID of Fig. 4.21 is obtained. It is limited to the top by the condition for elastic shakedown, Eq. (5.91).

In region 1 a pronounced ratcheting mechanism exists. However, the elastic-plastic strain in this region, as also in region 7, does not depend on temperature dependence of the yield stress. This is because the residual strains according to Eq. (5.102) do not contain a term $f_{y,\text{hot}}$. Accordingly, the anomalous behavior of a seemingly negative hardening modulus (Fig. 5.9) is also not possible here, in contrast to regions 4 and 6. However, the range of validity of region 1 as expressed by Eq. (5.91) is restricted compared to temperature-independent yield stress.

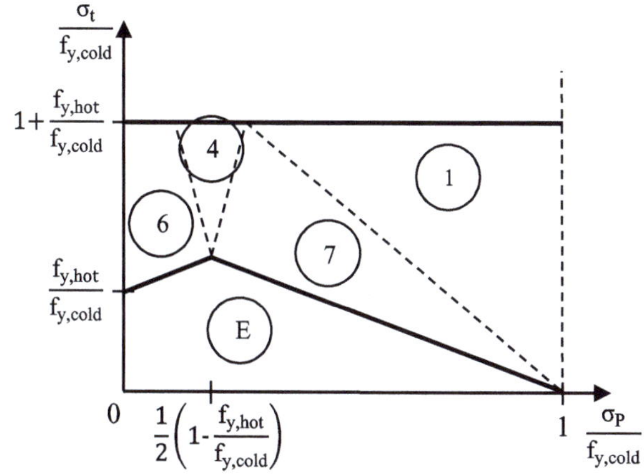

Fig. 5.20 Two-bar model: range of different solutions for the strain accumulation at elastic shakedown

The development of the stress–strain relationship with the load cycles from an incremental analysis in Fig. 5.21 for a parameter configuration in region 1 shows that the "hot" yield stress plays a role for the evolution of stresses and strains, but not for the state of shakedown, as is obvious by comparison with Fig. 4.23. Again, as in Fig. 5.8, seemingly different hardening modules can be recognizes here.

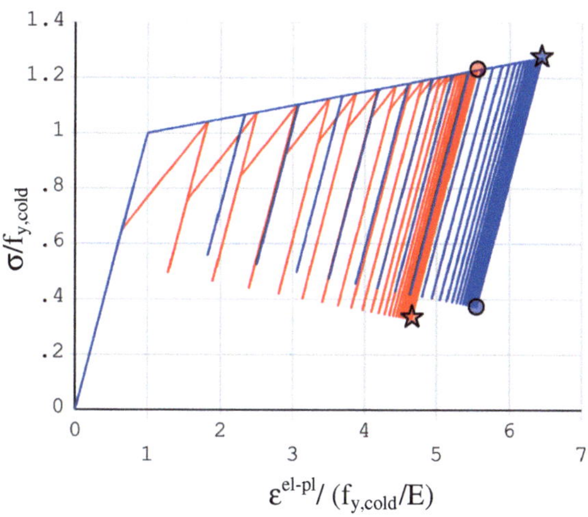

Fig. 5.21 Strain accumulation in the two-bar model with linear kinematic hardening: elastic shakedown ($\sigma_P/f_{y,cold} = 0.8$; $\sigma_t/f_{y,cold} = 0.9$; $E_t/E = 0.05$; $f_{y,hot}/f_{y,cold} = 0.6$); comparison of incremental solution with the STPZ (*circles* for minimum and *stars* for maximum load condition)

5.7.2 Multiaxial Ratcheting

The example of multiaxial ratcheting already treated in Sect. 4.5.2 with temperature-independent yield stress with respect to elastic shakedown brings no particular innovations if a temperature-dependent yield stress is taken into account. Because the temperature does not constitute a load in this example, it only plays the role of a parameter which modifies the yield stress with time.

Of course, some details are changed then. For example, Eqs. (4.141) and (4.142) remain valid to determine the coordinates of the vertex of the intersection area Ω of the yield surfaces, as do the results of the MEA in the form of residual stress in direction (2) and the residual strain in direction (1) in Eqs. (4.144) and (4.145) for the case that the projection of Y^* leads to the vertex of Ω. However, the parameters a and b to be inserted there are to be redefined according to Eqs. (5.89) and (5.90), resulting in longer terms, to be dispensed with here.

Analogous to Fig. 4.27, the evolution of the TIV is shown in the TIV-space in Fig. 5.22 by way of example as determined with an incremental analysis in the case where the temperature rises synchronously with the displacement-controlled loading, so that the yield stress decreases with increasing load. A good approximation of the state of shakedown is reached after 80 cycles in the incremental analysis. The stresses and strains present in the state of shakedown are, however, determined exactly only by the STPZ. So, for example, the elastic-plastic strain in the direction of the primary stress is $\varepsilon_1 = 2.315\ f_{y,\text{cold}}/E$ at both the minimum and the maximum loading condition because the Poisson's ratio $\nu = 0$.

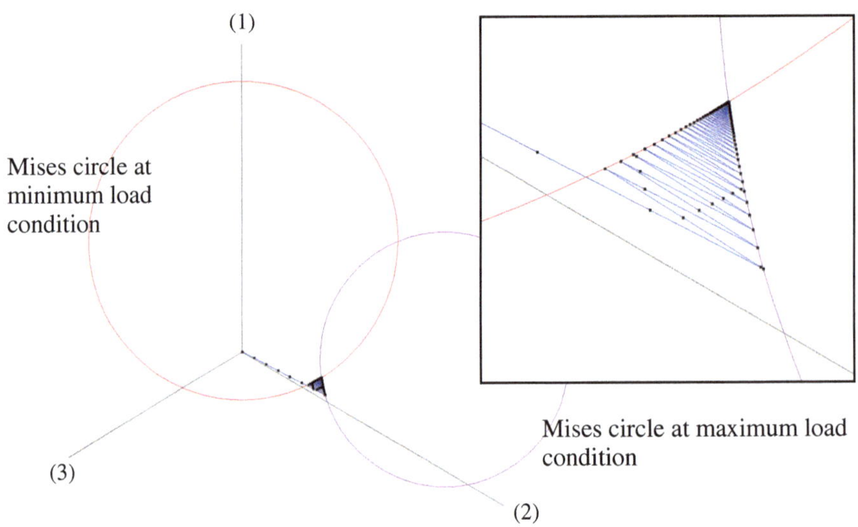

Fig. 5.22 Multiaxial ratcheting: trajectory of the TIV in the TIV-space at elastic shakedown with temperature-dependent yield stress ($\sigma_P/f_{y,\text{cold}} = 0.7$; $\sigma_t/f_{y,\text{cold}} = 1.5$; $E_t/E = 0.05$; $f_{y,\text{hot}}/f_{y,\text{cold}} = 0.8$; $\nu = 0$)

5.7.3 Bree Tube

It also applies to the axisymmetric model of the Bree tube that the results with temperature-dependent yield stress differ in detail from those with temperature-independent yield stress in Sect. 4.5.3. Because the inside surface is warmer than the outside surface, a lower yield stress is effective there. Anyway, a different yield stress is effective at the maximum load condition anywhere across the pipe wall because of the linear temperature distribution.

Figure 5.23 shows the stress–strain curve, after normalization to the "cold" yield stress or elastic limit strain, for the circumferential direction on the inside and the outside surface from an incremental analysis for the parameter combination

$$\frac{\sigma_P}{f_{y,\text{cold}}} = 0.75\ ;\ \frac{\sigma_t}{f_{y,\text{cold}}} = 1.4\ ;\ \nu = 0.3\ ;\ \frac{E_t}{E} = 0.02\ ;\ \frac{f_{y,\text{hot}}}{f_{y,\text{cold}}} = 0.6667\,. \qquad (5.133)$$

A linear dependence of the yield stress on temperature was assumed. In addition, the results of the STPZ for the shakedown state after five MEAs are given there by diamonds for the minimum and crosses for the maximum load condition. The quality of the approximation by the STPZ is regarded as quite good. It is certainly no worse than at temperature-independent yield stress in Fig. 4.31.

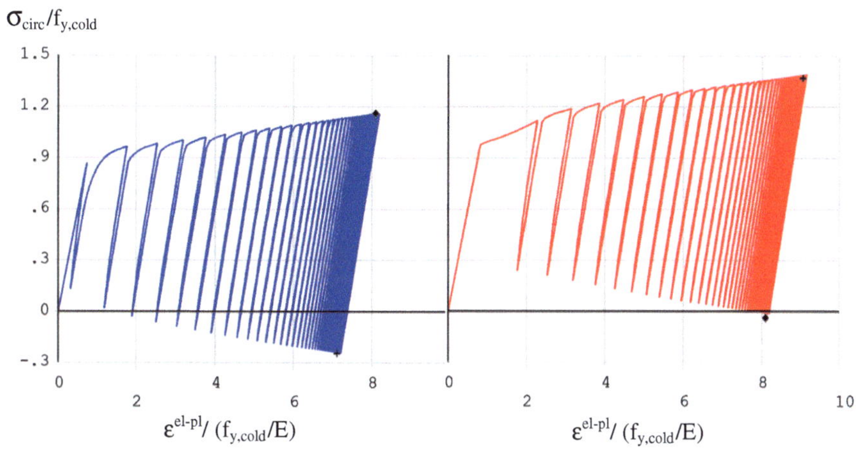

Fig. 5.23 Bree tube (axisymmetric model): stress–strain diagram for circumferential direction after incremental analysis, and stress–strain couples of the fifth MEA of the STPZ for the state of elastic shakedown (*diamonds* and *crosses*); *left* inside surface; *right* outside surface

5.8 Accumulated Strain at Plastic Shakedown

In contrast to calculating the plastic shakedown state with the STPZ at temperature-independent yield stress presented in Sect. 4.6, we now describe how temperature-dependent yield stress can be taken into account.

Whether it is useful in light of the discussion in Sect. 5.1 to take into account the temperature dependence of the elastic material parameters E and ν and of the hardening modulus E_t for calculating the accumulated strains, and how this can possibly be done, is currently still unclear and needs further research. It has already been pointed out in Sect. 5.4 that a separate consideration of strain ranges and accumulated strains at temperature-dependent E, ν, and E_t is not readily useful.

However, taking into account temperature-dependent yield stresses for calculating the accumulated strains at PS presents no particular difficulties. For this purpose, as with temperature-independent yield stress in Sect. 4.6, the stresses and strains are calculated for a mean load condition. It is necessary to use a prior calculation of the strain range with temperature-dependent yield stress according to Sect. 5.4.

The nature of the shakedown state, whether elastic or plastic shakedown occurs at temperature-dependent yield stress, has already been examined in Sect. 5.3. Plastic shakedown exists when

$$\Delta\sigma^{\mathrm{fel}}_{\mathrm{v}(\underline{x})} > \left(f_{y,\min} + f_{y,\max}\right)_{(\underline{x})} \quad \exists \underline{x} \in V. \tag{5.134}$$

As with temperature-independent yield stress discussed in Sect. 4.6, the plastic subvolume V_p is distinguished in the nth MEA according to where alternating plastic straining ($V_{p\Delta}$) is detected in the prior calculation of the strain range (Sect. 5.4) and where the plastic strains are virtually frozen ($V_{e\Delta}$):

$$V_p^{(n)} = V_{p\Delta}^{(n)} \cup V_{e\Delta}^{(n)} \tag{5.135}$$

$$V_{p\Delta}^{(n)} = \left\{ \underline{x} \mid \Delta\sigma_{\mathrm{v}}^{(n-1)} \geq f_{y,\min} + f_{y,\max} \right\} \tag{5.136}$$

$$V_{e\Delta}^{(n)} = \left\{ \underline{x} \mid \Delta\sigma_{\mathrm{v}}^{(n-1)} < f_{y,\min} + f_{y,\max} \wedge \left(\sigma_{\mathrm{v},\max}^{(n-1)} > f_{y,\max} \vee \sigma_{\mathrm{v},\min}^{(n-1)} > f_{y,\min} \right) \right\}. \tag{5.137}$$

The subvolume of the structure remaining elastic is thus defined by

$$V_e^{(n)} = \left\{ \underline{x} \,\middle|\, \sigma_{\mathrm{v},\min}^{(n-1)} < f_{y,\min} \quad \wedge \quad \sigma_{\mathrm{v},\max}^{(n-1)} < f_{y,\max} \right\}. \tag{5.138}$$

For a location of the structure in $V_{p\Delta}$, the TIV for the mean load condition is estimated by the middle of the clearance between the two yield surfaces of the minimum and maximum load conditions (Fig. 5.24):

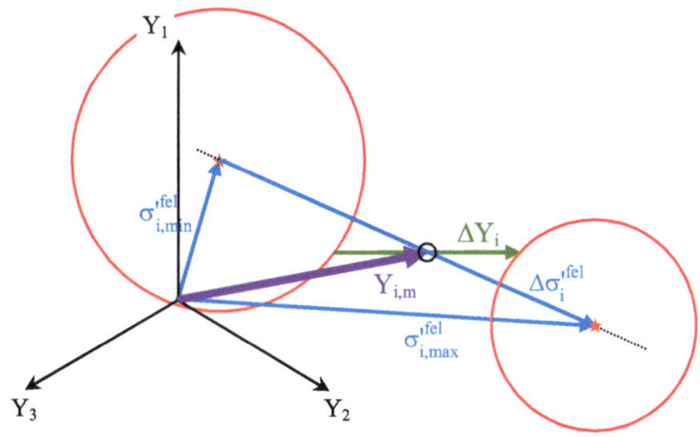

Fig. 5.24 Estimation of the TIV for the mean state of loading at PS for a location in $V_{p\Delta}$

$$Y_{i,m} = \sigma'^{\mathrm{fel}}_{i,\min} + \frac{1}{2}\Delta\sigma'^{\mathrm{fel}}_{i}\left(1 + \frac{f_{y,\min} - f_{y,\max}}{\Delta\sigma^{\mathrm{fel}}_{\mathrm{v}}}\right) \quad \forall \underline{x} \in V_{p\Delta}. \tag{5.139}$$

This mean state is given solely by fictitious elastic calculated stresses.

For a location of the structure in $V_{e\Delta}$, the TIV is estimated by means of a projection on the yield surface of one of the two extreme load conditions, the one where the equivalent stress exceeds the associated yield stress by a larger amount than at the other (see Sects. 4.6.2.2 and 4.6.2.3 for temperature-independent yield stress).

Thus, for

$$\sigma_{\mathrm{v,max}} - f_{y,\max} > \sigma_{\mathrm{v,min}} - f_{y,\min} \tag{5.140}$$

first

$$Y^*_{i,\max} = Y^*_{i,m} + \frac{1}{2}\Delta Y_i \tag{5.141}$$

is projected on the Mises circle of the maximum load condition (Fig. 5.25). Because

$$Y^*_{i,m} = -\rho'_{i,m} = -\frac{1}{2}\left(\rho'_{i,\max} + \rho'_{i,\min}\right), \tag{5.142}$$

and because the plastic strain range vanishes at locations in $V_{e\Delta}$,

$$\Delta Y_i = -\Delta\rho'_i = -\left(\rho'_{i,\max} - \rho'_{i,\min}\right), \tag{5.143}$$

we get

$$Y^*_{i,\max} = -\rho'_{i,\max}. \tag{5.144}$$

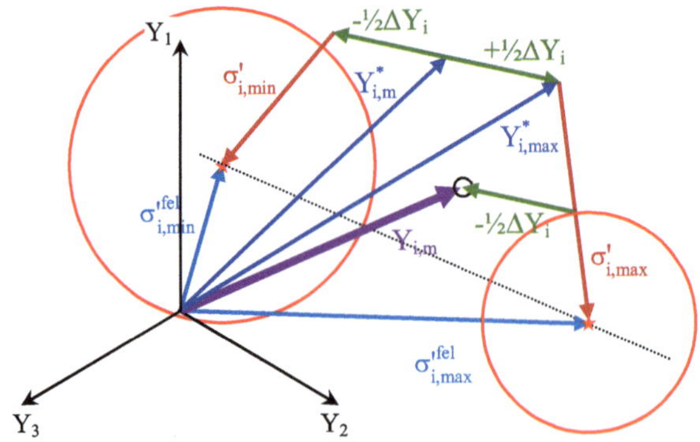

Fig. 5.25 Estimation of the TIV for the mean load condition at PS for a location in $V_{e\Delta}$, by projection on the maximum state

After the projection, half the TIV-range is to be subtracted again to get the mean state and we obtain as the TIV for the mean load condition

$$Y_{i,m} = \sigma_{i,\max}^{\prime\text{fel}} - \sigma_{i,\max}^{\prime} \cdot \frac{f_{y,\max}}{\sigma_{\text{v},\max}} + \frac{1}{2}\left(\Delta\sigma_i^{\prime} - \Delta\sigma_i^{\prime\text{fel}}\right). \tag{5.145}$$

Similarly we get for

$$\sigma_{\text{v},\max} - f_{y,\max} \leq \sigma_{\text{v},\min} - f_{y,\min} \tag{5.146}$$

through the projection of

$$Y_{i,\min}^{*} = -\rho_{i,\min}^{\prime} \tag{5.147}$$

on the Mises circle of the minimum load condition (Fig. 5.26) and subsequent addition of half the TIV-range

$$Y_{i,m} = \sigma_{i,\min}^{\prime\text{fel}} - \sigma_{i,\min}^{\prime} \cdot \frac{f_{y,\min}}{\sigma_{\text{v},\min}} - \frac{1}{2}\left(\Delta\sigma_i^{\prime} - \Delta\sigma_i^{\prime\text{fel}}\right). \tag{5.148}$$

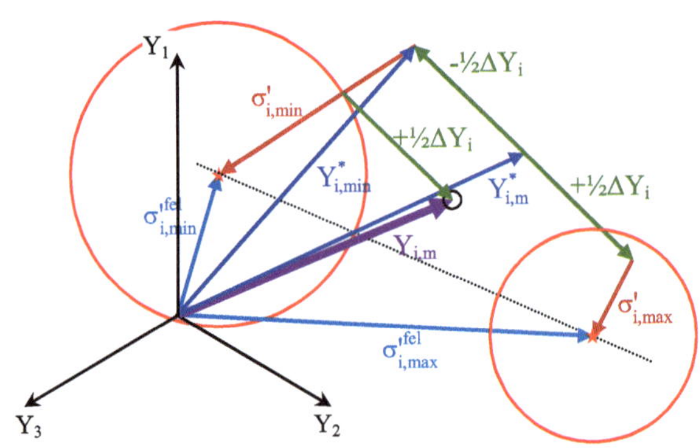

Fig. 5.26 Estimation of the TIV for the mean load condition at PS for a location in $V_{e\Delta}$, by projection on the minimum state

5.9 Examples for Accumulated Strains at Plastic Shakedown

5.9.1 Two-Bar Model

The procedure of Sect. 5.7.1 for the two-bar model at elastic shakedown, where regions of load levels are sought that correspond to given regions of residual stress fields, rather than otherwise looking for the plastic response of a structure to a given level of loading, is also examined here for investigating plastic shakedown. Four combinations can be distinguished, where the left bar in any case undergoes cyclic plastic straining, i.e., belongs to $V_{p\Delta}$. The right bar may also experience cyclic plastic straining ($V_{p\Delta}$) or may plasticize only monotonously ($V_{e\Delta}$), but in the considered parameter combinations only at the maximum, not the minimum load condition, or it can remain elastic (V_e) (Table 5.2).

5.9.1.1 Combination 10

For both bars, the mean value of the TIV is determined by Eq. (5.139). After adapting to the notation for uniaxial stress states we receive for positive secondary stresses σ_t

$$Y_{\text{left},m} = \sigma_P - \frac{1}{2}\sigma_t + \frac{1}{2}\left(f_{y,\text{hot}} - f_{y,\text{cold}}\right) \tag{5.149}$$

$$Y_{\text{right},m} = \sigma_P + \frac{1}{2}\sigma_t , \tag{5.150}$$

so that the modified elastic material laws are

$$\varepsilon^*_{\text{left},m} = \frac{1}{E_t}\rho_{\text{left},m} + \frac{1}{C}\left[\sigma_P - \frac{1}{2}\sigma_t + \frac{1}{2}\left(f_{y,\text{hot}} - f_{y,\text{cold}}\right)\right] \tag{5.151}$$

$$\varepsilon^*_{\text{right},m} = \frac{1}{E_t}\rho_{\text{right},m} + \frac{1}{C}\left(\sigma_P + \frac{1}{2}\sigma_t\right) . \tag{5.152}$$

For the equilibrium condition

$$\rho_{\text{right},m} = -\rho_{\text{left},m} \tag{5.153}$$

Table 5.2 Two-bar model: possible combinations of the states of the bars at PS

Combination	Left bar in	Right bar in
10	$V_{p\Delta}$	$V_{p\Delta}$
11	$V_{p\Delta}$	$V_{e\Delta}$ (plastic at max. load)
12	$V_{p\Delta}$	V_e

and the compatibility condition

$$\varepsilon^*_{\text{left},m} = \varepsilon^*_{\text{right},m} \tag{5.154}$$

we thus obtain as solution for the mean loading condition

$$\rho_{\text{left},m} = \left(1 - \frac{E_t}{E}\right)\left[\frac{1}{2}\sigma_t - \frac{1}{4}\left(f_{y,\text{hot}} - f_{y,\text{cold}}\right)\right] \tag{5.155}$$

$$\varepsilon^*_{\text{left},m} = \frac{1 - \frac{E_t}{E}}{E_t}\left[\sigma_P + \frac{1}{4}\left(f_{y,\text{hot}} - f_{y,\text{cold}}\right)\right] \tag{5.156}$$

as well as for the minimum and the maximum load condition

$$\rho_{\text{left,min}} = \rho_{\text{left},m} - \frac{1}{2}\Delta\rho_{\text{left}} \tag{5.157}$$

$$\rho_{\text{left,max}} = \rho_{\text{left},m} + \frac{1}{2}\Delta\rho_{\text{left}} \tag{5.158}$$

$$\varepsilon^*_{\text{left,min}} = \varepsilon^*_{\text{left},m} - \frac{1}{2}\Delta\varepsilon^*_{\text{left}} \tag{5.159}$$

$$\varepsilon^*_{\text{left,max}} = \varepsilon^*_{\text{left},m} + \frac{1}{2}\Delta\varepsilon^*_{\text{left}}\,, \tag{5.160}$$

with the residual stress range $\Delta\rho$ given by Eq. (5.45) and residual strain range $\Delta\varepsilon^*$ by Eq. (5.46). After superposition with the fictitious elastic solution we receive

$$\sigma_{\text{left,min}} = \left(1 - \frac{E_t}{E}\right)f_{y,\text{cold}} + \sigma_P \tag{5.161}$$

$$\sigma_{\text{left,max}} = \left(1 - \frac{E_t}{E}\right)\left(\sigma_t - \frac{1}{2}f_{y,\text{hot}} - \frac{1}{2}f_{y,\text{cold}}\right) + \sigma_P - \sigma_t \tag{5.162}$$

$$\varepsilon_{\text{left,min}} = \frac{\sigma_P}{E_t} \tag{5.163}$$

$$\varepsilon_{\text{left,max}} = \frac{\sigma_P}{E_t} - \frac{\sigma_t}{E} + \frac{1}{2}\frac{1 - \frac{E_t}{E}}{E_t}\left(f_{y,\text{hot}} - f_{y,\text{cold}}\right). \tag{5.164}$$

The range of validity of this solution is given by the requirement that the stress range must exceed twice the "cold" yield stress in the right bar,

$$\Delta\sigma_{\text{right}} = -\Delta\rho_{\text{left}} + \Delta\sigma_{\text{right}}^{\text{fel}} \geq 2f_{y,\text{cold}}\,, \tag{5.165}$$

leading by means of Eq. (5.45) to the condition

$$\sigma_t \geq \frac{3}{2} f_{y,\text{cold}} + \frac{1}{2} f_{y,\text{hot}} - \frac{1}{2}\frac{E}{E_t}\left(f_{y,\text{hot}} - f_{y,\text{cold}}\right). \tag{5.166}$$

5.9.1.2 Combination 11

If the right bar is experiencing only elastic strain ranges in the state of plastic shakedown, and if the stress at the maximum load is greater than at the minimum load, then the mean value of the TIV is determined according to Eq. (5.145). This can be written for the uniaxial stress state in the right bar as

$$Y_{\text{right},m} = \sigma_P + \sigma_t - \sigma_{\text{right,max}} \cdot \frac{f_{y,\text{cold}}}{\sigma_{\text{v,right,max}}} - \frac{1}{2}\Delta\rho_{\text{left}}\,. \tag{5.167}$$

After inserting Eq. (5.51), and because the stress in the right bar must be a tensile stress at the maximum load condition, we get

$$Y_{\text{right},m} = \sigma_P + \sigma_t - f_{y,\text{cold}} + \frac{1}{2}\frac{1 - \frac{E_t}{E}}{1 + \frac{E_t}{E}}\left(-\sigma_t + f_{y,\text{cold}} + f_{y,\text{hot}}\right). \tag{5.168}$$

For the left bar, Eq. (5.149) remains valid. Thus the modified elastic material laws are

$$\varepsilon^*_{\text{left},m} = \frac{1}{E_t}\rho_{\text{left},m} + \frac{1}{C}\left[\sigma_P - \frac{1}{2}\sigma_t + \frac{1}{2}\left(f_{y,\text{hot}} - f_{y,\text{cold}}\right)\right] \tag{5.169}$$

$$\varepsilon^*_{\text{right},m} = \frac{1}{E_t}\rho_{\text{right},m} + \frac{1}{C}\left(\sigma_P + \sigma_t - f_{y,\text{cold}} + \frac{1}{2}\frac{1 - \frac{E_t}{E}}{1 + \frac{E_t}{E}}\left(-\sigma_t + f_{y,\text{cold}} + f_{y,\text{hot}}\right)\right), \tag{5.170}$$

which provides a solution to the equilibrium and the compatibility condition for the mean load condition:

$$\rho_{\text{left},m} = \frac{1}{2}\frac{1 - \frac{E_t}{E}}{1 + \frac{E_t}{E}}\left[\left(1 + 2\frac{E_t}{E}\right)\sigma_t - \frac{E_t}{E}\left(f_{y,\text{hot}} + f_{y,\text{cold}}\right)\right] \tag{5.171}$$

$$\varepsilon^*_{\text{left},m} = \frac{1}{2}\frac{1}{E_t}\frac{1 - \frac{E_t}{E}}{1 + \frac{E_t}{E}}\left[2\left(1 + \frac{E_t}{E}\right)\sigma_P + \frac{E_t}{E}\sigma_t - \left(1 + 2\frac{E_t}{E}\right)f_{y,\text{cold}} + f_{y,\text{hot}}\right]. \tag{5.172}$$

This results in the stresses and strains at the minimum and the maximum load condition

$$\sigma_{\text{left,min}} = \frac{1}{2}\frac{1-\frac{E_t}{E}}{1+\frac{E_t}{E}}\left[2\frac{E_t}{E}\sigma_t + \left(1-\frac{E_t}{E}\right)\left(f_{y,\text{hot}} + f_{y,\text{cold}}\right)\right] + \sigma_P \tag{5.173}$$

$$\sigma_{\text{left,max}} = \left(1-\frac{E_t}{E}\right)\left(\sigma_t - \frac{1}{2}f_{y,\text{hot}} - \frac{1}{2}f_{y,\text{cold}}\right) + \sigma_P - \sigma_t \tag{5.174}$$

$$\varepsilon_{\text{left,min}} = \frac{1}{E_t}\frac{1-\frac{E_t}{E}}{1+\frac{E_t}{E}}\left[\frac{E_t}{E}\sigma_t + \frac{1}{2}\left(1-\frac{E_t}{E}\right)f_{y,\text{hot}} - \frac{1}{2}\left(1+3\frac{E_t}{E}\right)f_{y,\text{cold}}\right] + \frac{\sigma_P}{E_t} \tag{5.175}$$

$$\varepsilon_{\text{left,max}} = -\frac{\sigma_t}{E} + \frac{1}{2}\frac{1-\frac{E_t}{E}}{E_t}\left(f_{y,\text{hot}} - f_{y,\text{cold}}\right) + \frac{\sigma_P}{E_t}\,. \tag{5.176}$$

The range of validity of this solution is limited by the limits of validity of the other combinations.

5.9.1.3 Combination 12

In the event that the right bar remains fully elastic, the modified elastic material laws apply:

$$\varepsilon^*_{\text{left},m} = \frac{1}{E_t}\rho_{\text{left},m} + \frac{1}{C}\left[\sigma_P - \frac{1}{2}\sigma_t + \frac{1}{2}\left(f_{y,\text{hot}} - f_{y,\text{cold}}\right)\right] \tag{5.177}$$

$$\varepsilon^*_{\text{right},m} = \frac{1}{E}\rho_{\text{right},m} \tag{5.178}$$

with the solution of the equilibrium and compatibility conditions

$$\rho_{\text{left},m} = -\frac{1-\frac{E_t}{E}}{1+\frac{E_t}{E}}\left[\sigma_P - \frac{1}{2}\sigma_t + \frac{1}{2}\left(f_{y,\text{hot}} - f_{y,\text{cold}}\right)\right] \tag{5.179}$$

$$\varepsilon^*_{\text{left},m} = \frac{1}{E}\frac{1-\frac{E_t}{E}}{1+\frac{E_t}{E}}\left[\sigma_P - \frac{1}{2}\sigma_t + \frac{1}{2}\left(f_{y,\text{hot}} - f_{y,\text{cold}}\right)\right] \tag{5.180}$$

and the stresses and strains at the minimum and maximum load condition

$$\sigma_{\text{left,min}} = \frac{2\frac{E_t}{E}\sigma_P + f_{y,\text{cold}}\left(1-\frac{E_t}{E}\right)}{1+\frac{E_t}{E}} \tag{5.181}$$

$$\sigma_{\text{left,max}} = \frac{2\frac{E_t}{E}\left(\sigma_P - \sigma_t\right) - f_{y,\text{hot}}\left(1 - \frac{E_t}{E}\right)}{1 + \frac{E_t}{E}} \tag{5.182}$$

$$\varepsilon_{\text{left,min}} = \frac{\sigma_P}{E}\frac{2}{1 + \frac{E_t}{E}} - \frac{f_{y,\text{cold}}}{E}\frac{1 - \frac{E_t}{E}}{1 + \frac{E_t}{E}} \tag{5.183}$$

$$\varepsilon_{\text{left,max}} = \frac{2\left(\sigma_P - \sigma_t\right) + f_{y,\text{hot}}\left(1 - \frac{E_t}{E}\right)}{E\left(1 + \frac{E_t}{E}\right)}\,. \tag{5.184}$$

The condition that the yield stress may not be reached in the right bar at the maximum load condition leads to the limitation of the range of validity of combination 12 that is identical to Eq. (5.122).

5.9.1.4 Ratcheting Interaction Diagram

The solutions presented in the previous sections can be used to complete the ratcheting interaction diagram created already for the ES-region in Sect. 5.7.1.10 (Fig. 5.27).

For a parameter combination in region 11, Fig. 5.28 shows, as an example, how the stresses and strains develop with the number of loading cycles.

As long as the anomalous behavior with seemingly negative hardening produced by a very pronounced temperature dependence of the yield stress as described in Sect. 5.1.2 (Fig. 5.9) does not occur, the results obtained with the STPZ are accurate, for example for the parameter combination considered in Fig. 5.28. Otherwise, they are only approximate solutions. A variation of the parameter combination of

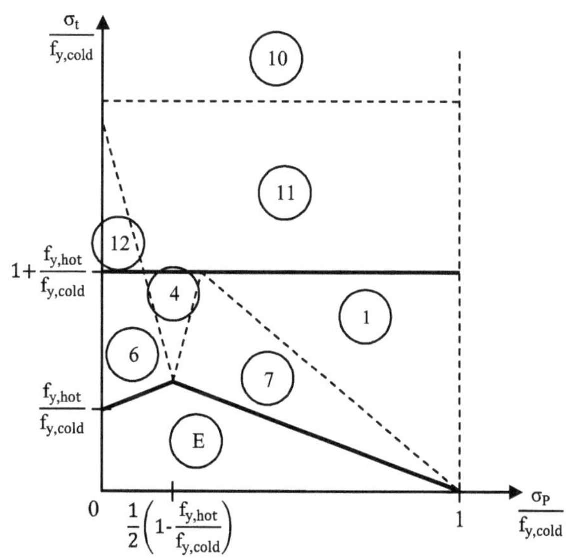

Fig. 5.27 Two-bar model: range of different solutions for the strain accumulation at elastic and plastic shakedown

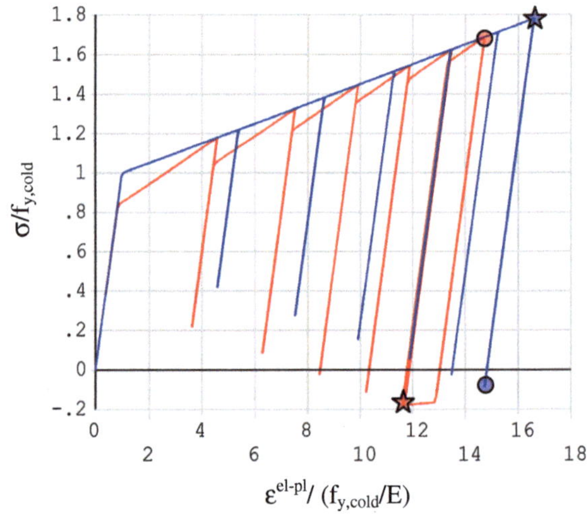

Fig. 5.28 Strain accumulation in the two-bar model with linear kinematic hardening ($\sigma_P/f_{y,\text{cold}} = 0.8$; $\sigma_t/f_{y,\text{cold}} = 2.5$; $E_t/E = 0.05$; $f_{y,\text{hot}}/f_{y,\text{cold}} = 0.8$); in the state of plastic shakedown the incremental solution is identical with the STPZ (*circles* for minimum and *stars* for maximum load condition)

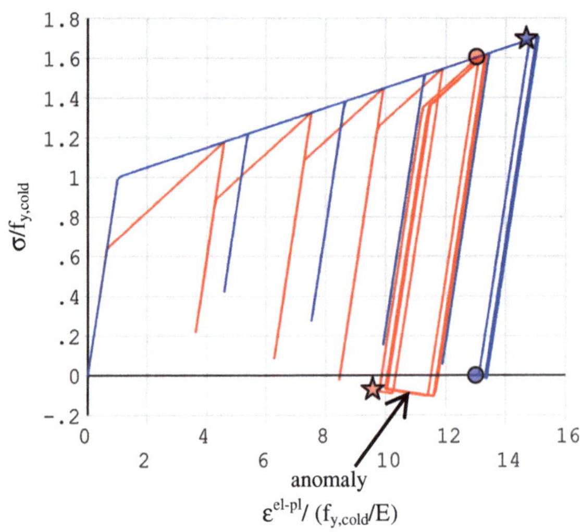

Fig. 5.29 Strain accumulation in the two-bar model with linear kinematic hardening ($\sigma_P/f_{y,\text{cold}} = 0.8$; $\sigma_t/f_{y,\text{cold}} = 2.5$; $E_t/E = 0.05$; $f_{y,\text{hot}}/f_{y,\text{cold}} = 0.6$); in the state of plastic shakedown the incremental solution and the STPZ (*circles* for minimum and *stars* for maximum load condition) are not identical

Fig. 5.28 may serve as an example of such a situation, when the ratio of the "hot" yield stress to the "cold" yield stress is reduced from 80 to 60 % (Fig. 5.29).

One recognizes the anomaly mentioned in the left bar shortly before reaching the maximum load condition where the compressive stress decreases despite increasing compressive plastic strain. Whereas the strain range is still exactly determined by the STPZ in the state of shakedown, the decrease in the compressive stress in the left bar is, for equilibrium reasons, responsible for the maximum stress in the right bar not being reached at maximum load. The strain accumulated in the state of shakedown is therefore not as expected on reaching the yield stress at maximum load, but at an intermediate load condition.

Table 5.3 provides a quantitative impression of the strains obtained with the STPZ compared to an incremental analysis for this example of anomalous behavior.

The cause of this anomaly can also be visualized in the TIV-space (Fig. 5.30). The correct values of the TIV are shown here in purple. The TIVs of the left bar are exactly on the edges of the yield surfaces associated with the minimum and the maximum load, which is indeed the condition for active plastic straining. Therefore both their range ΔY as well as their mean value Y_m are correctly identified by the STPZ. The TIVs of the right bar, however, are in the interior of the Mises circles associated with the minimum and the maximum load, which is the condition for elastic behavior. The STPZ correctly recognizes that the right bar is not plasticized cyclically and also is not plasticized at the minimum load condition, but assumes incorrectly that the TIV is exactly located on the edge of the yield surface at maximum load, and that the right bar is thus plasticized at the maximum load condition. However, in fact the plastic straining takes place at a load condition below the maximum load (green circle in Fig. 5.30). Because the maximum value of the TIV, Y_{max}, is therefore underestimated by the STPZ, but the range of the TIV,

Table 5.3 Strains obtained in the state of shakedown with the STPZ and incremental analysis in the case of anomalous behavior

	$\varepsilon\, E/f_{y,cold}$ in the left bar		$\varepsilon\, E/f_{y,cold}$ in the right bar	
	Incremental	STPZ	Incremental	STPZ
max ε	10.05	9.70	15.05	14.70
min ε	13.37	13.02	13.37	13.02
$\Delta\varepsilon$	−3.31	−3.31	1.69	1.69

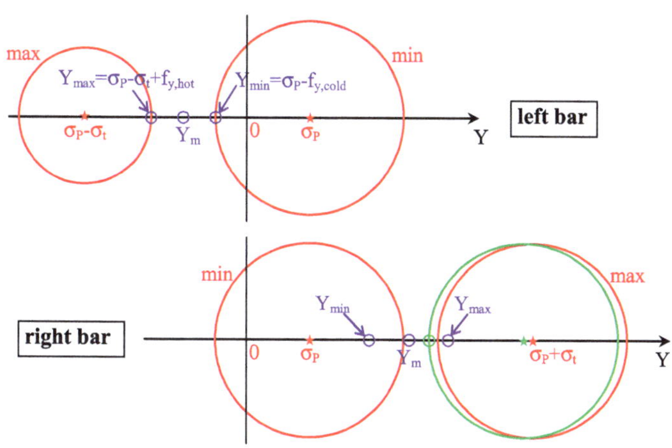

Fig. 5.30 Two-bar model with anomalous behavior: TIV-space in the state of plastic shakedown

ΔY, is correctly identified, the mean value of the TIV, Y_m, is underestimated by the STPZ.

Whether such a situation of anomalous behavior is present can be determined with the STPZ when the calculation is repeated with a slightly reduced level of cyclic loading. Anomalous behavior occurs if the accumulated strain is larger for a reduced load.

5.9.2 Multiaxial Ratcheting

In the example of multiaxial ratcheting already treated repeatedly, the plastic zone is immediately given and consists of the entire homogeneously stressed element. The TIV for the mean state is as in Eq. (5.139) with $v = 0$:

$$Y_{i,m} = \sigma_P \begin{pmatrix} \frac{2}{3} \\ -\frac{1}{3} \\ -\frac{1}{3} \end{pmatrix} + \frac{1}{2}\left(\sigma_t + f_{y,\text{cold}} - f_{y,\text{hot}}\right) \begin{pmatrix} -\frac{1}{3} \\ \frac{2}{3} \\ -\frac{1}{3} \end{pmatrix} . \tag{5.185}$$

As modified elastic material law for the mean loading state we get, if we just take into account the boundary conditions ($\varepsilon_2^* = 0$, $\rho_1 = \rho_3 = 0$),

$$\begin{pmatrix} \varepsilon_{1,m}^* \\ 0 \\ \varepsilon_{3,m}^* \end{pmatrix} = \frac{1}{E_t} \begin{pmatrix} 1 & -v^* & -v^* \\ -v^* & 1 & -v^* \\ -v^* & -v^* & 1 \end{pmatrix} \begin{pmatrix} 0 \\ \rho_{2,m} \\ 0 \end{pmatrix} + \frac{1}{C}\left[\frac{\sigma_P}{2}\begin{pmatrix} +2 \\ -1 \\ -1 \end{pmatrix} + \frac{\sigma_t + f_{y,\text{cold}} - f_{y,\text{hot}}}{4}\begin{pmatrix} -1 \\ +2 \\ -1 \end{pmatrix}\right] . \tag{5.186}$$

This leads to the solution

$$\rho_{2,m} = \frac{1}{2}\left(1 - \frac{E_t}{E}\right)\left(\sigma_P - \sigma_t - f_{y,\text{cold}} + f_{y,\text{hot}}\right) . \tag{5.187}$$

By superposition of the fictitious elastic solution for the mean state and subtraction or addition of half the range quantities given by Eqs. (4.42) and (4.43), where, according to Sect. 5.5.2, the arithmetic mean of the "cold" and "hot" yield stresses is to be used for the yield stress f_y, we obtain for the two extreme states of loading:

$$\frac{\sigma_{2,\min}}{f_{y,\text{cold}}} = \left(1 - \frac{E_t}{E}\right)\left(\frac{1}{2}\frac{\sigma_P}{f_{y,\text{cold}}} - 1\right) \tag{5.188}$$

$$\frac{\sigma_{2,\max}}{f_{y,\text{cold}}} = \left(1 - \frac{E_t}{E}\right)\left(\frac{1}{2}\frac{\sigma_P}{f_{y,\text{cold}}} + \frac{f_{y,\text{hot}}}{f_{y,\text{cold}}}\right) + \frac{E_t}{E}\frac{\sigma_t}{f_{y,\text{cold}}} \tag{5.189}$$

$$\frac{E}{f_{y,\text{cold}}}\varepsilon_{1,\min} = \frac{\sigma_P}{f_{y,\text{cold}}}\left(\frac{3}{4}\frac{E}{E_t} + \frac{1}{2} - \frac{1}{4}\frac{E_t}{E}\right) - \frac{1}{2}\left(1 - \frac{E_t}{E}\right) \tag{5.190}$$

$$\frac{E}{f_{y,\text{cold}}}\varepsilon_{1,\max} = \frac{\sigma_P}{f_{y,\text{cold}}}\left(\frac{3}{4}\frac{E}{E_t} + \frac{1}{2} - \frac{1}{4}\frac{E_t}{E}\right) + \frac{1}{2}\left(1 - \frac{E_t}{E}\right)\left(\frac{f_{y,\text{hot}}}{f_{y,\text{cold}}} - \frac{\sigma_t}{f_{y,\text{cold}}}\right). \quad (5.191)$$

These results are exact, even if anomalous behavior occurs according to Sect. 5.1.2 (Fig. 5.9).

5.9.3 Bree Tube

The Bree tube has already been discussed several times. As in Sect. 5.7.3 for elastic shakedown, plastic shakedown is now considered for the axisymmetric model of the Bree tube.

For the parameter combination

$$\frac{\sigma_P}{f_{y,\text{cold}}} = 0.8\ ;\ \frac{\sigma_t}{f_{y,\text{cold}}} = 1.9\ ;\ \nu = 0.3\ ;\ \frac{E_t}{E} = 0.02\ ;\ \frac{f_{y,\text{hot}}}{f_{y,\text{cold}}} = 0.6667 \quad (5.192)$$

the stress–strain curves for the circumferential direction on the inside and the outside surface, normalized to the “cold” yield stress or the elastic limit strain, are shown in Fig. 5.31 from an incremental analysis of linear dependence of yield stress on temperature. The results of the STPZ for the state of shakedown are also marked after five MEAs. Cyclic plastic straining in the shakedown state can be clearly seen on the inside surface, whereas, on the outside surface, the change of stress and strain is almost purely elastic. Unlike at temperature-independent yield stress (Fig. 4.42), a drop of the maximum stress is observed towards the end of the

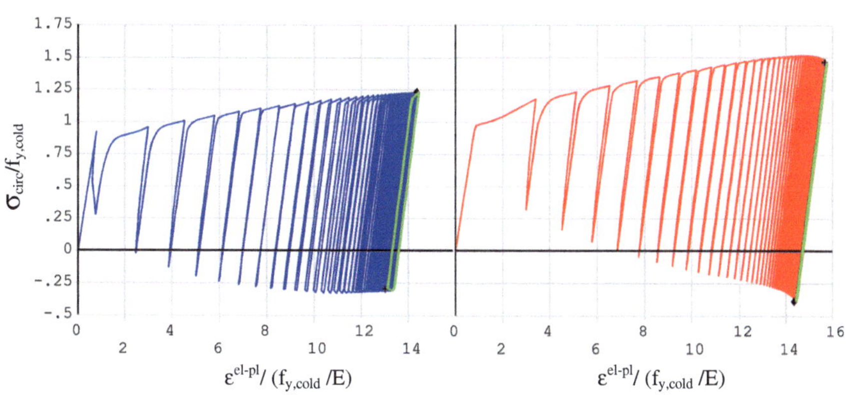

Fig. 5.31 Bree tube (axisymmetric model): stress–strain diagram for the circumferential direction according to an incremental analysis, stress–strain couples of the fifth MEA of the STPZ (*diamonds* and *crosses*); *left* inside surface; *right* outside surface; hysteresis loop in the shakedown state highlighted in *green*

Table 5.4 Development of circumferential strain on the outside surface with the MEAs of the STPZ as compared with an incremental analysis at maximum load level in the state of shakedown

STPZ, MEA	$E/f_{y,\text{cold}}{}^{*}\varepsilon_z$
1	2.22
2	6.14
3	15.37
4	15.70
5	15.67
6	15.64
7	15.62
8	15.61
Incremental	15.77

shakedown process on the outside surface. Despite this somewhat unusual behavior, the quality of the approximation by the STPZ must be regarded as very good.

As an example of the development of strains with the number of MEAs of the STPZ in the state of plastic shakedown, the circumferential strain component on the outside surface is listed in Table 5.4 for the maximum load level. After three MEAs the result changes only slightly.

5.9.4 Thick-Walled Cylinder Loaded by a Temperature Transient and Internal Pressure

In Sect. 4.7.4 the accumulated strains were calculated for a thick-walled cylinder with temperature-independent material data, which is exposed to a constant internal pressure and a temperature transient of the fluid moving through. Now, the yield stress is regarded as a function of temperature. The geometry and material data, as well as the thermal and the fictitious elastic analysis of Sect. 4.3.4 are adopted.

So far, only a linear dependence of the yield stress of the temperature was used in the examples. Now a nonlinear dependence is selected. Table 5.5 provides some reference points, between which linear interpolations can be made. A linear dependence between the lowest and the highest temperature would result in the mean value of 100 N/mm^2 as used in Sect. 4.7.4.

Table 5.5 Nonlinear temperature dependence of the yield stress

Temperature (°C)	Yield stress (N/mm^2)
0	130
100	105
200	89
300	78
400	70

Figure 5.32 represents the results of an incremental analysis performed only for the pair of transient times 330 and 525 s in the form of the strain distribution across the wall thickness at the two extreme loading times in the state of shakedown. For a good approximation to the state of shakedown, about the same number of cycles were to be calculated as at temperature-independent yield stress (about 50 cycles).

As examples of the initial strains or initial stresses to be applied as modified loading in the STPZ, the distributions of initial stresses across the wall thickness are shown for the first MEA to calculate the range quantities and for the first MEA to calculate the mean quantities in Figs. 5.33 and 5.34.

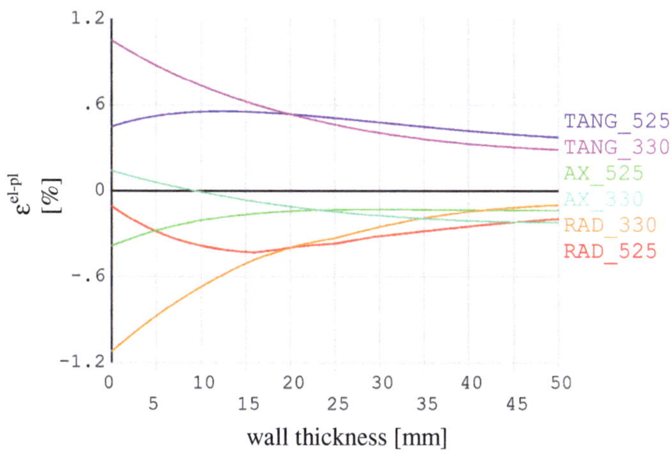

Fig. 5.32 Distribution of elastic-plastic strain components across the wall thickness at the state of plastic shakedown (incremental cyclic analysis only for transient times 330 and 525 s)

Fig. 5.33 Distribution of initial stress components across the wall thickness with respect to calculating range quantities (first MEA)

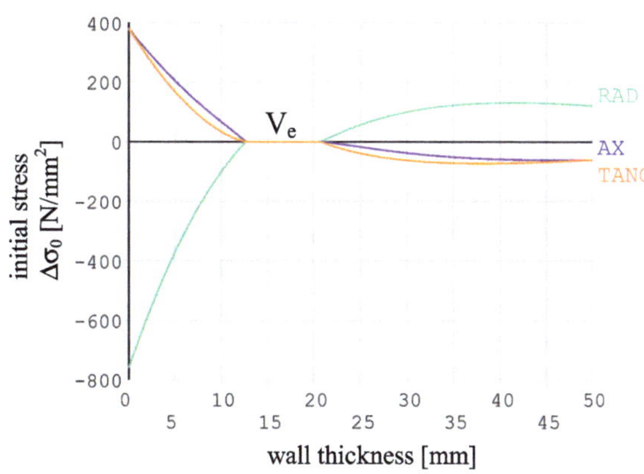

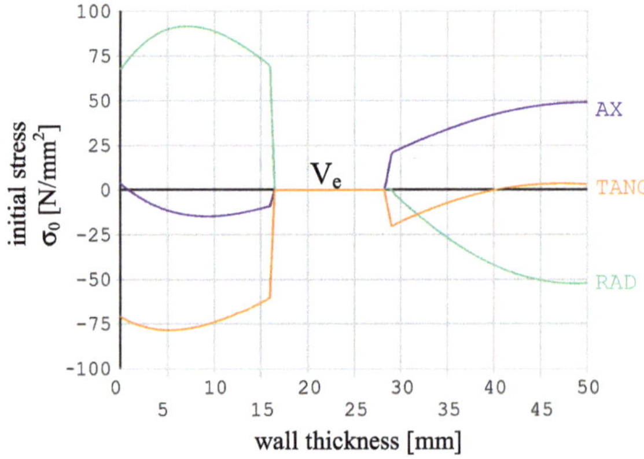

Fig. 5.34 Distribution of initial stress components across the wall thickness with respect to calculating mean quantities (first MEA)

A comparison of the results of the STPZ shown in Fig. 5.35 in the form of the distribution of the elastic-plastic strain components after the first and the second MEA with the incremental results in Fig. 5.32 reveals that after two linear analyses a good approximation to the incremental analysis is reached.

Figures 5.36 and 5.37 represent the circumferential stresses and strains in the state of plastic shakedown obtained with the STPZ along with the strain histogram on the inside and outside surfaces and the stress–strain diagram. There, the approximate character of the STPZ is visible, because the strains of the incremental analysis are somewhat underestimated by the STPZ.

The development of circumferential strains on the inside surface in the state of shakedown with the MEAs is shown in Table 5.6 for the two extreme loading states at the transient times 330 and 525 s.

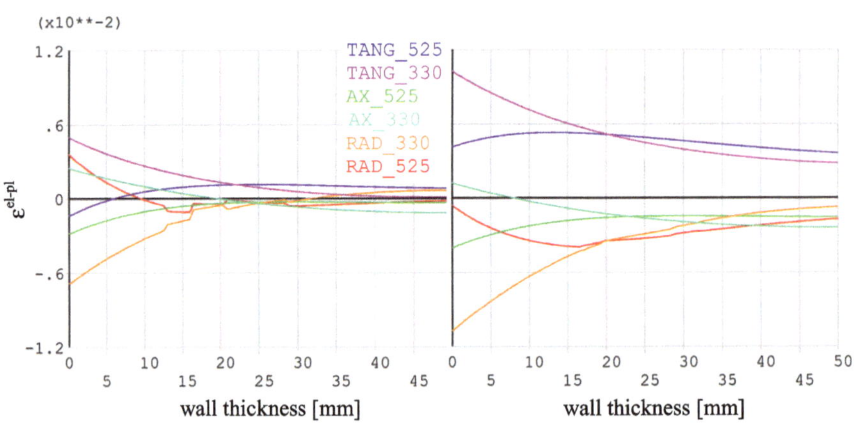

Fig. 5.35 Distribution of elastic-plastic strain components across the wall thickness for the transient times of 330 and 525 s after one MEA (*left*) or two MEAs (*right*) of the STPZ

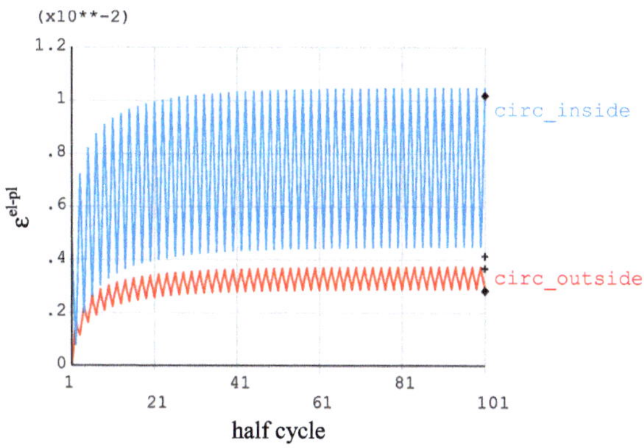

Fig. 5.36 Histogram of the circumferential strain on the inner and outer surface from incremental calculation for the pair of loading at the transient times 330 and 525 s and results after four MEAs of the STPZ at the state of plastic shakedown (*diamonds* for transient time 330 s, *crosses* for 525 s)

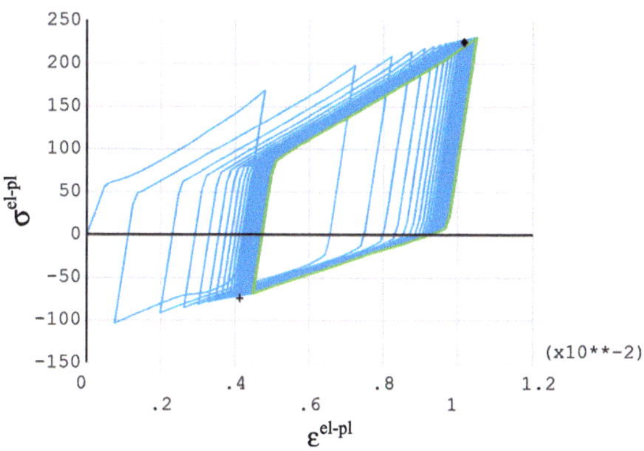

Fig. 5.37 Stress–strain diagram for the circumferential direction on the inner surface from incremental calculation for the pair of loading at the transient times 330 and 525 s (last cycle highlighted in *green*) and results after four MEAs of the STPZ at the state of plastic shakedown (*diamonds* for transient time 330 s, *crosses* for 525 s)

Table 5.6 Development of circumferential strain on the inner surface with the MEAs of the STPZ, compared with an incremental analysis in the shakedown condition at the two extreme load states 330 and 525 s

STPZ, MEA	ε_z (%) at 330 s	ε_z (%) at 525 s
1	0.496	−0.139
2	1.025	0.416
3	1.021	0.414
4	1.019	0.414
5	1.018	0.413
6	1.018	0.413
Incremental	1.051	0.448

Overall, it can be concluded that the results of the STPZ with temperature-dependent yield stress are no worse than those with temperature-independent yield stress in Sect. 4.7.4.

References

1. Ponter, A.R.S., Cocks, A.C.F.: The Anderson-Bishop problem– thermal ratchetting of a polycrystalline metals. In: Spiliopoulos, K., Weichert, D. (eds.) Direct Methods for Limit States in Structures and Materials, pp 243–255. Springer Science + Business Media, Dordrecht (2014). doi:10.1007/978-94-007-6827-7
2. Hübel, H.: Basic conditions for material and structural ratcheting. Nucl. Eng. Des. **162**, 55–65 (1996)
3. ANSYS Release 14.5, ANSYS Inc. Canonsburg, USA (2012)
4. Hübel, H.: Simplified theory of plastic zones for cyclic loading and multilinear hardening. Int. J. Press. Vess. Pip. **129–130**, 19–31 (2015). doi:10.1016/j.ijpvp.2015.03.002
5. Sicherheitstechnische Regel des KTA, KTA 3201.2. Komponenten des Primärkreises von Leichtwasserreaktoren, Teil 2: Auslegung, Konstruktion und Berechnung. Fassung 6/96 (including correction from BAnz. Nr. 129, 13.07.2000). Office of the KTA c/o Bundesamt für Strahlenschutz, Salzgitter (2000)

Overlay Model

6

Before the STPZ can be expanded in terms of nonlinear kinematic hardening in Chap. 7, some preliminary work is required, because a suitable material model must be provided first. So far, only linear kinematic hardening, Prager–Ziegler hardening, has been considered. Now an extension to multilinear kinematic hardening is made.

6.1 Multilinear Kinematic Hardening

Multilinear kinematic hardening can be modeled by multiple yield surfaces, which are coupled to each other, so that their movements in the stress space depend on each other. If they could move independently, an overlap of the yield surfaces in multiaxial stress states would not be ruled out. However, this is thermodynamically not readily admissible because Drucker's postulates for stable material behavior may then be violated. Otherwise it would be possible that, with increasing load, starting in the elastic range with plasticizing beginning, the trilinear branch of the stress–strain diagram would be activated first and then, with further increase in load, the bilinear branch. Therefore, for example, a simple addition of multiple yield surfaces with Prager–Ziegler hardening fails.

In a trilinear stress–strain diagram, the two yield surfaces are coupled in the Mróz model [1] by the fact that the smaller yield surface, belonging to the first breakpoint of the stress–strain curve, moves in the stress space in the direction of the stress point on the edge of the larger yield surface, where the larger yield surface has the same normal as at the stress point on the smaller yield surface. This means that the second yield surface has an influence on the plastic behavior, even if it is not active, i.e., if the stress point is only inside and not on the edge of the greater yield surface.

H. Hübel, *Simplified Theory of Plastic Zones*,
DOI 10.1007/978-3-319-29875-7_6

Stresses and strains in the bilinear branch of the stress–strain diagram, i.e., below the second breakpoint, are then influenced by the stress–strain curve above the second breakpoint. A consequence of this is in turn that degeneration to a bilinear stress–strain behavior (when the tangent modules in the bilinear and trilinear branches are the same) is not unique, but that the level of the second yield stress plays a role. Only when, in addition to identical tangent modulus, the second yield stress is chosen to be infinitely high, does the Mróz model reduce to the bilinear behavior of the Prager–Ziegler hardening.

The Mróz model often predicts severe ratcheting. Despite the unlimited kinematic hardening, even infinite ratcheting, i.e., a constant strain increment in each load cycle, is possible. The Mróz model is unsuitable as a basis for the STPZ, because the centers of the yield surfaces are not uniquely correlated with the plastic strains, because the yield surface is not moving in the direction of the normal to the yield surface in the current stress point.

Instead, reference is made to the overlay model of Besseling [2]. In this, an elementary volume of material consists of several layers with different elastic-plastic properties. By coupling these layers to satisfy the kinematic and static conditions at the elementary volume of material, complex material behavior can be simulated, even if the behavior of each of the individual layers is simple.

6.2 Layer Without Hardening

With two layers of non-hardening material, linear kinematic hardening, i.e., bilinear stress–strain behavior, and thereby the Bauschinger effect, can be described. Linear kinematic hardening, however, is then limited. One layer less than the number of segments in a multilinear stress–strain diagram is always needed if the last segment has no more hardening. For any multilinear behavior of the total state with kinematic hardening, the equations for the Young's moduli and the yield stresses of the individual linear elastic-perfectly plastic layers are available in the literature (for example in [3]). Weighting factors w_k are defined for the kth layer:

$$w_k = \frac{E - E_{tk}}{E - \frac{1-2\nu}{3} E_{tk}} - \sum_{i=1}^{k-1} w_i \,, \tag{6.1}$$

where E_{tk} is the slope of the kth segment of the multilinear stress–strain diagram. The Young's modulus E_k of layer k is obtained by multiplying the Young's modulus E of the total state by the weighting factor of this layer:

$$E_k = w_k E \,. \tag{6.2}$$

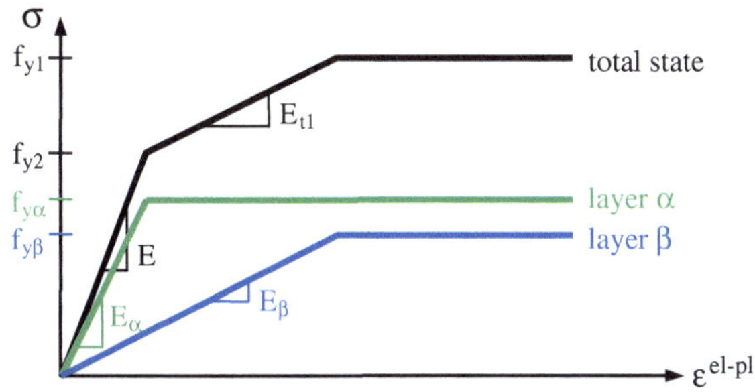

Fig. 6.1 Besseling model: two layers with linear elastic-perfectly plastic behavior to describe trilinear stress–strain behavior with kinematic hardening (except in the last segment)

If the breakpoints of the stress–strain curve are given by pairs of values σ_k, ε_k, the yield stress of layer k follows from

$$f_{yk} = \frac{w_k}{2(1+v)}\left[3E\varepsilon_k - (1-2v)\sigma_k\right]. \tag{6.3}$$

In all layers the same Poisson's ratio v is used as in the total state.

So all the parameters necessary to describe the behavior of the individual linear elastic-perfectly plastic layers are identified.

For a trilinear stress–strain curve without hardening in the third segment (Fig. 6.1), for example, we have for the two layers α and β:

for layer α (k = 1):

$$w_\alpha = \frac{E - E_{t1}}{E - \frac{1-2v}{3}E_{t1}} \tag{6.4}$$

$$E_\alpha = w_\alpha E \tag{6.5}$$

$$f_{y\alpha} = w_\alpha f_{y1} \tag{6.6}$$

for layer β (k = 2):

$$w_\beta = 1 - w_\alpha \tag{6.7}$$

$$E_\beta = w_\beta E \tag{6.8}$$

$$f_{y\beta} = f_{y2} - f_{y\alpha}\,. \tag{6.9}$$

6.3 Layer with Linear Kinematic Hardening

In the following we want to describe each of the individual layers with linear kinematic hardening with regard to the STPZ. For example, two layers are necessary for a trilinear stress–strain behavior of the total state with unlimited hardening in the last segment, denoted in Fig. 6.2 with α and β. We wonder, therefore, how the three material parameters of each layer, thus with two layers E_α, $E_{t\alpha}$, $f_{y\alpha}$, E_β, $E_{t\beta}$, and $f_{y\beta}$, can be deduced from the known material parameters of a given multilinear description of the stress–strain diagram of the total state, in the trilinear case E, E_{t1}, f_{y1}, E_{t2}, f_{y2}, and Poisson's ratio ν. If the hardening of the material is limited, the slope in the last segment must be chosen as 0.

6.3.1 Basic Equations for Determining the Material Parameters

In an overlay model, an elementary volume of material is imagined that is composed of several layers, each of which has the same volume, but different elastic-plastic material properties and different yield surfaces. From this immediately follows that the strains of all layers and of the total state are the same. The layers are coupled kinematically with each other, via their elastic-plastic strain.

This is unlike other multi-surface models, where the yield surfaces are coupled to each other via plastic strains, such as the Mróz model. The yield surfaces of the layers of the overlay model are therefore related only indirectly with each other. An overlap of different yield surfaces is therefore not excluded in contrast to the Mróz model, but is thermodynamically harmless here. This can be justified for the reason that an equivalent reformulation of the behavior of n layers with hardening is possible to $n-1$ layers without hardening (and therefore no overlap between the flow surfaces), and the nth layer with unlimited elastic behavior (which at the end

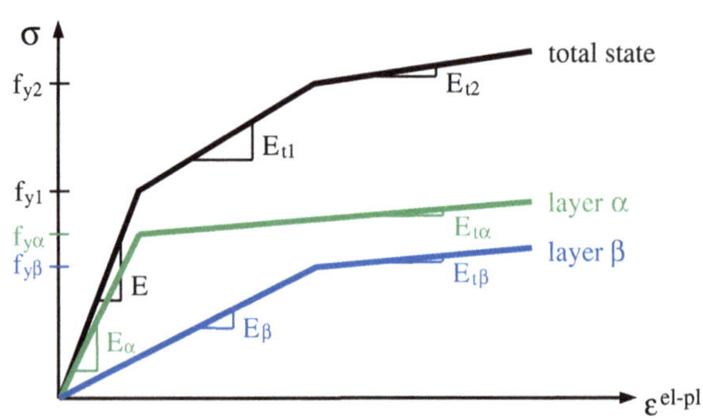

Fig. 6.2 Besseling model: two layers each with linear kinematic hardening to describe trilinear stress–strain behavior with kinematic hardening

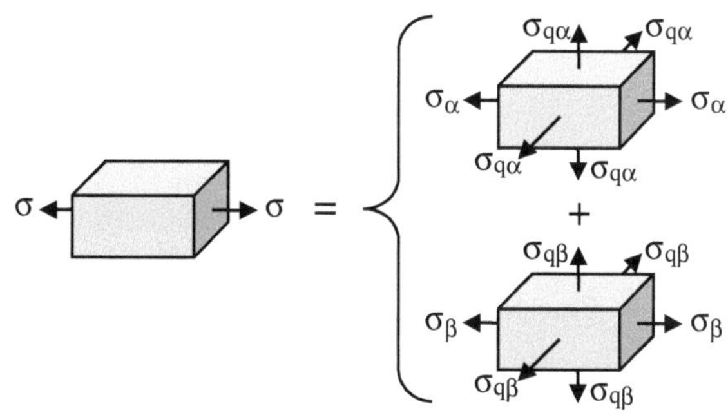

Fig. 6.3 Two layers with each multiaxial stress state for describing a uniaxial total stress state

this chapter is more evident than here already). For incremental analyses this would indeed be sufficient, but not for the STPZ.

For the derivation of the basic equations for determining the material parameters of each layer, an elementary volume of material is considered, which is subjected to monotonic uniaxial tensile stress. In the layers a triaxial stress state can arise as a result of constraints between the different layers (Fig. 6.3).

Occasionally readings can be such that the layers could be interpreted to consist of different volumes. This is at the least inaccurate, because then there would be no need to request the same strain perpendicular to these layers in all layers. The author believes that this is also the reason for the incorrectness in ANSYS already mentioned in Sect. 2.3 in treating a plane stress state with `PLANE182`- and `PLANE183`-elements.

Because multiaxial stress states occur in the individual layers because of a uniaxial stress state of the total state, equivalent stresses are to be provided. These arise both after Mises and after the Tresca hypothesis in each layer k to give

$$\sigma_{\mathrm{v}} = \sigma \tag{6.10}$$

$$\sigma_{\mathrm{vk}} = \sigma_k - \sigma_{qk}; \quad k = \alpha, \beta, \gamma, \ldots \tag{6.11}$$

Because the components of the respective deviatoric stress vector

$$\sigma'_{i,k} = \frac{\sigma_k - \sigma_{qk}}{3} \begin{pmatrix} 2 \\ -1 \\ -1 \end{pmatrix} \tag{6.12}$$

are proportional to each other in each layer, independently of the stress level, the loading is radial in the deviatoric stress space, so that the Prandtl–Reuss flow law, Eq. (1.18), can be integrated into the finite flow law Eq. (1.19). Using Eq. (6.11) we can realize that the equivalent plastic strain $\varepsilon_{\mathrm{v}}^{\mathrm{pl}}$ in each layer k is the same as its plastic strain component in axial direction, $\varepsilon_k^{\mathrm{pl}}$:

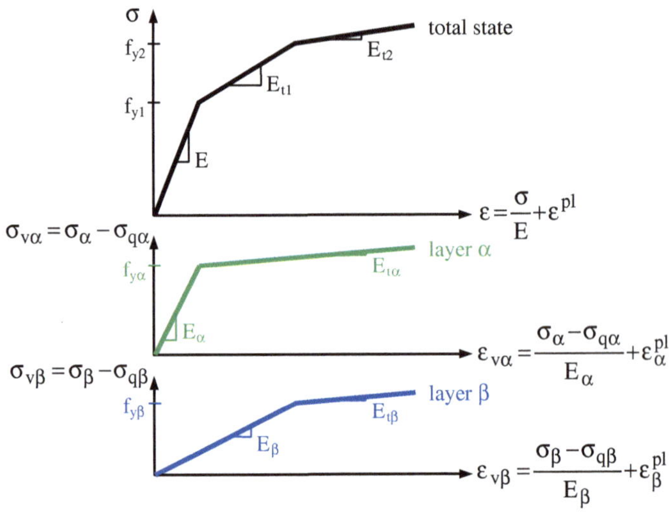

Fig. 6.4 Relationship between equivalent stress and equivalent strain in the layers

$$\varepsilon_{\mathrm{vk}}^{\mathrm{pl}} = \varepsilon_k^{\mathrm{pl}} \,. \tag{6.13}$$

Thus, the material behavior can be represented as a relationship between equivalent stress and equivalent strain for the two layers of the trilinear material behavior outlined in Fig. 6.2, for example, as in Fig. 6.4. It is worth noting that the breakpoints of the stress–strain curve no longer necessarily have to lie exactly one above the other, because at a given load level, although the elastic-plastic strains of the layers and of the total state are equal in the longitudinal direction, the elastic-plastic equivalent strains are not necessarily equal.

Furthermore, in the individual layers, just as in the overall element, use is made of the additivity theorem

$$\varepsilon = \varepsilon^{\mathrm{el}} + \varepsilon^{\mathrm{pl}} \tag{6.14}$$

$$\varepsilon_k = \varepsilon_k^{\mathrm{el}} + \varepsilon_k^{\mathrm{pl}} \tag{6.15}$$

and of the incompressibility of plastic strains. The elastic Poisson's ratio ν is the same in all layers as in the total state.

The equilibrium condition in the longitudinal direction requires that the sum of the stresses in the layers corresponds to the stress of the total state:

$$\sigma = \sum_k \sigma_k \,. \tag{6.16}$$

In the transverse direction, however, the total stress must vanish, so that the equilibrium condition is

$$\sum_k \sigma_{qk} = 0\,. \tag{6.17}$$

It follows then from Eqs. (6.10) and (6.11) that the equivalent stresses of the layers also add up to the equivalent stress of the total state:

$$\sigma = \sigma_{\mathrm{v}} = \sum_k \sigma_{\mathrm{vk}}\,. \tag{6.18}$$

The compatibility condition in the longitudinal direction requires that the strain of the total state, ε, is identical with the longitudinal strain ε_k in each layer k:

$$\varepsilon = \varepsilon_k \tag{6.19}$$

and likewise in the transverse direction:

$$\varepsilon_q = \varepsilon_{qk}\,. \tag{6.20}$$

With the abbreviations for the total state

$$C_1 = \frac{E\,E_{t1}}{E - E_{t1}} \tag{6.21}$$

$$C_2 = \frac{E\,E_{t2}}{E - E_{t2}} \tag{6.22}$$

etc., for other segments, as well as

$$C_k = \frac{E_k\,E_{tk}}{E_k - E_{tk}} \tag{6.23}$$

for each layer, the material laws for the total state and for the layers in the respective segments can be written by inserting the equivalent stresses from Eqs. (6.10) and (6.11):

material law for the total state:

first segment:

$$\text{in the longitudinal direction}: \ \varepsilon = \frac{\sigma}{E} \tag{6.24}$$

$$\text{in the transverse direction}: \ \varepsilon_q = -\nu \frac{\sigma}{E} \tag{6.25}$$

second segment:

$$\text{in the longitudinal direction}: \ \varepsilon = \frac{\sigma}{E} + \frac{\sigma - f_{y1}}{C_1} \tag{6.26}$$

$$\text{in the transverse direction}: \ \varepsilon_q = -\nu\frac{\sigma}{E} - \frac{1}{2}\frac{\sigma - f_{y1}}{C_1} \tag{6.27}$$

third segment:

$$\text{in the longitudinal direction}: \ \varepsilon = \frac{\sigma}{E} + \frac{f_{y2} - f_{y1}}{C_1} + \frac{\sigma - f_{y2}}{C_2} \tag{6.28}$$

$$\text{in the transverse direction}: \ \varepsilon_q = -\nu\frac{\sigma}{E} - \frac{1}{2}\frac{f_{y2} - f_{y1}}{C_1} - \frac{1}{2}\frac{\sigma - f_{y2}}{C_2} \tag{6.29}$$

material law for each layer k ($k = \alpha, \beta, \gamma, \ldots$):
in the elastic range:

$$\text{in the longitudinal direction}: \ \varepsilon_k = \frac{\sigma_k - 2\nu\sigma_{qk}}{E_k} \tag{6.30}$$

$$\text{in the transverse direction}: \ \varepsilon_{qk} = \frac{-\nu\sigma_k + \sigma_{qk}(1-\nu)}{E_k} \tag{6.31}$$

in the hardening range:

$$\text{in the longitudinal direction}: \ \varepsilon_k = \frac{\sigma_k - 2\nu\sigma_{qk}}{E_k} + \frac{\sigma_k - \sigma_{qk} - f_{yk}}{C_k} \tag{6.32}$$

$$\text{in the transverse direction}: \ \varepsilon_{qk} = \frac{-\nu\sigma_k + \sigma_{qk}(1-\nu)}{E_k} - \frac{1}{2}\frac{\sigma_k - \sigma_{qk} - f_{yk}}{C_k}. \tag{6.33}$$

Because of Eqs. (6.19) and (6.20),

$$\varepsilon + 2\varepsilon_q = \varepsilon_k + 2\varepsilon_{qk}\,, \tag{6.34}$$

and for the transverse stresses in all layers in all segments of the total state

$$\sigma_{qk} = \frac{1}{2}\left(\sigma\frac{E_k}{E} - \sigma_k\right) \tag{6.35}$$

proves true.

6.3.2 First Segment of the Total State

Without plastic straining in the total state, none of the layers can plasticize. In the compatibility conditions for each layer, Eqs. (6.19) and (6.20), therefore the material laws Eqs. (6.24) and (6.25) as well as Eqs. (6.30) and (6.31) are used. The only viable solution for the transverse stress in all layers is

$$\sigma_{qk} = 0\,. \tag{6.36}$$

After inserting the material laws in Eq. (6.19), they are rephrased for σ_k and inserted in the equilibrium condition Eq. (6.16), resulting in

$$E = \sum_k E_k\,. \tag{6.37}$$

No further equations are available from which a division of Young's modulus to the individual layers could be derived. For the time being still infinitely many possibilities exist for this.

6.3.3 Second Segment of the Total State

In the second segment of the total state, layer α is in the hardening state, whereas all the other layers still behave linearly elastic.

At the transition from the linear to the hardening behavior of the total state ($\sigma = f_{y1}$) the following applies:

$$\varepsilon = \frac{f_{y1}}{E}\,. \tag{6.38}$$

From the compatibility condition Eq. (6.19) it follows for layer α from the just still valid Eq. (6.30) together with the also just still valid Eq. (6.36) that

$$\sigma_\alpha = f_{y1}\frac{E_\alpha}{E}\,. \tag{6.39}$$

Because of Eq. (6.11) for layer α together with Eq. (6.36) and the requirement

$$\sigma_{v\alpha} = f_{y\alpha} \tag{6.40}$$

the transition condition from the first to the second segment is available:

$$f_{y\alpha} = f_{y1}\frac{E_\alpha}{E}\,. \tag{6.41}$$

Insertion of material laws Eq. (6.26) for the total state and Eq. (6.30) for layer β in the continuity equation $\varepsilon = \varepsilon_\beta$ returns, taking into account Eq. (6.35),

$$\sigma_\alpha = \sigma\left[1 - \frac{1}{1+v}\left(1 - \frac{E_\alpha}{E}\right)\left(\frac{E}{E_{t1}} + v\right)\right] + f_{y1}\frac{1}{1+v}\left(1 - \frac{E_\alpha}{E}\right)\left(\frac{E}{E_{t1}} - 1\right) \quad (6.42)$$

and, similarly, also

$$\frac{\sigma_\beta}{E_\beta}(1+v) = \frac{1}{E_{t1}}\left[\sigma\left(1 + v\frac{E_{t1}}{E}\right) - f_{y1}\left(1 - \frac{E_{t1}}{E}\right)\right]. \quad (6.43)$$

Inserting into Eq. (6.35) yields

$$\sigma_{q\alpha} = \frac{1}{2(1+v)}(\sigma - f_{y1})\left(\frac{E}{E_{t1}} - 1\right)\left(1 - \frac{E_\alpha}{E}\right), \quad (6.44)$$

showing that transverse stresses are generated in the layers because of the plasticizing of the total state, so that the uniaxial stress state of the total state is associated with multiaxial stress states in the layers!

The equivalent stress in layer α arises via Eqs. (6.11) and (6.35) to

$$\sigma_{v\alpha} = \sigma\left[\frac{3}{2} - \frac{1}{2}\frac{E_\alpha}{E} - \frac{\frac{3}{2}}{1+v}\left(1 - \frac{E_\alpha}{E}\right)\left(\frac{E}{E_{t1}} + v\right)\right] + f_{y1}\frac{\frac{3}{2}}{1+v}\left(1 - \frac{E_\alpha}{E}\right)\left(\frac{E}{E_{t1}} - 1\right), \quad (6.45)$$

and, with the aid of Eq. (6.35),

$$\varepsilon_\alpha^{pl} = (1+v)\left(\frac{\sigma_\beta}{E_\beta} - \frac{\sigma_\alpha}{E_\alpha}\right) = \frac{1}{E_\alpha}\left(\frac{E}{E_{t1}} - 1\right)(\sigma - f_{y1}) \quad (6.46)$$

can be specified. From

$$\sigma_{v\alpha} = f_{y\alpha} + \varepsilon_{v\alpha}^{pl} C_\alpha \quad (6.47)$$

we obtain by Eqs. (6.13), (6.45), and (6.46)

$$\frac{E_{t\alpha}}{E_\alpha} = \frac{a\,E_{t1} - (E - E_\alpha)}{a\,E - (E - E_\alpha)} \quad (6.48)$$

and, employing the abbreviation a for an expression of quantities known a priori,

$$a = \frac{1+v}{\frac{3}{2} - \left(\frac{1}{2} - v\right)\frac{E_{t1}}{E}}. \quad (6.49)$$

It is evident from Eq. (6.42) that the slope of the stress–strain diagram in the total state ($=E_{t1}$ in the second segment) is in general not equal to the sum of the slopes in the layers. This would only happen in the case $a = 1$, i.e., with elastic incompressible material ($\nu = 0.5$).

With Eqs. (6.37), (6.41), and (6.48), three equations for the desired material parameters are available as a function of the known material parameters E, f_{y1}, E_{t1}, and ν. Other determining equations cannot be deduced.

With two layers, for example, the parameters E_α, $E_\beta, f_{y\alpha}$, $E_{t\alpha}$ are required. One of these parameters can therefore be chosen freely. If, for example, $E_{t\alpha} = 0$ is chosen, as in the Besseling model with non-hardening layers considered in Sect. 6.2, Eq. (6.48) leads to E_α as given by Eqs. (6.5) and (6.41) to $f_{y\alpha}$ as given by Eq. (6.6).

6.3.4 Third Segment of the Total State

In the third segment of the total state, layer β is also in its hardening range in addition to layer α.

At the transition from the second to the third segment of the total state ($\sigma = f_{y2}$) the following applies:

$$\varepsilon = \frac{f_{y1}}{E} + \frac{f_{y2} - f_{y1}}{E_{t1}} \,. \tag{6.50}$$

From the compatibility condition $\varepsilon = \varepsilon_\beta$, and because Eq. (6.30) is just still valid, we obtain for layer β with the aid of Eq. (6.44)

$$\sigma_\beta = f_{y1}\frac{E_\beta}{E} + \left(f_{y2} - f_{y1}\right)\frac{E_\beta}{E_{t1}}\left[1 - \frac{\nu}{1+\nu}\left(1 - \frac{E_{t1}}{E}\right)\right]. \tag{6.51}$$

Because of Eq. (6.11) for layer β along with Eq. (6.44) and the requirement

$$\sigma_{v\beta} = f_{y\beta} \tag{6.52}$$

the transition condition is available:

$$f_{y\beta} = f_{y1}\frac{E_\beta}{E} + \left(f_{y2} - f_{y1}\right)\frac{E_\beta}{a\,E_{t1}} \,. \tag{6.53}$$

Inserting the material laws Eq. (6.28) for the total state and Eq. (6.32) for layer α in $\varepsilon = \varepsilon_\alpha$ we get, with the aid of Eq. (6.35),

$$\sigma_\alpha = \frac{2}{3}\frac{E_\alpha}{E}\frac{\frac{1}{2}\sigma\left[\left(2\frac{E}{E_{t2}}+2\nu-1\right)\frac{E_{t\alpha}}{E_\alpha}+1\right]+f_{y1}\left(1-\frac{E_{t\alpha}}{E_\alpha}\frac{E}{E_{t1}}\right)+f_{y2}\frac{E_{t\alpha}}{E_\alpha}\frac{E}{E_{t1}}\left(1-\frac{E_{t1}}{E_{t2}}\right)}{1-\frac{1-2\nu}{3}\frac{E_{t\alpha}}{E_\alpha}}. \tag{6.54}$$

Inserting the material laws Eqs. (6.28) and (6.32) for layer β in $\varepsilon = \varepsilon_\beta$, we get with the aid of Eqs. (6.53) and (6.35),

$$\frac{\sigma_\beta}{E_\beta} = \frac{\frac{\sigma}{E}\left(\frac{E}{E_{t2}}+\nu+\frac{1}{2}\frac{E_\beta}{C_\beta}\right)}{\frac{3}{2}\frac{E_\beta}{E_{t\beta}}-\frac{1}{2}+\nu} + \frac{\frac{E_\beta}{C_\beta(1+\nu)}\left\{\frac{f_{y1}}{E}\frac{3}{2}\left(1-\frac{E}{E_{t1}}\right)+\frac{f_{y2}}{E}\left[\frac{3}{2}\frac{E}{E_{t1}}-\frac{1}{2}+\nu\right]\right\}+\frac{f_{y1}}{E}\left(1-\frac{E}{E_{t1}}\right)+\frac{f_{y2}}{E}\left(\frac{E}{E_{t1}}-\frac{E}{E_{t2}}\right)}{\frac{3}{2}\frac{E_\beta}{E_{t\beta}}-\frac{1}{2}+\nu}. \tag{6.55}$$

The equations are becoming more and more complex, so that hereinafter only the solution to a trilinear material behavior of the total state, i.e., with two layers, which are both plasticized, is given and the reproduction of the equations in case of four and more segments, which would require additional layers remaining elastic in the third segment, is dispensed with.

Substituting Eqs. (6.54) and (6.55) in the equilibrium condition Eq. (6.16), $\sigma = \sigma_\alpha + \sigma_\beta$, leads to a long expression that can be reformulated so that, in addition to the known material data of the trilinear total state, only two of the six material parameters of the two layers appear, and the total stress σ. Because the two as yet unknown material parameters have to be independent of the stress level σ, the coefficients of all σ-terms together must evaluate to null. This requirement leads to the following expression for $E_{t\beta}$:

$$\frac{E_{t\beta}}{E_\beta} = \frac{1-\frac{E_{t\alpha}}{E_\alpha}\left[\frac{E}{E_{t2}}-\frac{E_\beta}{E}\left(\frac{E}{E_{t2}}-\frac{1-2\nu}{3}\right)\right]}{\frac{1-2\nu}{3}\left(1-\frac{E_{t\alpha}}{E_\alpha}\frac{E}{E_{t2}}\right)+\frac{E_\beta}{E}\left(\frac{E}{E_{t2}}-\frac{1-2\nu}{3}\right)}. \tag{6.56}$$

As $E_{t\alpha}/E_\alpha$ can be expressed by E_β (=$E - E_\alpha$) according to Eq. (6.48), Eq. (6.56) ultimately provides $E_{t\beta}$ as a function of E_β.

Alternatively, and with the same result, advantage could be taken from the fact that the terms in Eqs. (6.54) and (6.55), which are independent of σ, must add up to zero.

Only in special cases, such as, for example, when $\nu = 0.5$, $E_{t\alpha}$ and $E_{t\beta}$ complement each other to E_{t2}:

$$E_{t\alpha} + E_{t\beta} = E_{t2} \quad \text{for } \nu = 0.5\,. \tag{6.57}$$

For linear elastic-perfectly plastic material behavior in the third segment of the total state, i.e., if $E_{t2} = 0$, the only solution to Eqs. (6.56) and (6.48) is $E_{t\beta} = E_{t\alpha} = 0$ if neither for $E_{t\beta}$ nor for $E_{t\alpha}$ are negative values admitted, and we obtain the same values for E_α, $f_{y\alpha}$, E_β, $f_{y\beta}$ as given by Eqs. (6.4)–(6.9) in Sect. 6.2.

6.3.5 Material Parameters at Elastic Incompressibility

If the governing equations for all segments of the total state are developed, they can be used to identify the three material parameters per layer. For more than two layers, and thus, for example, quadrilinear stress–strain behavior (three layers), quite extensive algebra is generally required.

For some special cases, however, simple correlations can be found. So we get for an elastic incompressible material ($\nu = 0.5$)

$$E_\alpha + E_\beta + E_\gamma + \cdots = E \tag{6.58}$$

$$E_{t\alpha} + E_\beta + E_\gamma + \cdots = E_{t1} \tag{6.59}$$

$$E_{t\alpha} + E_{t\beta} + E_\gamma + \cdots = E_{t2} \tag{6.60}$$

$$E_{t\alpha} + E_{t\beta} + E_{t\gamma} + \cdots = E_{t3} \tag{6.61}$$

etc.

However, these equations are insufficient for a unique identification, but, apart from the first layer, one value can be chosen freely for each layer. For example, if the additional conditions

$$\frac{E_{t\alpha}}{E_\alpha} = \frac{E_{t\beta}}{E_\beta} = \frac{E_{t\gamma}}{E_\gamma} = \cdots \tag{6.62}$$

are chosen, we get for the special case of $\nu = 0.5$, if E_{tm} indicates the tangential slope in the final segment of the total state,

$$\frac{E_{t\alpha}}{E_\alpha} = \frac{E_{t\beta}}{E_\beta} = \frac{E_{t\gamma}}{E_\gamma} = \cdots = \frac{E_{tm}}{E} \tag{6.63}$$

$$E_\alpha = E\frac{E - E_{t1}}{E - E_{tm}} \tag{6.64}$$

$$E_\beta = E\frac{E_{t1} - E_{t2}}{E - E_{tm}} \tag{6.65}$$

$$E_\gamma = E\frac{E_{t2} - E_{t3}}{E - E_{tm}} \tag{6.66}$$

etc., and

$$f_{y\alpha} = E_\alpha \frac{f_{y1}}{E} \tag{6.67}$$

$$f_{y\beta} = E_\beta \left[\frac{f_{y1}}{E} + \frac{f_{y2} - f_{y1}}{E_{t1}} \right] \tag{6.68}$$

$$f_{y\gamma} = E_\gamma \left[\frac{f_{y1}}{E} + \frac{f_{y2} - f_{y1}}{E_{t1}} + \frac{f_{y3} - f_{y2}}{E_{t2}} \right] \tag{6.69}$$

etc.

6.3.6 Material Parameters for Trilinear Material Law

For any values of Poisson's ratio, the equations Eqs. (6.37), (6.41), (6.48), (6.53), and (6.56), can be used to identify the material parameters of two layers, each with linear kinematic hardening, of a given trilinear stress–strain diagram of the total state with hardening in the third segment.

To determine the six unknowns E_α, $E_{t\alpha}$, $f_{y\alpha}$, E_β, $E_{t\beta}$, and $f_{y\beta}$ from the known material data of the trilinear total state, only five equations are available. So a sixth condition can be freely selected. Two more equations are available with each additional segment of the total state, but a further layer is required, and three new unknown material parameters arise (E_k, E_{tk}, f_{yk}). That is, with each additional layer, an additional condition may be chosen freely.

For trilinear behavior of the total state, with respect to the usefulness in the STPZ, chosen freely as the sixth condition is

$$\frac{E_{t\alpha}}{E_\alpha} = \frac{E_{t\beta}}{E_\beta} \,. \tag{6.70}$$

The only sensible solution of Eq. (6.56) is then

$$\frac{E_{t\alpha}}{E_\alpha} = \frac{E_{t\beta}}{E_\beta} = \frac{E_{t2}}{E} \,, \tag{6.71}$$

leading to

$$a = \frac{1 + \nu}{\frac{3}{2} - \left(\frac{1}{2} - \nu\right) \frac{E_{t1}}{E}} \tag{6.72}$$

$$E_\beta = a\, E \, \frac{E_{t1} - E_{t2}}{E - E_{t2}} \tag{6.73}$$

$$E_\alpha = E - E_\beta \tag{6.74}$$

$$E_{t\alpha} = \frac{E_{t2}}{E} E_\alpha \tag{6.75}$$

$$E_{t\beta} = \frac{E_{t2}}{E} E_\beta \tag{6.76}$$

$$f_{y\alpha} = f_{y1} \frac{E_\alpha}{E} \tag{6.77}$$

$$f_{y\beta} = f_{y1} \frac{E_\beta}{E} + \left(f_{y2} - f_{y1}\right) \frac{E_\beta}{a\, E_{t1}} \, . \tag{6.78}$$

One consequence of the choice of Eq. (6.70) is that $E_{t\alpha}$ and $E_{t\beta}$ complement each other to E_{t2}:

$$E_{t\alpha} + E_{t\beta} = E_{t2} \, . \tag{6.79}$$

If, instead of Eq. (6.70), a different choice had been made, we would have obtained other material parameters for the layers, which however would ultimately be equivalent. The choice here has been made only for the sake of expediency in regard to the application of the STPZ.

As a numerical example a trilinear stress–strain curve with pure kinematic hardening is considered with the material data for the total state:

E = 16,103
E_{t1} = 4,318
E_{t2} = 1,666
f_{y1} = 10.00
f_{y2} = 12.50.

Equations (6.73)–(6.78) lead to the material parameters of the two layers specified in Table 6.1 for three different values of Poisson's ratio.

When visualizing the stress–strain diagrams of both layers in Fig. 6.5, it can be seen that the breakpoint of layer β occurs at a different strain than in the total state when the Poisson's ratio is not 0.5. This is attributable to the multiaxial stress state

Table 6.1 Material parameters of layers α and β

	ν = 0.0	ν = 0.3	ν = 0.5
E_α	13,937	13,444	13,145
E_β	2,166	2,659	2,958
$E_{t\alpha}$	1,442	1,391	1,360
$E_{t\beta}$	224	275	306
$f_{y\alpha}$	8.655	8.349	8.163
$f_{y\beta}$	3.057	3.364	3.550

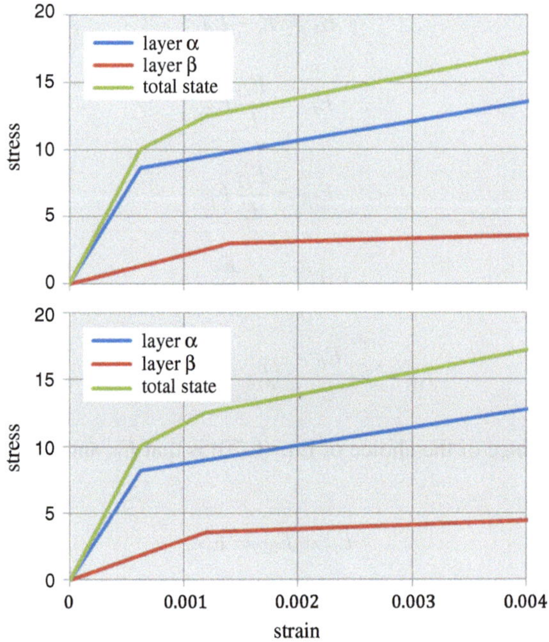

Fig. 6.5 Relationship between equivalent stress and equivalent strain in the layers with $\nu = 0$ (*top*) and $\nu = 0.5$ (*bottom*)

developed in the layers, which has the consequence that in Fig. 6.5 it is not the stress and strain components in the direction of the uniaxial stress of the total state, σ_β, that are shown, but rather the equivalent stress $\sigma_{v\beta}$ and the equivalent strain $\varepsilon_{v\beta}$ (see Fig. 6.4).

6.3.7 Material Parameters for Quadrilinear Material Law

For the identification of the material parameters of each layer for a quadrilinear material law of the total state, longer mathematical derivations and expressions are required. Therefore, only the results are reported here.

Three linear kinematic hardening layers are required to describe the behavior of quadrilinear behavior. These require a total of nine elastic-plastic material parameters, when the Poisson's ratio is assumed the same in all layers and identical to the total state. Two conditions can be chosen freely for their determination. Similar to trilinear behavior of the total state, in view of the usefulness for the STPZ, the following is chosen:

$$\frac{E_{t\alpha}}{E_\alpha} = \frac{E_{t\beta}}{E_\beta} = \frac{E_{t\gamma}}{E_\gamma}\,. \tag{6.80}$$

With the abbreviations

$$a = \frac{1+\nu}{\frac{3}{2} - \left(\frac{1}{2} - \nu\right)\frac{E_{t1}}{E}} \tag{6.81}$$

$$b = \frac{1}{2} - \frac{1}{2}\nu - \nu^2 \tag{6.82}$$

$$c = 1 - \frac{5}{2}\nu + \nu^2 \tag{6.83}$$

we get

$$\frac{E_\gamma}{E} = \frac{\left(\frac{E}{E_{t3}} - \frac{E}{E_{t2}}\right)\left(\frac{E}{E_{t3}}\frac{3}{2}(1+\nu) - b\right)}{\left(\frac{E}{E_{t3}}\right)^2\left[\frac{9}{4}\frac{E}{E_{t2}} - \frac{3}{2}\left(\frac{1}{2} - \nu\right)\right] - \frac{E}{E_{t3}}\left[\frac{E}{E_{t2}}3\left(1 - \frac{\nu}{2}\right) - c\right] + \frac{E}{E_{t2}}b - \nu\left(\frac{1}{2} - \nu\right)} \tag{6.84}$$

$$E_{t\gamma} = E_\gamma \frac{E_{t3}}{E} \tag{6.85}$$

$$E_\alpha = E\frac{\frac{E_{t3}}{E}(a-1) - a\frac{E_{t1}}{E} + 1}{1 - \frac{E_{t3}}{E}} \tag{6.86}$$

$$E_{t\alpha} = E_\alpha \frac{E_{t3}}{E} \tag{6.87}$$

$$E_\beta = E - E_\alpha - E_\gamma \tag{6.88}$$

$$E_{t\beta} = E_\beta \frac{E_{t3}}{E} \tag{6.89}$$

$$f_{y\alpha} = f_{y1}\frac{E_\alpha}{E} \tag{6.90}$$

$$f_{y\beta} = \frac{E_\beta}{E}\left[f_{y1} + (f_{y2} - f_{y1})\frac{E}{a\,E_{t1}}\right] \tag{6.91}$$

$$f_{y\gamma} = \frac{E_\gamma}{\frac{2}{3}(1+\nu)}\left[\frac{f_{y1}}{E}\left(1 - \frac{E}{E_{t1}}\right) + \frac{f_{y2}}{E}\left(\frac{E}{E_{t1}} - \frac{E}{E_{t2}}\right) + \frac{f_{y3}}{E}\left(\frac{E}{E_{t2}} - \frac{2}{3}\left(\frac{1}{2} - \nu\right)\right)\right]. \tag{6.92}$$

6.4 Trilinearization of Real Stress–Strain Curves

In Sects. 6.3.6 and 6.3.7 we assumed that a tri- or quadrilinear stress–strain diagram is given, and obtained the corresponding material parameters for each layer of the overlay model.

It is still undecided, however, how a tri- or quadrilinear approximation to a given cyclic saturated stress–strain curve, usually represented as stress range versus strain range at half the fatigue life, see, e.g., [4], can be made. If a temperature-dependent yield stress is to be considered, a set of cyclic saturated stress–strain diagrams obtained for different temperatures under isothermal conditions, is required.

Because trilinear stress–strain curves should usually be sufficient with respect to applying the STPZ in a practical engineering environment, a way is shown in the following how real stress–strain curves can be trilinearized.

The scattering of the saturated stress–strain curves of a specific material is generally high, in particular when several batches of steel are investigated or even different laboratories are commissioned with the investigations. Considering this fact, the author believes that a trilinearization could even be carried out by eye. The disadvantage of this would be less the poor accuracy but more the fact that the judgment is not reproducible and cannot be automated.

Instead, the following procedure is suggested in accordance with the specific requirements of the STPZ that E, E_{t1}, and E_{t2} must be temperature-independent and only f_{y1} and f_{y2} may be temperature-dependent.

In the case of a temperature loading, first a mean temperature $\overline{T}$ is evaluated according to Eq. (5.39) and the stress–strain curve $\sigma(\varepsilon, \overline{T})$ is determined by linear interpolation between the stress–strain curves of the two adjacent temperatures. Through the initial slope, the modulus of elasticity is already determined (all quantities specified for the temperature $\overline{T}$ are indicated by an overbar):

$$\overline{E} = \frac{\mathrm{d}\sigma(\varepsilon = 0, \overline{T})}{\mathrm{d}\varepsilon} . \tag{6.93}$$

Then a maximum occurring strain is preliminarily roughly estimated (ε_s). At that strain value, the trilinear stress–strain curve $\sigma_{\text{tri}}(\varepsilon)$ should be the same as that of the original stress–strain diagram (Fig. 6.6):

$$\sigma_{\text{tri}}(\varepsilon_s, \overline{T}) = \sigma(\varepsilon_s, \overline{T}) . \tag{6.94}$$

The transition from the second to the third segment of the trilinear stress–strain curve should be located at half the estimated strain.

The slope E_{t2}, and thus automatically also f_{y2}, is determined from the requirement that the specific strain energy should be the same, between ½ε_s and ε_s (the olive area in Fig. 6.6 is supposed to be zero):

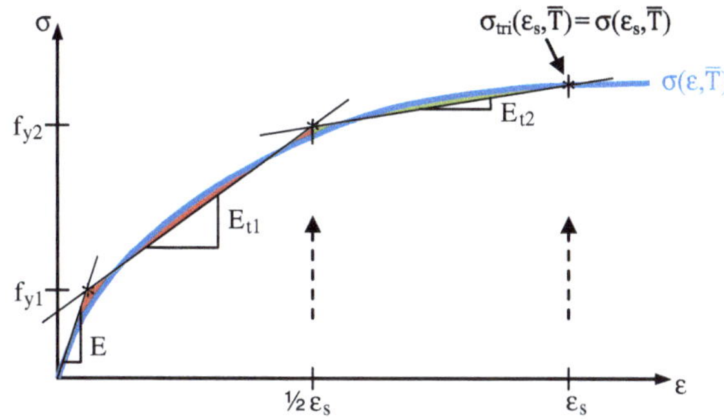

Fig. 6.6 Schematic diagram for trilinearizing a stress–strain curve at the temperature $\bar{T}$

$$\int_{\varepsilon_s/2}^{\varepsilon_s} \sigma_{\text{tri}}(\varepsilon, \overline{T})\,\mathrm{d}\varepsilon = \int_{\varepsilon_s/2}^{\varepsilon_s} \sigma(\varepsilon, \overline{T})\,\mathrm{d}\varepsilon\,. \tag{6.95}$$

Similarly, E_{t1}, and thus automatically also f_{y1}, is obtained by the requirement that the specific strain energy should also be the same between 0 and $½\varepsilon_s$ (the red area in Fig. 6.6 should be zero):

$$\int_{0}^{\varepsilon_s/2} \sigma_{\text{tri}}(\varepsilon, \overline{T})\,\mathrm{d}\varepsilon = \int_{0}^{\varepsilon_s/2} \sigma(\varepsilon, \overline{T})\,\mathrm{d}\varepsilon\,. \tag{6.96}$$

The integrals on the right hand side of Eqs. (6.95) and (6.96) possibly need to be solved numerically, depending on the mathematical description of the curve $\sigma(\varepsilon, \bar{T})$. If, however, $\sigma(\varepsilon, \bar{T})$ is given in the form of the Ramberg–Osgood law, such as for many cyclic stress–strain curves provided in [4],

$$\varepsilon = \frac{\sigma}{E} + A\sigma^n\,, \tag{6.97}$$

then the following hold:

$$\int_{\varepsilon_s/2}^{\varepsilon_s} \sigma(\varepsilon)\,\mathrm{d}\varepsilon = \frac{n}{1+n}\varepsilon_s\sigma_{(\varepsilon_s)} + \frac{1-n}{1+n}\frac{1}{2E}\sigma^2_{(\varepsilon_s)} - \int_{0}^{\varepsilon_s/2} \sigma(\varepsilon)\,\mathrm{d}\varepsilon \tag{6.98}$$

$$\int_{0}^{\varepsilon_s/2} \sigma(\varepsilon)\,\mathrm{d}\varepsilon = \frac{n}{1+n}\frac{\varepsilon_s}{2}\sigma_{(½\varepsilon_s)} + \frac{1-n}{1+n}\frac{1}{2E}\sigma^2_{(½\varepsilon_s)}\,. \tag{6.99}$$

However, the stresses at the strain values ½ε_s and ε_s must be evaluated numerically, e.g., by a Newton iteration. The integrals on the left hand side of Eqs. (6.95) and (6.96) can be solved very easily and we get

$$\overline{f}_{y2} = 4\frac{\int_{\varepsilon_s/2}^{\varepsilon_s} \sigma(\varepsilon, \overline{T})\,\mathrm{d}\varepsilon}{\varepsilon_s} - \sigma(\varepsilon_s, \overline{T}) \tag{6.100}$$

$$\overline{E}_{t2} = 2\frac{\sigma(\varepsilon_s, \overline{T}) - \overline{f}_{y2}}{\varepsilon_s} \tag{6.101}$$

$$\overline{f}_{y1} = \frac{2 \cdot \int_0^{\varepsilon_s/2} \sigma(\varepsilon, \overline{T})\,\mathrm{d}\varepsilon - \frac{1}{2}\overline{f}_{y2}\,\varepsilon_s}{\frac{1}{2}\varepsilon_s - \frac{\overline{f}_{y2}}{\overline{E}}} \tag{6.102}$$

$$\overline{E}_{t1} = \frac{\overline{f}_{y2} - \overline{f}_{y1}}{\frac{1}{2}\varepsilon_s - \frac{\overline{f}_{y1}}{\overline{E}}}\,. \tag{6.103}$$

By Eqs. (6.93) and (6.100)–(6.103), all five parameters to describe the trilinear stress–strain diagram at temperature $\overline{T}$ are now determined.

At other temperatures T, the values of E, E_{t1}, and E_{t2} are maintained and only f_{y1} and f_{y2} are adjusted, each by retaining the specific strain energies analogous to Eqs. (6.95) and (6.96), so that beside

$$E(T) = \overline{E} \tag{6.104}$$

$$E_{t1}(T) = \overline{E}_{t1} \tag{6.105}$$

$$E_{t2}(T) = \overline{E}_{t2} \tag{6.106}$$

with the abbreviations

$$a_1 = \frac{\mathrm{d}\sigma(\varepsilon = 0, T)}{\mathrm{d}\varepsilon} \tag{6.107}$$

$$a_2 = 4\frac{\int_{\varepsilon_s/2}^{\varepsilon_s} \sigma(\varepsilon, T)\,\mathrm{d}\varepsilon}{\varepsilon_s} - \sigma(\varepsilon_s, T) \tag{6.108}$$

$$a_3 = 2\frac{\sigma(\varepsilon_s, T) - a_2}{\varepsilon_s} \tag{6.109}$$

$$a_4 = \frac{2 \cdot \int_0^{\varepsilon_s/2} \sigma(\varepsilon, T)\,\mathrm{d}\varepsilon - \frac{1}{2} a_2 \varepsilon_s}{\frac{1}{2}\varepsilon_s - \frac{a_2}{a_1}} \tag{6.110}$$

$$a_5 = \frac{a_2 - a_4}{\frac{1}{2}\varepsilon_s - \frac{a_4}{a_1}} \tag{6.111}$$

then

$$f_{y1}(T) = \overline{E}\frac{\varepsilon_s}{2} - \sqrt{\left(\overline{E}\frac{\varepsilon_s}{2}\right)^2 + \frac{\overline{E}}{\frac{\overline{E}_{t1}}{\overline{E}} - 1}\left[\frac{a_4^2}{a_1} + \frac{a_2^2 - a_4^2}{a_5} - \overline{E}_{t1}\left(\frac{\varepsilon_s}{2}\right)^2\right]} \tag{6.112}$$

and, with further abbreviations,

$$a_6 = f_{y1}(T)\left(1 - \frac{\overline{E}_{t1}}{\overline{E}}\right) + \overline{E}_{t1}\frac{\varepsilon_s}{2} \tag{6.113}$$

$$a_7 = a_6 + \left(\overline{E}_{t2} - a_3\right)\frac{\varepsilon_s}{4} - a_2 \tag{6.114}$$

$$a_8 = \sqrt{\left(\frac{\varepsilon_s}{2}\right)^2 - \frac{\varepsilon_s}{\overline{E}_{t2} - \overline{E}_{t1}} a_7} \tag{6.115}$$

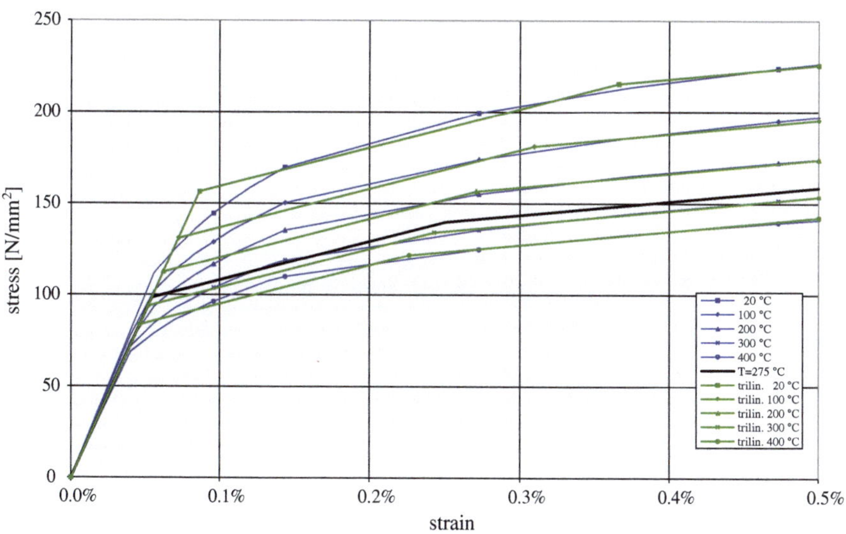

Fig. 6.7 Example of trilinearization of temperature-dependent stress–strain curves

then

$$f_{y2}(T) = f_{y1}(T)\left(1 - \frac{\overline{E}_{t1}}{\overline{E}}\right) + \overline{E}_{t1}(\varepsilon_s - a_8) \tag{6.116}$$

are determined.

Figure 6.7 gives an impression of this kind of trilinearization (green curves) for a mean temperature $\bar{T} = 275\,°\mathrm{C}$ at a pre-estimated maximum strain $\varepsilon_s = 0.5\ \%$.

At the end of an analysis carried out either incrementally or with the STPZ by adopting the trilinearized stress–strain diagram, it can be checked whether the maximum occurring strain has actually been estimated reasonably or grossly wrongly by ε_s.

References

1. Mróz, Z.: On the description of anisotropic workhardening. J. Mech. Phys. Solids **15**, 163–175 (1967)
2. Besseling, J.F.: A theory of elastic, plastic, and creep deformations of an initially isotropic material showing anisotropic strain-hardening, creep recovery, and secondary creep. ASME J. Appl. Mech. **25**, 529–536 (1958)
3. ANSYS Release 14.5, ANSYS Inc. Canonsburg, USA (2012)
4. Boller, C., Seeger, T., Vormwald, M.: Materials Database for Cyclic Loading, Fachgebiet Werkstoffmechanik, TU Darmstadt (2008). http://www.wm.tu-darmstadt.de/mat-db-html/index.html

STPZ at Multilinear Hardening

7

In Chap. 6 only a uniaxial stress state with multilinear hardening, largely independent of the STPZ, needed to be considered to determine the material parameters for each layer of the overlay model. In the present chapter, the STPZ is developed for the overlay model, taking into account multiaxial stress states.

7.1 Theory

As has already emerged from Chap. 3, the STPZ as the Zarka method is based on the proportionality of the backstress vector ξ_i and the plastic strain vector ε_i^{pl}, which requires a constant hardening parameter C in the Prager–Ziegler hardening law (Sect. 1.2.6). In the following, an extension is described to take account of multiple, segment-wise constant hardening parameters of a multilinear kinematic hardening material; see also [1]. Zarka has already dealt with this question in [2].

If the hardening is limited, i.e., if the slope of the stress–strain curve vanishes in the last segment ($C = 0$), this cannot be taken into account readily in the STPZ. This is because then the initial strains are no longer defined as finite quantities (C is in the denominator of Eq. (3.14)), and the Young's modulus of the modified elastic material law in V_p becomes $E^* = E_t = 0$. In this case, the limitation of the hardening should be ignored when applying the STPZ, and a strain limit should be introduced instead, marking the limit of validity of the results.

Taking into account multilinear instead of linear hardening is important with respect to ratcheting. On the one hand the larger strains occurring in the presence of ratcheting require a better approximation of the stress–strain curve over the entire domain of stresses and strains, and on the other hand the fact can properly be reflected that the strain ranges are naturally associated with smaller strains and thus steeper tangent modules than the strains accumulated in plastic shakedown (Fig. 7.1).

At a multilinear approximation of the stress–strain behavior, the lowest yield stress is always less than the yield stress f_y at a bilinear approximation, so that generally a larger portion of the structure becomes plastic. So the effort for the

H. Hübel, *Simplified Theory of Plastic Zones*,
DOI 10.1007/978-3-319-29875-7_7

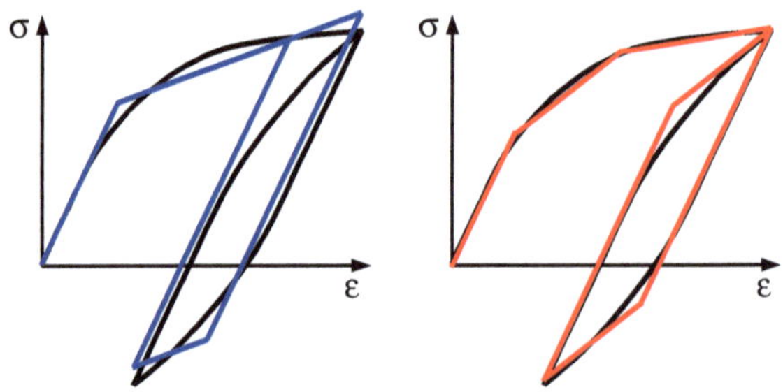

Fig. 7.1 Schematic diagram of the difference between bilinear (*blue*) and trilinear (*red*) approximation of the curved stress–strain diagram (*black*)

calculations with the STPZ increases at the local level, namely for the determination of initial strains to be applied in the plastic zone as modified loads.

Moreover, it now is no longer sufficient to determine only the geometry of the plastic zone V_p, but it has to be determined the geometry of the plastic zone associated with each segment of the multilinear stress–strain diagram individually (V_{p1}, V_{p2}, V_{p3}, …). In addition, the hardening parameters in the individual layers become naturally smaller the more layers taken into account, which is not beneficial to the numerical stability of an application of the STPZ.

This has the consequence that more iterations are often needed to satisfy the field equations at multilinear behavior, so that more modified elastic analyses (MEAs) are required than with bilinear behavior. In addition, more often plastic instead of elastic shakedown is achieved because of the lower yield stress, so that apart from the accumulated elastic-plastic state, the elastic-plastic strain range must also be determined.

Overall, it is thus to be expected that the calculation cost of the STPZ increases at multilinear compared to bilinear behavior, and that it is greater with a quadrilinear than with a trilinear stress–strain curve.

7.1.1 Behavior of the Individual Layers

Because only a bilinear stress–strain diagram is present in the individual layers, the same relationships hold there as described in the previous chapters for a bilinear total state. Thus for each layer k ($k = \alpha, \beta, \gamma, \ldots$) the modified elastic material law applies:

$$\varepsilon^*_{i,k} = \left(E_{ij}\right)^{-1}_k \rho_{j,k} \quad \text{in } V_{ek} \tag{7.1}$$

$$\varepsilon^*_{i,k} = \left(E^*_{ij}\right)^{-1}_k \rho_{j,k} + \varepsilon_{i,0k} \quad \text{in } V_{pk} \tag{7.2}$$

(summation rule applies only to j, not to k) with the modified elasticity matrix

$$E^*_{ij,k} = \frac{E^*_k}{(1-2\nu^*_k)(1+\nu^*_k)} \begin{pmatrix} 1-\nu^*_k & \nu^*_k & \nu^*_k & 0 & 0 & 0 \\ \nu^*_k & 1-\nu^*_k & \nu^*_k & 0 & 0 & 0 \\ \nu^*_k & \nu^*_k & 1-\nu^*_k & 0 & 0 & 0 \\ 0 & 0 & 0 & \frac{1}{2}(1-2\nu^*_k) & 0 & 0 \\ 0 & 0 & 0 & 0 & \frac{1}{2}(1-2\nu^*_k) & 0 \\ 0 & 0 & 0 & 0 & 0 & \frac{1}{2}(1-2\nu^*_k) \end{pmatrix} \tag{7.3}$$

$$\text{with} \quad E^*_k = E_{tk}; \quad \nu^*_k = 0.5 - (0.5 - \nu)\frac{E_{tk}}{E_k} \tag{7.4}$$

and the initial strains

$$\varepsilon_{i,0k} = \frac{3}{2}\frac{1}{E_{tk}}\left(1 - \frac{E_{tk}}{E_k}\right)Y_{i,k}. \tag{7.5}$$

In Chap. 6 it was mentioned that certain free values could be chosen to determine the elastic-plastic material parameters in the layers of the overlay model, and that in regard to the use of the STPZ the conditions given by Eqs. (6.70) and (6.80) are useful. One reason for this is that in this way the slopes of the different segments of the tri- or quadrilinear stress–strain diagram of the total state can be split evenly over the individual layers, which benefits the numeric stability. Another reason we see just at this point: because of

$$\frac{E_{t\alpha}}{E_\alpha} = \frac{E_{t\beta}}{E_\beta} \tag{7.6}$$

or

$$\frac{E_{t\alpha}}{E_\alpha} = \frac{E_{t\beta}}{E_\beta} = \frac{E_{t\gamma}}{E_\gamma} = \ldots. \tag{7.7}$$

the modified Poisson's ratios are, because of Eq. (7.4), the same in all layers and the modified elasticity matrices all proportional to each other!

7.1.2 Total State

The individual layers are now assembled into a total state. If only layer α is plastic, then the location of the structure under consideration is in the second segment, or the first plastic segment, of the total state. The sum of all these locations constitutes the first plastic zone, V_{p1}. Accordingly, all locations where the layers α and β are plastic belong to the third segment, or the second plastic segment, of the total state, and constitute the second plastic zone, V_{p2}. The locations where the layers α, β, and γ are plastic are in the fourth segment of the total state and form the third plastic zone V_{p3}, etc.

7.1.2.1 Locations in V_{p2}

If a location of the structure is attributed at the nth iteration to V_{p2}, i.e., if the total state is in the third segment of a multilinear stress–strain diagram, then layers α and β are plastic. For the total state the modified elastic material law applies:

$$\varepsilon_{i,2}^{*} = \left(E_{ij}^{*}\right)_{2}^{-1} \rho_j + \varepsilon_{i,02} \quad \text{in } V_{p2} \tag{7.8}$$

$$E_{ij,2}^{*} = \frac{E_2^{*}}{\left(1-2\nu_2^{*}\right)\left(1+\nu_2^{*}\right)} \begin{pmatrix} 1-\nu_2^{*} & \nu_2^{*} & \nu_2^{*} & 0 & 0 & 0 \\ \nu_2^{*} & 1-\nu_2^{*} & \nu_2^{*} & 0 & 0 & 0 \\ \nu_2^{*} & \nu_2^{*} & 1-\nu_2^{*} & 0 & 0 & 0 \\ 0 & 0 & 0 & \frac{1}{2}\left(1-2\nu_2^{*}\right) & 0 & 0 \\ 0 & 0 & 0 & 0 & \frac{1}{2}\left(1-2\nu_2^{*}\right) & 0 \\ 0 & 0 & 0 & 0 & 0 & \frac{1}{2}\left(1-2\nu_2^{*}\right) \end{pmatrix} \tag{7.9}$$

$$E_2^{*} = E_{t2} \tag{7.10}$$

$$\nu_2^{*} = 0.5 - (0.5 - \nu)\frac{E_{t2}}{E}. \tag{7.11}$$

Because of the choice of free values in Eq. (6.70) it turns out that:

$$\nu_2^{*} = \nu_\beta^{*} = \nu_\alpha^{*}, \tag{7.12}$$

according to which the modified Poisson's ratios in all layers are not only the same but also identical with the modified Poisson's ratio of the total state. Therefore the following also applies:

$$E_{ij,\alpha}^{*} = E_{ij,2}^{*} \frac{E_{t\alpha}}{E_{t2}} \tag{7.13}$$

$$E_{ij,\beta}^{*} = E_{ij,2}^{*} \frac{E_{t\beta}}{E_{t2}}. \tag{7.14}$$

Because of the compatibility conditions of the overlay model, known from Sect. 6.3.1, according to which each layer has the same deformations as the total state,

$$\varepsilon_{i,\beta}^{*} = \varepsilon_{i,\alpha}^{*} = \varepsilon_{i,2}^{*} \tag{7.15}$$

and the equilibrium condition

$$\rho_{i,2} = \rho_{i,\alpha} + \rho_{i,\beta} \tag{7.16}$$

it follows that the initial strains of the overall condition in Eq. (7.8) consist of the sum of the weighted initial strains of the two layers:

$$\varepsilon_{i,02} = \frac{E_{t\alpha}}{E_{t2}}\,\varepsilon_{i,0\alpha} + \frac{E_{t\beta}}{E_{t2}}\,\varepsilon_{i,0\beta}. \tag{7.17}$$

As usual, the following apply to the initial strains in the layers (see Eq. (7.5)):

$$\varepsilon_{i,0\alpha} = \frac{3}{2}\frac{1}{E_{t\alpha}}\left(1 - \frac{E_{t\alpha}}{E_{\alpha}}\right) Y_{i,\alpha} \tag{7.18}$$

$$\varepsilon_{i,0\beta} = \frac{3}{2}\frac{1}{E_{t\beta}}\left(1 - \frac{E_{t\beta}}{E_{\beta}}\right) Y_{i,\beta}, \tag{7.19}$$

so that Eq. (7.17) can also be written as

$$E_{t2}\,\varepsilon_{i,02} = \frac{3}{2}\left(1 - \frac{E_{t2}}{E}\right)\left(Y_{i,\alpha} + Y_{i,\beta}\right). \tag{7.20}$$

The initial strains for the total state can be converted into initial stresses:

$$\sigma_{i,02} = -E^{*}_{ij,2}\,\varepsilon_{j,02} \tag{7.21}$$

or for each layer:

$$\sigma_{i,0\alpha} = -\frac{E_{\alpha}}{E}\,E^{*}_{ij,2}\,\varepsilon_{j,0\alpha} \tag{7.22}$$

$$\sigma_{i,0\beta} = -\frac{E_{\beta}}{E}\,E^{*}_{ij,2}\,\varepsilon_{j,0\beta}. \tag{7.23}$$

7.1.2.2 Locations in V_{px}

For locations of the structure, which are at the nth iteration in V_{p3}, V_{p4}, etc., referred to generally as V_{px}, i.e., in the fourth or higher segment of the total state of a multilinear stress–strain diagram, so that then the layers α, β and γ, etc., are plastic, the modified elastic material law for the total state applies analogous to V_{p2} (no summation over index x):

$$\varepsilon^{*}_{i,x} = \left(E^{*}_{ij}\right)^{-1}_{x}\,\rho_j + \varepsilon_{i,0x} \quad \text{in } V_{px} \tag{7.24}$$

$$E_{ij,x}^* = \frac{E_x^*}{(1-2\nu_x^*)(1+\nu_x^*)} \begin{pmatrix} 1-\nu_x^* & \nu_x^* & \nu_x^* & 0 & 0 & 0 \\ \nu_x^* & 1-\nu_x^* & \nu_x^* & 0 & 0 & 0 \\ \nu_x^* & \nu_x^* & 1-\nu_x^* & 0 & 0 & 0 \\ 0 & 0 & 0 & \frac{1}{2}(1-2\nu_x^*) & 0 & 0 \\ 0 & 0 & 0 & 0 & \frac{1}{2}(1-2\nu_x^*) & 0 \\ 0 & 0 & 0 & 0 & 0 & \frac{1}{2}(1-2\nu_x^*) \end{pmatrix} \tag{7.25}$$

$$E_x^* = E_{tx} \tag{7.26}$$

$$\nu_x^* = 0.5 - (0.5-\nu)\frac{E_{tx}}{E} \tag{7.27}$$

$$\nu_x^* = \nu_\alpha^* = \nu_\beta^* = \nu_\gamma^* = \ldots \tag{7.28}$$

$$E_{ij,\alpha}^* = E_{ij,x}^* \frac{E_{t\alpha}}{E_{tx}} \tag{7.29}$$

$$E_{ij,\beta}^* = E_{ij,x}^* \frac{E_{t\beta}}{E_{tx}} \tag{7.30}$$

$$E_{ij,\gamma}^* = E_{ij,x}^* \frac{E_{t\gamma}}{E_{tx}} \tag{7.31}$$

etc.

Because of the compatibility and the equilibrium condition of the overlay model

$$\varepsilon_{i,x}^* = \varepsilon_{i,\alpha}^* = \varepsilon_{i,\beta}^* = \varepsilon_{i,\gamma}^* = \ldots \tag{7.32}$$

$$\rho_{i,x} = \rho_{i,\alpha} + \rho_{i,\beta} + \rho_{i,\gamma} + \ldots \tag{7.33}$$

we get the initial strains

$$\varepsilon_{i,0x} = \frac{E_{t\alpha}}{E_{tx}}\varepsilon_{i,0\alpha} + \frac{E_{t\beta}}{E_{tx}}\varepsilon_{i,0\beta} + \frac{E_{t\gamma}}{E_{tx}}\varepsilon_{i,0\gamma} + \ldots \tag{7.34}$$

$$\varepsilon_{i,0\alpha} = \frac{3}{2}\frac{1}{E_{t\alpha}}\left(1 - \frac{E_{t\alpha}}{E_\alpha}\right) Y_{i,\alpha} \tag{7.35}$$

$$\varepsilon_{i,0\beta} = \frac{3}{2}\frac{1}{E_{t\beta}}\left(1 - \frac{E_{t\beta}}{E_\beta}\right) Y_{i,\beta} \tag{7.36}$$

$$\varepsilon_{i,0\gamma} = \frac{3}{2}\frac{1}{E_{t\gamma}}\left(1 - \frac{E_{t\gamma}}{E_\gamma}\right) Y_{i,\gamma} \tag{7.37}$$

etc.,

so that Eq. (7.34) can also be written as

$$E_{tx}\,\varepsilon_{i,0x} = \frac{3}{2}\left(1 - \frac{E_{tx}}{E}\right)\left(Y_{i,\alpha} + Y_{i,\beta} + Y_{i,\gamma} + \ldots\right), \tag{7.38}$$

and the initial stresses

$$\sigma_{i,0x} = -E^*_{ij,x}\,\varepsilon_{i,0x} = -\frac{E_{tx}}{1+\nu^*}\varepsilon_{i,0x} \tag{7.39}$$

$$\sigma_{i,0\alpha} = -\frac{E_\alpha}{E}\,E^*_{ij,x}\,\varepsilon_{j,0\alpha} \tag{7.40}$$

$$\sigma_{i,0\beta} = -\frac{E_\beta}{E}\,E^*_{ij,x}\,\varepsilon_{j,0\beta} \tag{7.41}$$

$$\sigma_{i,0\gamma} = -\frac{E_\gamma}{E}\,E^*_{ij,x}\,\varepsilon_{j,0\gamma} \tag{7.42}$$

etc.

7.1.2.3 Modified Elastic Analysis

According to the previous sections, the TIVs $Y_{i,\alpha}$, $Y_{i,\beta}$, etc., are required in the respective plasticizing layers. They are calculated individually and separately in the layers according to the usual procedures for bilinear stress–strain behavior, such as described in Chap. 3 for monotonic loading and in Chap. 4 for cyclic loading, with the addition in Chap. 5 for temperature-dependent material data, so that no special explanations are necessary here (but see also the examples following in the next sections).

To be able to perform the projections on the respective yield surfaces, the fictitious elastic and elastic-plastic stress shares attributable to each layer are required.

For elastic and thus also fictitious elastic behavior with the corresponding material law Eq. (1.3) we get from the equilibrium and compatibility conditions of the overlay model Eqs. (6.16) and (6.19) for the kth layer (in the following equations summation rule does not apply to x and k)

$$\sigma^{\mathrm{fel}}_{i,k} = \frac{E_k}{E}\sigma^{\mathrm{fel}}_i \tag{7.43}$$

and with the modified elastic material law Eq. (7.24) for the kth layer, i.e., under the assumption that this layer is actually plasticized,

$$\rho'_{i,k} = \frac{E_{tk}}{E_{tx}} \rho'_i + \frac{E_{tk}}{E_{tx}} E^*_{ij,x} \left(\varepsilon_{j,0x} - \varepsilon_{j,0k} \right) \quad \text{in } V_{px}, \tag{7.44}$$

resulting in the elastic-plastic stresses

$$\sigma'_i = \sigma'^{\text{fel}}_i + \rho'_i \tag{7.45}$$

$$\sigma'_{i,k} = \frac{E_k}{E} \left[\sigma'_i + E^*_{ij,x} \left(\varepsilon_{j,0k} - \varepsilon_{j,0k} \right) \right] \quad \text{in } V_{px} \tag{7.46}$$

or with initial stresses instead of initial strains

$$\sigma'_{i,k} = \frac{E_k}{E} \left[\sigma'_i - \sigma_{i,0x} \right] + \sigma_{i,0k} \quad \text{in } V_{px}. \tag{7.47}$$

With an MEA, the residual stresses ρ'_i are calculated for the total state and can then be partitioned to the individual layers according to Eq. (7.44).

Because only a bilinear stress–strain diagram is present in each layer, the projection of the respective negative deviatoric residual stress of the layer

$$Y^*_{i,k} = -\rho'_{i,k} \tag{7.48}$$

on the respective Mises circle is then made according to the rules which are familiar from the previous chapters.

The MEAs are therefore carried out only for the total state. For each layer, only local calculations are required to determine the initial strains, etc.

7.1.2.4 Alternating Between Segments

It is clear from Sect. 3.1.2 that there are generally two options to formulate the modified elastic material law for the non-plasticizing subvolume V_e: either by Eq. (3.18) with the same modified elastic material parameters E^* and ν^* as in V_p and initial strains according to Eq. (3.19) or by Eq. (3.20) with the actual elastic material parameters E and ν without initial strains. These two possibilities are also available to take into account the layers. Accordingly, there is:

1. The possibility to consider all layers in accordance with the existing number of segments of the material behavior of the total state directly, even if they behave purely elastic (this corresponds to Eqs. (3.18) and (3.19)), or
2. The possibility to consider only as many layers at a specific location of the structure that all plasticize, or to assign the entire location to the elastic subvolume V_e (this corresponds to Eq. (3.20)).

Option 1 allows one to deal with the same number of layers and thus always with the same modified elastic material data throughout all MEAs at one location of the structure. If in the course of the iteration process the affiliation of a layer is interchanged between V_e and V_p, this has no serious consequences except that the initial strains are determined differently. Because in each layer there is a bilinear material model, the procedures described in Chaps. 3–5 are sufficient for determining the TIV in the layers. After an MEA is performed, the residual stresses in the layers, however, cannot always be determined according to Eq. (7.44). If single layers remain elastic, the transverse stresses of the overlay model are captured correctly only at elastic incompressible material. There is also evidence that option 1 is associated with only moderate convergence speeds.

By contrast, option 2 is indeed effective, but may require a variable number of layers at the same location of a structure in two consecutive MEAs and thus modified elastic material parameters changing during the iteration process, which involves a certain complication of the calculation process to be described hereinafter.

So far it has been assumed in Sects. 7.1.1 and 7.1.2 that the considered location of the structure is clearly assignable to a certain segment of the total state. At this location, so many layers have been provided that they all are plasticizing, and then the material parameters of these layers have been identified. That, thereby, different numbers of layers are used at different locations of the structure does not matter. However, it is not usually known a priori which segment of the total state is to be assigned to a specific location of the structure, but this is discovered in the course of an iteration process after several MEAs.

For a change of the segment allocation of one MEA to another, and therefore an iterative change of the number of layers considered at one location of the structure, and for the determination of their residual stresses, certain relationships are to be observed. On one hand, only the total states are required to perform an MEA, and thus only the stresses and strains for the entirety of all layers considered at any location of the structure. On the other hand, it is necessary for the projections required for estimating the TIV in the layers, to know how the residual stress ρ_i determined in the previous MEA for the total state on the basis of a different number of layers is to be distributed among the current number of layers to determine their $\rho_{i,k}$.

From the modified elastic material law for a plastic layer k according to Eq. (7.2) and the modified elastic material law for the xth plastic segment of the total state according to Eq. (7.24) ($x \geq 1$), there arises from the compatibility condition $\varepsilon^*_{i,k} = \varepsilon^*_{i,x}$:

$$\rho'_{i,k} = E^*_{ij,k}\left(E^*_{ij,x}\right)^{-1}\rho'_j + E^*_{ij,k}\left(\varepsilon_{j,0x} - \varepsilon_{j,0k}\right). \tag{7.49}$$

Making use of the reformulation

$$E^*_{ij,k}\left(E^*_{ij,x}\right)^{-1}\rho'_j = \frac{E_{tk}}{E_{tx}}\frac{1+\nu^*_x}{1+\nu^*_k}\rho'_i \tag{7.50}$$

it can be seen that Eq. (7.49) is only then equivalent to Eq. (7.44) if all layers existing in the xth plastic segment are indeed plastic (only then does Eq. (7.28) hold).

At the previous MEA, initial strains $\varepsilon_{i,0k}$ and material data E_{tk}, ν_k^*, $E_{ij,k}^*$ were used in layer k to calculate the residual total stress state $\rho_i^{'}$ of the xth plastic segment. For a different segment y, where the material data of the individual layers do not match those in segment x, the residual stress in layer k arises from

$$\rho_{i,k}^{'} = \frac{E_{tk}}{E_{ty}} \frac{1+\nu_y^*}{1+\nu_k^*} \rho_j^{'} + E_{ij,k}^* \left(\varepsilon_{j,0y} - \varepsilon_{j,0k}\right), \tag{7.51}$$

where

$$\nu_k^* = \nu_x^* \tag{7.52}$$

$$E_{ij,k}^* = \frac{E_{tk}}{E_{tx}} E_{ij,x}^*. \tag{7.53}$$

If, for example, in the $(n-1)$th MEA, a location of the structure belonged to V_{p1}, so that there thus existed only one layer, but belonging to V_{p2} in the nth MEA, then the residual stress and the elastic-plastic stress of the one layer of V_{p1} is to be converted first to the two layers of V_{p2} before the TIV in the two layers of V_{p2} can be determined. Because by definition no initial stresses exist in layer β at the $(n-1)$th MEA because of the assignment of this location to V_{p1}, we have

$$\sigma_{i,01}^{(n-1)} = \sigma_{i,0\alpha}^{(n-1)} = \sigma_{i,02}^{(n-1)}, \tag{7.54}$$

and we thus get for layer α of V_{p2} the elastic-plastic stress that is equivalent to the stress in V_{p1}:

$$\sigma_{i,\alpha}^{'(n-1)} = \frac{E_\alpha}{E} \left[\sigma_i^{'\text{fel}} + \frac{E_{t2}}{E_{t1}} \frac{1+\nu_1^*}{1+\nu_2^*} \left(-\sigma_i^{'\text{fel}} + \sigma_i^{'(n-1)} - \sigma_{i,01}^{(n-1)} \right) \right] + \sigma_{i,01}^{(n-1)}. \tag{7.55}$$

In determining the elastic-plastic strain ranges, stresses are to be replaced by stress ranges in Eq. (7.55). This is not only important for the iterative improvement of the results of a calculation of strain ranges, but also because they serve as a basis for calculating accumulated strains in the state of plastic shakedown. This is because the strain ranges are typically in a lower segment as the accumulated strains. This was already shown in the right sketch of Fig. 7.1, where the accumulated state is in the third segment, the strain range, however, being in the second segment. So if the strain range was determined with a different number of layers, as is necessary for the subsequent calculation of accumulated strains, an equivalent transition must be ensured.

If $\Delta\sigma_{i,\alpha}^{'}$ is the elastic-plastic stress range as a result of the last iteration of a calculation of range values, converted to layer α according to Eq. (7.55) if necessary, and if the location of the structure under consideration remains in the subvolume V_{p2} in two consecutive iterations for determining the mean state, then the elastic-plastic stress state of layer α can be calculated, for example, for the minimum state of a cyclic loading, based on the previous MEA for the mean state, according to

$$\sigma_{i,\alpha,\min}^{\prime(n-1)} = \frac{E_\alpha}{E}\left[\frac{1}{2}\left(\sigma_{i,\min}^{\prime(n-1)} + \sigma_{i,\max}^{\prime(n-1)}\right) - \sigma_{i,02}^{(n-1)}\right] + \sigma_{i,0\alpha}^{(n-1)} - \frac{1}{2}\Delta\sigma_{i,\alpha}^{\prime}. \tag{7.56}$$

If, however, this location still belongs to V_{p1} in the $(n-1)$th iteration, then the elastic-plastic stress state of layer α, needed as prerequisite for the projections in the nth iteration, is determined by

$$\sigma_{i,\alpha,\min}^{\prime(n-1)} = \frac{E_\alpha}{E}\left[\sigma_{i,m}^{\prime\text{fel}} + \frac{E_{t2}}{E_{t1}}\frac{1+\nu_1^*}{1+\nu_2^*}\left[\frac{1}{2}\left(\sigma_{i,\min}^{\prime(n-1)} + \sigma_{i,\max}^{\prime(n-1)}\right) - \sigma_{i,m}^{\prime\text{fel}} - \sigma_{i,01}^{(n-1)}\right]\right] + \sigma_{i,0\alpha}^{(n-1)} \\ - \frac{1}{2}\Delta\sigma_{i,\alpha}^{\prime} \tag{7.57}$$

The mean fictitious elastic stress state is given there by

$$\sigma_{i,m}^{\prime\text{fel}} = \frac{1}{2}\left(\sigma_{i,\min}^{\prime\text{fel}} + \sigma_{i,\max}^{\prime\text{fel}}\right). \tag{7.58}$$

7.1.3 Uniaxial Stress State

In uniaxial stress states, considerable simplifications arise because the TIV in the various layers must be collinear and the modified Poisson's ratios, which differ in different segments, are not relevant. For instance, the initial strains of the total state given by Eq. (7.38) may be expressed for monotonic loading, with recourse to Eq. (3.67), by

$$E_{tx}\,\varepsilon_{0x} = \left(1 - \frac{E_{tx}}{E}\right)\left[\sigma^{\text{fel}} - \left(f_{y\alpha} + f_{y\beta} + f_{y\gamma} + \ldots\right)\ \text{sgn}\left(\sigma^{\text{fel}}\right)\right], \tag{7.59}$$

which can also be written after inserting the yield stresses of the individual layers according to Eqs. (6.67)–(6.69) as

$$E_{tx}\,\varepsilon_{0x} = \sigma^{\text{fel}}\left\{\left(1 - \frac{E_{tx}}{E}\right) - \frac{1}{\sigma_{\text{v}}^{\text{fel}}}\frac{E_{tx}}{E}\left[f_{y1}\left(\frac{E}{E_{t1}} - 1\right) + f_{y2}\left(\frac{E}{E_{t2}} - \frac{E}{E_{t1}}\right) + f_{y3}\left(\frac{E}{E_{t3}} - \frac{E}{E_{t2}}\right) + \ldots\right]\right\}. \tag{7.60}$$

7.2 Examples

7.2.1 Two-Bar Model

The two-bar model often considered in this book (see, e.g., Figure 4.3) is now investigated at a multilinear stress–strain diagram with 11 segments. The material parameters of the total state are independent of the temperature and shown in Fig. 7.2.

The material parameters of each layer can be determined according to Sect. 6.3.5, because transverse effects are irrelevant because of the uniaxial stress state. Because a different number of segments is addressed, they are for the calculation of strain range different than for the calculation of the accumulated state, in the left bar unlike in the right bar, and vary from one to another MEA.

The load level is given by the constant primary stress $\sigma_p = 50$ and the fictitious elastic calculated cyclically occurring secondary stress $\sigma_t = 600$ as a result of the temperature change in the left bar. Plastic shakedown is achieved.

Stresses and strains in the state of shakedown are listed in Table 7.1.

Figure 7.3 shows as a result of an incremental analysis, how the stresses and strains develop with the number of loading cycles. The state of shakedown is highlighted in green.

The STPZ provides the exact stresses and strains at the load reversals in the state of shakedown (Table 7.1 and Fig. 7.3). Five MEAs are required to determine the strain range, and additional four MEAs for the subsequent determination of the accumulated strains. The expectation expressed in Sect. 7.1, that the calculation effort increases with increasing number of layers, is evident here. The iteration sequence of the residual stresses is given in Table 7.2.

segment	tangent modulus	yield stress
1	$E = 1e5$	$f_{y1} = 50$
2	$E_{t1} = E*0.5$	$f_{y2} = 80$
3	$E_{t2} = E*0.3$	$f_{y3} = 100$
4	$E_{t3} = E*0.2$	$f_{y4} = 120$
5	$E_{t4} = E*0.1$	$f_{y5} = 130$
6	$E_{t5} = E*0.06$	$f_{y6} = 140$
7	$E_{t6} = E*0.04$	$f_{y7} = 150$
8	$E_{t7} = E*0.03$	$f_{y8} = 160$
9	$E_{t8} = E*0.02$	$f_{y9} = 170$
10	$E_{t9} = E*0.01$	$f_{y10} = 180$
11	$E_{t10} = E*0.005$	any high value

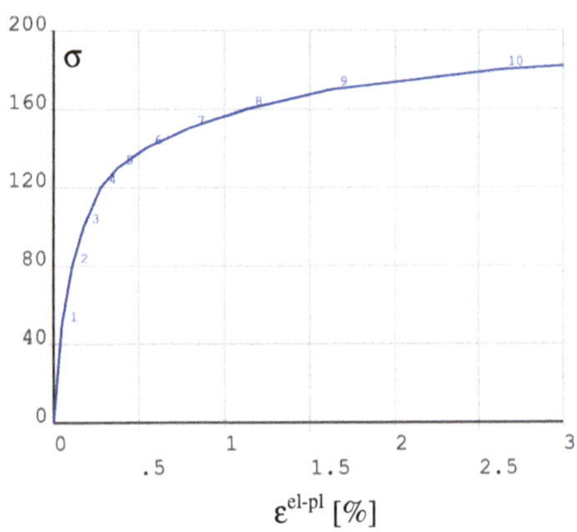

Fig. 7.2 Multilinear stress–strain diagram

Table 7.1 Stresses and strains in the state of shakedown

	Max. loading condition		Min. loading condition	
	Stress	Strain (%)	Stress	Strain (%)
Left bar	−75.556	0.9822	169.111	1.5822
Right bar	175.556	2.1822	−69.111	1.5822

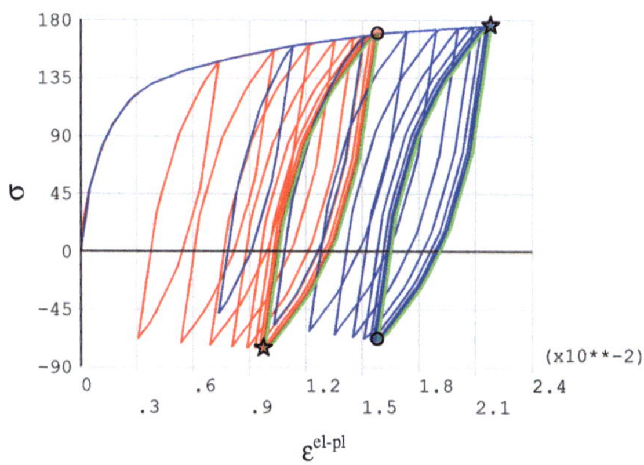

Fig. 7.3 Strain accumulation in the two-bar model with multilinear kinematic hardening (11 segments); in the state of plastic shakedown, incremental solution and STPZ (*circles* for minimum and *stars* for maximum load) are identical

Table 7.2 Iteration sequence of residual stresses of the STPZ

	Residual stress range		Residual stress range	
	Left bar	Right bar	Left bar	Right bar
1. MEA	−336.73	336.73	252.47	−252.47
2. MEA	−286.93	286.93	298.54	−298.54
3. MEA	−260.53	260.53	297.00	−297.00
4. MEA	−250.80	250.80	296.78 = exact	−296.78 = exact
5. MEA	−244.67 = exact	244.67 = exact		

Because of the uniaxial stress state, these iterations are not needed for determining the TIV, but only for determining the plastic zones of the different layers, and thus the correct segment in the plastic zone of the total state (the fifth segment for calculating the ranges, the eighth segment on the left side, the ninth segment in the right bar for calculating the accumulated state).

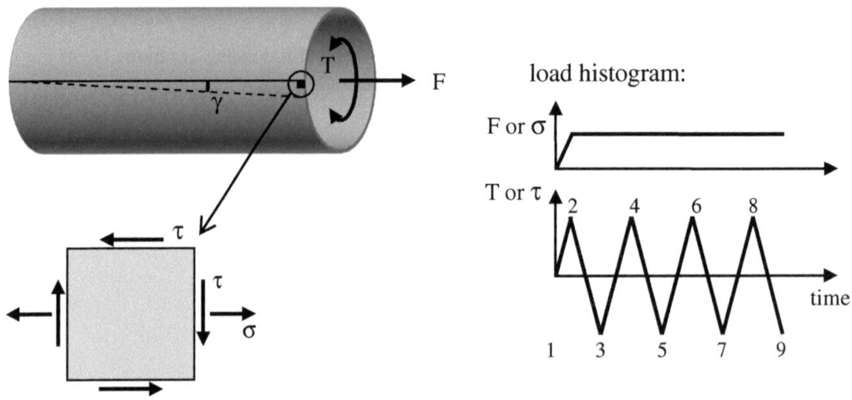

Fig. 7.4 Tribout-example of multiaxial ratcheting: thin-walled tube under axial force and stress-controlled torsion

7.2.2 Multiaxial Ratcheting

So far, a homogeneously stressed element has been considered several times in this book as a simple example of the application of the STPZ in multiaxial stress states, which has been loaded in one direction by a constant stress, in a different direction by a cyclically applied strain. In Sect. 2.5.2.1, such a loading has been described as typical of a thin-walled tube under constant internal pressure and displacement-controlled elongation. In Sect. 2.5.2.2 it was noted that similar behavior occurs when a thin-walled tube is subjected to an axial force and displacement-controlled twist (Fig. 2.18).

As an illustrative example of the application of the STPZ with multilinear material behavior, a modification thereof is now resorted to, which has already been considered with the Zarka method by Tribout et al. in [3], namely applying the torsion stress-controlled instead of displacement-controlled (Fig. 7.4).

In [3], Zarka's method was applied differently compared to the STPZ, namely subsequently to an incremental analysis for the first cycles of a load histogram, in order to conduct an extrapolation to the shakedown condition. In the following, however, the STPZ serves as usual as a substitute for an incremental analysis.

A plane stress state is examined, where in one direction a constant stress $\sigma = 9$ and then a cyclic shear stress of $\tau = \pm 5, 54$ is applied.[1] The structural system is shown in Fig. 7.5 using unit lengths.

[1]As elsewhere in this book, here also units are not provided, because they play no role for the procedure of the calculation. The figures used here are provided with the units hbar for stresses in [3].

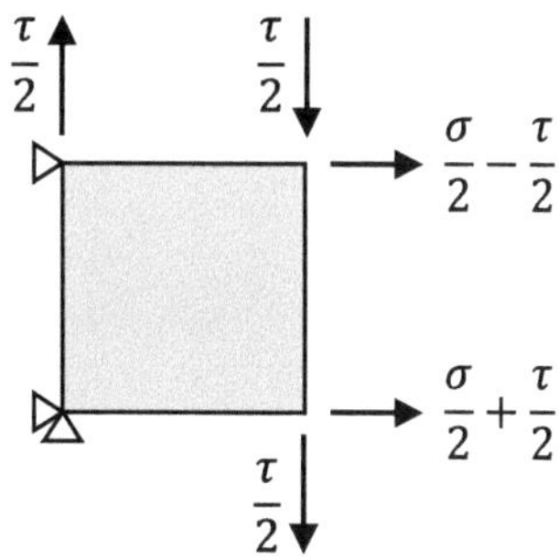

Fig. 7.5 Structural system for Tribout sample

Both loads are therefore stress-controlled, so that from the outset it is known that the residual stresses to be determined with the MEAs are zero. This is, however, only true for the total state, and the residual stresses in the layers can well be nonzero, and here they are!

It is also noteworthy that the principal axes rotate heavily during each loading cycle because of the cyclic shear stresses.

A trilinear stress–strain curve is used with pure kinematic hardening with constant, i.e., temperature-independent, material data for the total state:

$$\begin{aligned} E &= 16{,}103 \\ E_{t1} &= 4{,}318 \\ E_{t2} &= 1{,}666 \\ f_{y1} &= 10.00 \\ f_{y2} &= 12.50. \end{aligned} \tag{7.61}$$

The Poisson's ratio is not specified in [3]. It appears, however, that the results presented there are valid only for $\nu = 0.5$.

Because the fictitious elastic equivalent stress range is smaller than twice the yield stress,

$$\Delta\sigma_{\mathrm{v}}^{\mathrm{fel}} = 2\sqrt{3}\tau_{\max} = 19.19 \quad < \quad 2f_{y1} = 20.0, \tag{7.62}$$

it comes to elastic shakedown (ES).

As a result of an incremental analysis it is indicated in [3] that, after 20 cycles, the plastic strain component in the direction of the constant stress is $\varepsilon^{\mathrm{pl}} = 0.192\ \%$. In fact, the shakedown condition has almost been reached after 20 cycles. After further cycles, the plastic strain increases but still up to $\varepsilon^{\mathrm{pl}} = 0.1948\ \%$.

The decomposition of the given trilinear stress–strain diagram for the total state into the two bilinear stress–strain diagrams of the two layers is carried out according to Sect. 6.3.6. The numerical example indicated there is identical to the material data in Eq. (7.61). The material data of the two layers derived from there are given

Table 7.3 Material parameters of layers α and β

	According to [3]	For $\nu = 0.5$ from Table 6.1
E_α	12,618	13,145
E_β	3,485	2,958
$E_{t\alpha}$	833	1,360
$E_{t\beta}$	833	306
$f_{y\alpha}$	7.836	8.163
$f_{y\beta}$	4.182	3.550

for different numbers of Poisson's ratio in Table 6.1. A comparison with [3] reveals that there the material data of the layers must have been obtained from the total state in a different way (Table 7.3), namely by choosing as a free value, instead of Eq. (6.70), that both layers should have the same hardening modulus. In addition, the material parameters of the total state consist of pure addition of the material parameters of the layers, which according to Sect. 6.3 is only true for incompressible material, i.e., $\nu = 0.5$ (see Eq. (6.57)).

In the following the data of the right column in Table 7.3 are used. Four stress components are to be considered, consisting of the three deviatoric direct stresses and the shear stress. The shares of the fictitious elastic deviatoric stresses of the minimum and maximum load state attributable to the two layers amount to, according to Eq. (7.43):

$$\sigma'^{\text{fel}}_{i,\min,\alpha} = \frac{E_\alpha}{E} \sigma'^{\text{fel}}_{i,\min} = \begin{pmatrix} +4.8978 \\ -2.4489 \\ -2.4489 \\ -4.5223 \end{pmatrix} \tag{7.63}$$

$$\sigma'^{\text{fel}}_{i,\max,\alpha} = \frac{E_\alpha}{E} \sigma'^{\text{fel}}_{i,\max} = \begin{pmatrix} +4.8978 \\ -2.4489 \\ -2.4489 \\ +4.5223 \end{pmatrix} \tag{7.64}$$

$$\sigma'^{\text{fel}}_{i,\min,\beta} = \frac{E_\beta}{E} \sigma'^{\text{fel}}_{i,\min} = \begin{pmatrix} +1.1022 \\ -0.5511 \\ -0.5511 \\ -1.0177 \end{pmatrix} \tag{7.65}$$

$$\sigma'^{\text{fel}}_{i,\max,\beta} = \frac{E_\beta}{E} \sigma'^{\text{fel}}_{i,\max} = \begin{pmatrix} +1.1022 \\ -0.5511 \\ -0.5511 \\ +1.0177 \end{pmatrix}. \tag{7.66}$$

The Y-spaces of the two layers are shown in Fig. 7.6. The graphical representation here is not in the four-dimensional deviatoric TIV-space but in the Y_σ-$\sqrt{3}Y_\tau$-plane,

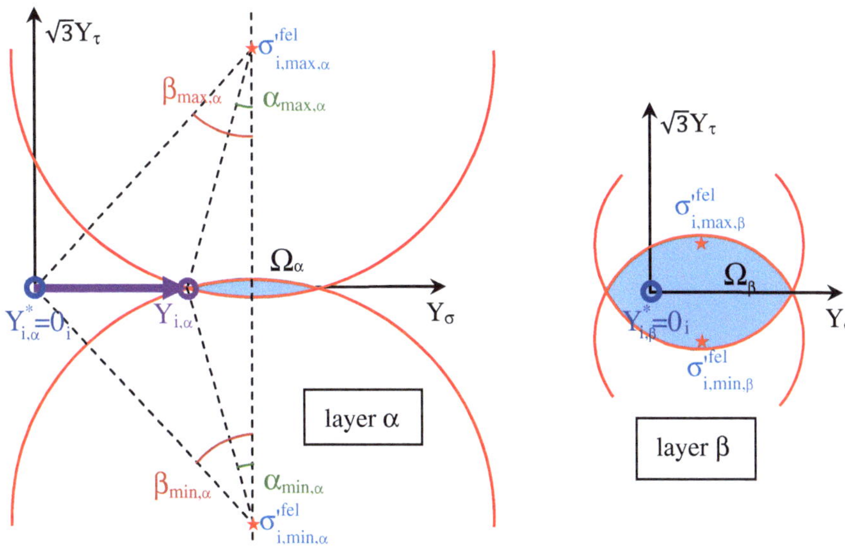

Fig. 7.6 Mises yield surfaces and intersection areas Ω of the two layers in the respective Y-spaces at the first MEA

similar to the yield surface according to Eq. (2.10), shown in the σ-$\sqrt{3}\tau$-plane (Fig. 2.19), with

$$Y_\sigma = \frac{3}{2} Y'_x; \quad Y_\tau = Y_{xy}. \tag{7.67}$$

The vertices of the intersection area Ω_α in layer α lie at the angles $\alpha_{\min,\alpha}$ and $\alpha_{\max,\alpha}$ in Fig. 7.6. Because the yield stresses are temperature-independent, both angles are the same and amount, according to Eq. (4.63), to

$$\alpha_{\min,\alpha} = \alpha_{\max,\alpha} = 16.35^\circ. \tag{7.68}$$

Because, by definition, no residual stresses are available before the first MEA, the quantity $Y^*_{i,\alpha}$ (see Eq. (7.48)) in layer α to be projected on the intersection area Ω_α is at the first MEA

$$Y^{*(1)}_{i,\alpha} = -\rho'^{(0)}_{i,\alpha} = \begin{pmatrix} 0 \\ 0 \\ 0 \\ 0 \end{pmatrix}. \tag{7.69}$$

Thus, the elastic-plastic stresses are identical to the fictitious elastic stresses. According to Eqs. (4.64) and (4.65), the negative deviatoric residual stresses in Fig. 7.6 are inclined under the angles

$$\beta_{\min,\alpha} = \beta_{\max,\alpha} = 43.17^\circ \tag{7.70}$$

relative to the connection of the two circle centers. Thus

$$\beta_{\min,\alpha} > \alpha_{\min,\alpha} \tag{7.71}$$

$$\beta_{\max,\alpha} > \alpha_{\max,\alpha}, \tag{7.72}$$

and, according to Eq. (4.70), $Y^*_{i,\alpha}$ is projected to the nearest vertex of Ω_α. With a and b of Eqs. (4.73) and (4.74),

$$a = 0.31280 \tag{7.73}$$

$$b = 0.34360, \tag{7.74}$$

Eq. (4.71) results in

$$Y^{(1)}_{i,\alpha} = \begin{pmatrix} +3.3658 \\ -1.6829 \\ -1.6829 \\ 0 \end{pmatrix} \tag{7.75}$$

(see Fig. 7.6).

In layer β, as is clear from Fig. 7.6, $Y^*_{i,\beta}$ according to Eq. (7.48),

$$Y^{*(1)}_{i,\beta} = -\rho'^{(0)}_{i,\beta} = \begin{pmatrix} 0 \\ 0 \\ 0 \\ 0 \end{pmatrix} \tag{7.76}$$

lies within the area Ω_β, because

$$\sigma_{\mathrm{v,min},\beta} = 2.417 \; < \; f_{y\beta,\min} = 4.182 \tag{7.77}$$

and

$$\sigma_{\mathrm{v,max},\beta} = 2.417 \; < \; f_{y\beta,\max} = 4.182. \tag{7.78}$$

According to Eq. (4.60), layer β is assigned to the subvolume V_e at the first MEA, so that

$$Y_{i,\beta}^{(1)} = \begin{pmatrix} 0 \\ 0 \\ 0 \\ 0 \end{pmatrix} \tag{7.79}$$

and thus no initial strains are applied. This is noteworthy because the equivalent stresses both at the minimum and the maximum load exceed the second yield stress,

$$\sigma_{\mathrm{v,min}} = \sigma_{\mathrm{v,max}} = 13.16 \quad > \quad f_{y2} = 12.50 \tag{7.80}$$

which actually requires plastic straining of both layers.

The initial strains in layer α are, according to Eq. (7.18),

$$\varepsilon_{i,0\alpha}^{(1)} = \begin{pmatrix} +0.33282\% \\ -0.16641\% \\ -0.16641\% \\ 0 \end{pmatrix} \tag{7.81}$$

or for the total state of both layers, according to Eq. (7.20),

$$\varepsilon_{i,02}^{(1)} = \begin{pmatrix} +0.27169\ \% \\ -0.13585\ \% \\ -0.13585\ \% \\ 0 \end{pmatrix}. \tag{7.82}$$

For the MEA, the modified elastic material parameters for the second segment of the total state are to be evaluated according to Eqs. (7.4) and (7.11):

$$E_2^* = 1666; \quad \nu_2^* = 0.5. \tag{7.83}$$

Because layer β was still classified as elastic in this first MEA, the initial strains of the total state could instead have been determined as with bilinear material behavior according to Sect. 4.4:

$$\varepsilon_{i,01}^{(1)} = \begin{pmatrix} +0.10483\ \% \\ -0.05241\ \% \\ -0.05241\ \% \\ 0 \end{pmatrix}. \tag{7.84}$$

The MEA would then have to be performed with modified elastic material parameters for the first segment of the total state

$$E_1^* = 4318; \quad \nu_1^* = 0.5. \tag{7.85}$$

In both cases, the result of the first MEA is

$$\rho_i^{(1)} = \rho_i^{'(1)} = \begin{pmatrix} 0 \\ 0 \\ 0 \\ 0 \end{pmatrix}, \tag{7.86}$$

which was predictable because the original actual load is stress-controlled. The division of the deviatoric residual stresses on the two layers according to Eq. (7.44) is possible only under the condition that all existing layer are plasticizing, i.e., in the present case, where only layer α is plasticizing, only on the basis of the initial strains $\varepsilon_{i,01}$ in Eq. (7.84). Equation (7.44) then becomes

$$\rho'_{i,\alpha} = \frac{E_{t\alpha}}{E_{t1}} \rho'_i + \frac{E_{t\alpha}}{E_{t1}} E^*_{ij,1} \left(\varepsilon_{j,01} - \varepsilon_{j,0\alpha} \right), \tag{7.87}$$

and the deviatoric residual stresses are partitioned amongst the two layers on the basis of initial strains $\varepsilon_{i,02}$ because of

$$E^*_{ij,1} \, \varepsilon_{j,01} = E^*_{ij,2} \, \varepsilon_{j,02} \tag{7.88}$$

with $\nu_1^* = \nu_2^*$, leading to

$$\rho'_{i,\alpha} = \frac{E_{t\alpha}}{E_{t2}} \rho'_i + \frac{E_{t\alpha}}{E_{t2}} E^*_{ij,2} \left(\frac{E_{t2}}{E_{t1}} \varepsilon_{j,02} - \varepsilon_{j,0\alpha} \right) \tag{7.89}$$

with $E^*_{ij,2}$ of Eq. (7.9). In both cases, we obtain by satisfying the equilibrium condition ($\rho_{i,\alpha} + \rho_{i,\beta} = 0$)

$$\rho_{i,\alpha}^{'(1)} = \begin{pmatrix} -2.0672 \\ +1.0336 \\ +1.0336 \\ 0 \end{pmatrix}; \quad \rho_{i,\beta}^{'(1)} = \begin{pmatrix} +2.0672 \\ -1.0336 \\ -1.0336 \\ 0 \end{pmatrix}. \tag{7.90}$$

Thus, all data are available, needed for the second MEA:

$$Y_{i,\alpha}^{*(2)} = -\rho_{i,\alpha}^{'(1)} = \begin{pmatrix} +2.0672 \\ -1.0336 \\ -1.0336 \\ 0 \end{pmatrix}; \quad Y_{i,\beta}^{*(2)} = -\rho_{i,\beta}^{'(1)} = \begin{pmatrix} -2.0672 \\ +1.0336 \\ +1.0336 \\ 0 \end{pmatrix}, \tag{7.91}$$

see Fig. 7.7.

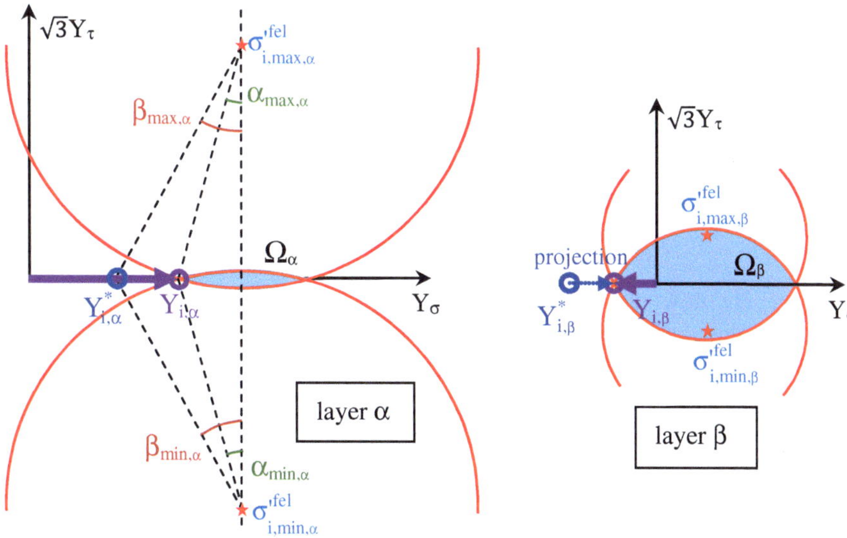

Fig. 7.7 Mises yield surfaces and intersection areas Ω of the two layers in the respective Y-spaces at the second MEA

For layer α, the angles now arising in the second MEA are

$$\beta_{\min,\alpha} = \beta_{\max,\alpha} = 28.46^\circ. \tag{7.92}$$

Accordingly, in the second MEA the projection remains, as already made for the first MEA, on the vertex of area Ω_α (see Fig. 7.7):

$$Y_{i,\alpha}^{(2)} = \begin{pmatrix} +3.3658 \\ -1.6829 \\ -1.6829 \\ 0 \end{pmatrix}. \tag{7.93}$$

In contrast, Fig. 7.7 shows that $Y_{i,\beta}^*$ for layer β is no longer located within Ω_β, so that layer β now becomes plastic and $Y_{i,\beta}^*$ is to be projected on Ω_β:

$$\alpha_{\min,\beta} = \alpha_{\max,\beta} = 60.23^\circ \tag{7.94}$$

$$\beta_{\min,\beta} = \beta_{\max,\beta} = 69.66^\circ. \tag{7.95}$$

With a and b of Eqs. (4.73) and (4.74)

$$a = 0.64807 \tag{7.96}$$

$$b = 0.17596, \tag{7.97}$$

Eq. (4.71) results in

$$Y_{i,\beta}^{(2)} = \begin{pmatrix} -0.9518 \\ +0.4759 \\ +0.4759 \\ 0 \end{pmatrix}. \tag{7.98}$$

There are now also initial strains to be applied in layer β. According to Eq. (7.19) they amount to

$$\varepsilon_{i,0\beta}^{(2)} = \begin{pmatrix} +0.41830\% \\ -0.20915\% \\ -0.20915\% \\ 0 \end{pmatrix} \tag{7.99}$$

and the initial strains of the total state according to Eq. (7.20)

$$\varepsilon_{i,02}^{(2)} = \begin{pmatrix} +0.1948\ \% \\ -0.0974\ \% \\ -0.0974\ \% \\ 0 \end{pmatrix}. \tag{7.100}$$

The second MEA, carried out with modified elastic material parameters as in Eq. (7.83) for the second segment of the total state, provides of course again vanishing residual stresses for the total state and therefore the residual stresses in the individual layers according to Eq. (7.44)

$$\rho_{i,\alpha}^{'(2)} = \begin{pmatrix} -1.2510 \\ +0.6255 \\ +0.6255 \\ 0 \end{pmatrix}; \quad \rho_{i,\beta}^{'(2)} = \begin{pmatrix} +1.2510 \\ -0.6255 \\ -0.6255 \\ 0 \end{pmatrix}. \tag{7.101}$$

In the illustration of Fig. 7.7, $Y_{i,\alpha}^{*}$ moves again a bit further to the left and $Y_{i,\beta}^{*}$ slightly further to the right, without any change in the projections on the vertices of the areas Ω_α and Ω_β. Thus, a further MEA is not required, and the end of the iterative improvement is reached.

The corresponding plastic strains of the total state are the same here as its initial strains, because the residual stresses of the total state are zero. The plastic strain in the direction of the constant stress is thus

$$\varepsilon^{\mathrm{pl}} = 0.1948\ \%, \tag{7.102}$$

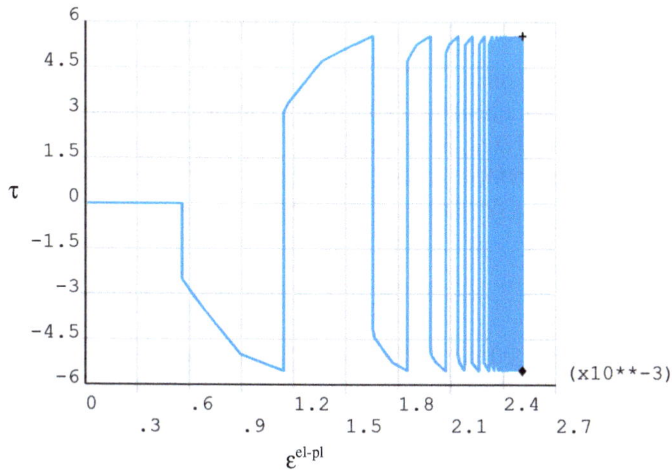

Fig. 7.8 Development of shear stress vs axial strain for the Tribout sample from the incremental analysis, and results of the STPZ for the extreme loading states (*cross* and *diamond*)

which is identical to the above result of the incremental analysis for the shakedown state. Figure 7.8 shows how the axial strain, i.e., the strain component acting in the direction of the constant direct stress, develops as a function of the cyclic shear stress until shakedown is ultimately achieved, along with the result of the STPZ in the shakedown state.

How the TIV develop in the two layers according to the incremental analysis, is shown in Fig. 7.9 (limited to the load reversal points).

The final state was determined correctly with the STPZ after two MEAs. The result is the same as the one obtained with Zarka's method in [3], but was achieved in other ways, with different material data of the layers, and also much faster.

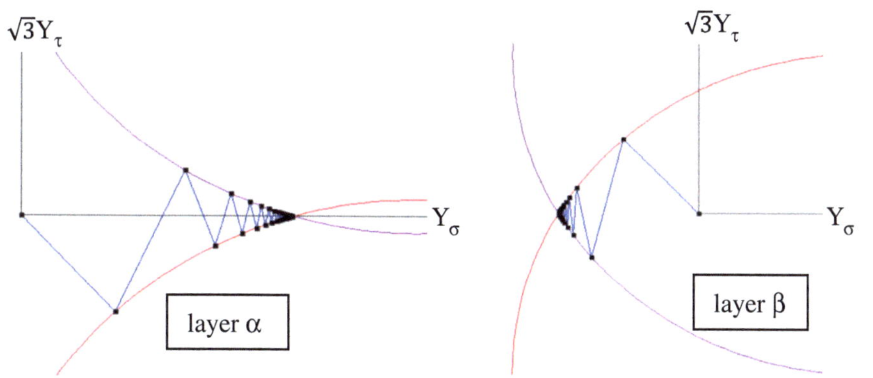

Fig. 7.9 Trajectories of the TIV in the two layers (the two layers are not represented at the same scale)

7.2.3 Bree Tube

The axisymmetric model of the thin-walled Bree tube under constant internal pressure and cyclic temperature gradient through the thickness of the wall addressed recently in Sect. 5.9.3 is now analyzed with a trilinear material model. The elastic-plastic material parameters are temperature-independent and indicated in Table 7.4.

The load consists of a constant internal pressure p, causing the equivalent stress

$$\sigma_p = 72 \frac{\mathrm{N}}{\mathrm{mm}^2} \tag{7.103}$$

if cap forces are accounted for. All temperatures are equal to zero at the minimum load condition. A cyclic linear temperature distribution over the wall thickness is applied, so that the inner surface is warmer than the outside surface, and the fictitious elastic calculation results in a cyclical equivalent stress range at both the inner and the outer surface:

$$\sigma_t = 360 \frac{\mathrm{N}}{\mathrm{mm}^2}. \tag{7.104}$$

Figure 7.10 presents the results of an incremental analysis in the form of the stress distribution across the wall thickness at the two extreme load times in the

Table 7.4 Material parameters of the trilinear material model

Material data	
E	1×10^5 N/mm^2
ν	0.3
E_{t1}	0.1×10^5 N/mm^2
E_{t2}	0.02×10^5 N/mm^2
f_{y1}	75 N/mm^2
f_{y2}	120 N/mm^2

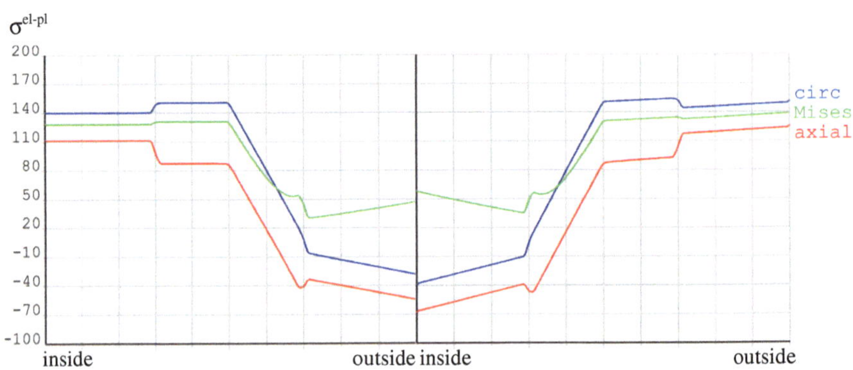

Fig. 7.10 Distribution of stresses across the wall thickness in the plastic shakedown state (incremental cyclic analysis) at minimum (*left*) and maximum (*right*) load condition

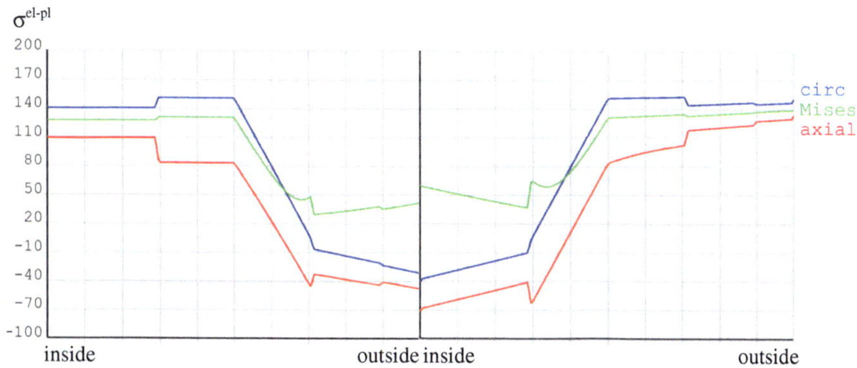

Fig. 7.11 Distribution of stresses across the wall thickness after the fifth MEA of the STPZ at minimum (*left*) and maximum (*right*) load condition

shakedown state. For a good approximation to the shakedown state, about 50 cycles were to be calculated.

Comparison with the results of the STPZ given in Fig. 7.11 shows that a good approximation to the incremental analysis is already reached after a few linear analyses.

Figures 7.12 and 7.13 allow a comparison of the stresses and strains obtained by the STPZ for the shakedown state with the results of an incremental analysis in terms of a histogram of circumferential and axial strains on the inside and outside surface, and of the stress–strain diagram for the circumferential direction.

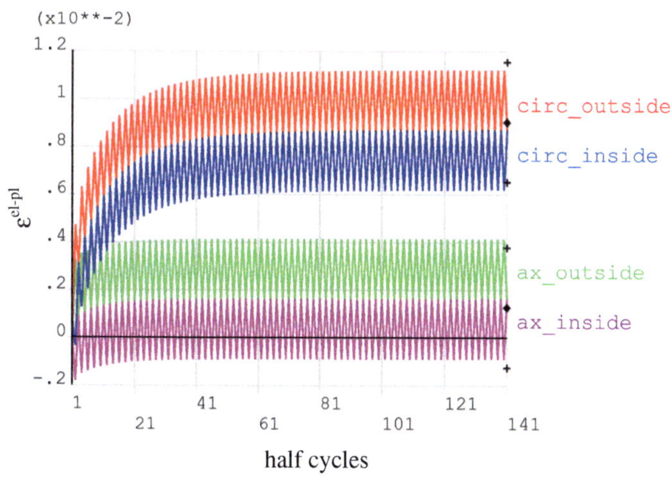

Fig. 7.12 Histogram of circumferential and axial strains on the inside and outside surface according to an incremental analysis, and results after the fifth MEA of the STPZ for the state of plastic shakedown (*diamonds* for minimum, *crosses* for maximum loading condition)

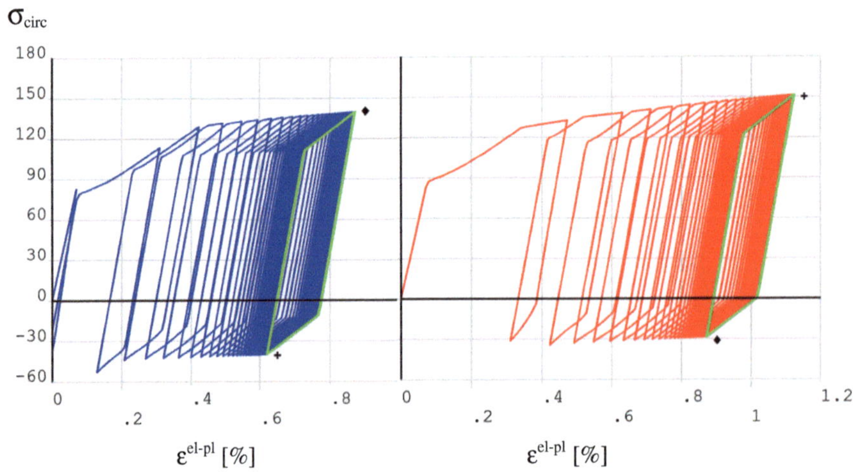

Fig. 7.13 Stress–strain diagram for the circumferential direction from an incremental analysis and stress–strain couples after the fifth MEA of the STPZ for the state of plastic shakedown (*diamonds* and *crosses*); *left* inside surface; *right* outside surface; hysteresis loop in the state of shakedown highlighted in *green*

It is evident that the range is characterized by bilinear behavior, and the strain accumulation is strongly influenced by the trilinear segment of the material behavior. The range is represented by the STPZ in very good quality, although the accumulated strains are somewhat overestimated.

The development of the circumferential strains on the outside surface with the number of MEAs of the STPZ can be found in Table 7.5.

Table 7.5 Development of the circumferential strains on the outside surface with the number of MEAs of the STPZ compared with an incremental analysis for the two extreme loading conditions in the shakedown state

STPZ, MEA	ε_z (%) min. loading condition	ε_z (%) max. loading condition
1	0.145	0.397
2	0.501	0.753
3	0.894	1.146
4	0.890	1.142
5	0.901	1.153
6	0.902	1.154
Incremental	0.872	1.123

7.2.4 Nozzle

7.2.4.1 Geometry

Several simplified methods for determining elastic-plastic strain ranges were examined in [4] on the basis of practical examples from power plant technology in terms of fatigue usage, where the STPZ has consistently performed very well.

In the following, a practical example of power plant technology is also covered, the original specification of which was provided by the company AREVA NP GmbH in Erlangen. The author would like to thank them for the permission to use this sample here. However, a number of simplifications concerning the geometry, the material behavior, and the loading is introduced for this book.

A liquid medium moves through a pipe into a large vessel. The connection between pipe and vessel is manufactured by an axisymmetric nozzle (Fig. 7.14). Pipe and nozzle are made of an austenitic steel (grayed out in Fig. 7.14). As in the example of the thick-walled cylinder in Sect. 4.3.4, the loading consists of a change in the fluid temperature, namely a down-transient of 350 to 50 °C, followed after a holding time by a faster up-transient back to 350 °C, as already depicted in Fig. 4.6. Because of the compact geometry, a possibly existing internal pressure constitutes here no ratcheting mechanism and is therefore left in the following disregarded, as well as internal forces that are initiated from the connected pipeline.

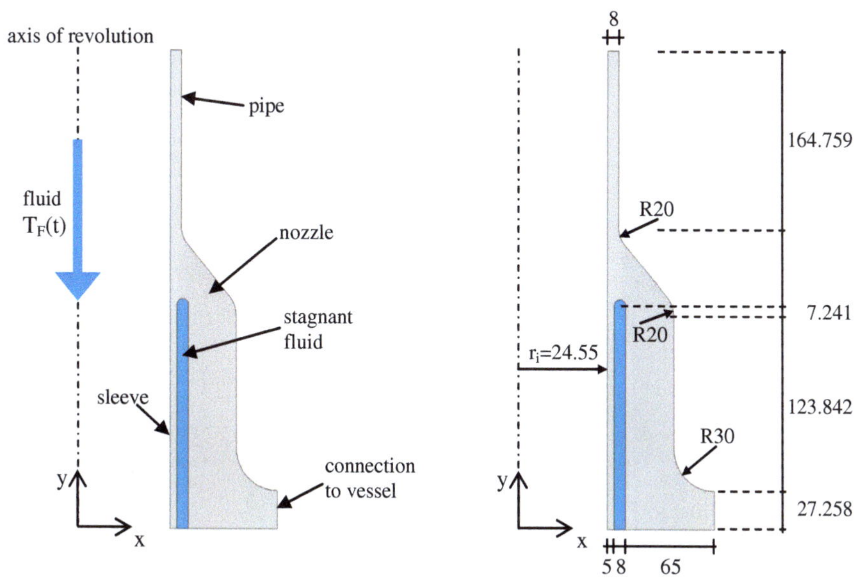

Fig. 7.14 Nozzle as a transition between pipe and vessel

7.2.4.2 Thermal Analysis

When the fluid changes its temperature, a convective heat transfer takes place on the inner surface of the structure. Now by thermal conduction not only does a radial temperature gradient occur, as in Sect. 4.3.4, but also an axial temperature gradient.

To reduce the thermal stresses in the thick part of the structure, this is protected by a sleeve, through which an area of nearly stagnant fluid is produced (shown in blue in Fig. 7.14), which has a damping effect on the temperature of the vessel surface. To take into account the low convection in this area, the thermal conductivity of the fluid is increased in the calculation of the temperature fields. The thermal material data are temperature-independent and can be seen in Table 7.6.

On the inner surface of the tube and the sleeve, a convective heat transfer takes place. The bottom of the connection to the vessel ($y = 0$) is held constant at 350 °C. The top edge of the pipe and the entire right outer contour (at the pipe, at the nozzle, at the connection to the vessel) exhibit adiabatic boundary conditions.

The temperature distributions in Fig. 7.15 are exemplary for the thermal analysis for the temperature transient of Fig. 4.6 at the end of the down-transient (time 330 s)

Table 7.6 Thermal material data

	Steel	Fluid (stagnant area)
Density ρ	7.93×10^{-9} N s^2/mm^4	0.80×10^{-9} N s^2/mm^4
Thermal conductivity k	17 N/(s K)	9 N/(s K)
Specific heat c	0.49×10^9 mm^2/(K s^2)	5×10^9 mm^2/(K s^2)
Heat transfer coefficient h_f	30 N/(K s mm)	

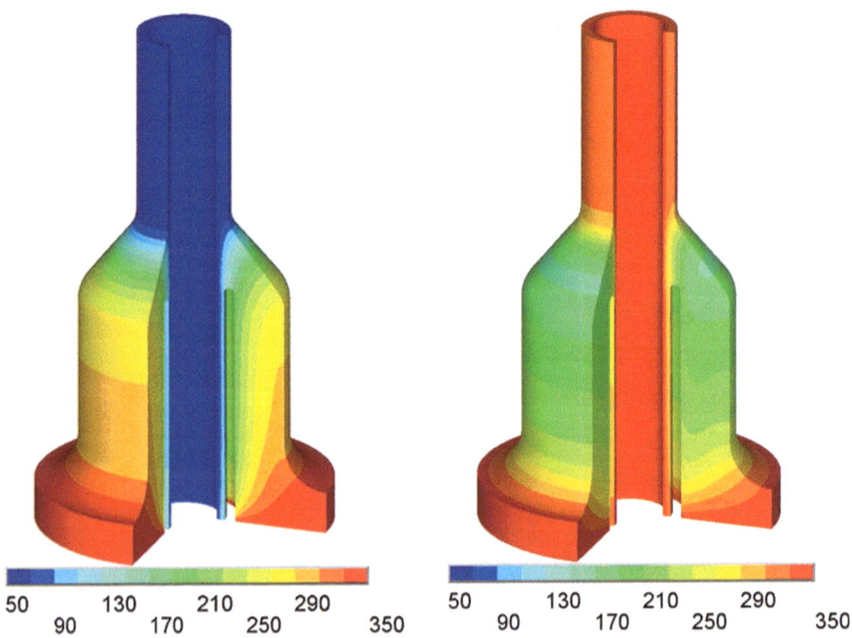

Fig. 7.15 Temperature distributions; *left* at the end of the down-transient (time 330 s); *right* at the end of the up-transient (525 s)

and at the end of the up-transient (time 525 s). Radial and axial temperature gradients are each clearly visible.

7.2.4.3 Fictitious Elastic Analysis

The fictitious elastic calculations are made with the following material data:

$$E = 180{,}750\ \mathrm{N/mm^2}$$
$$\nu = 0.3$$
$$\alpha_T = 2.5 \times 10^{-5}/\mathrm{K}\ .$$

The bottom of the connection to the vessel ($y = 0$), but not the sleeve, are held against vertical displacement.

Figure 7.16 shows the temperatures on the deformed structure in 80-fold magnification for the transient times 0, 330, and 525 s. From these it is already evident that, mainly as a result of the axial temperature gradient, axial bending stresses as well as circumferential membrane and, because of transverse strain constraints, circumferential bending stresses occur. They also give an impression at which locations the highest stresses are to be expected. Unlike the thick-walled cylinder in Sect. 4.3.4, shear stresses also occur in addition to the three direct stresses.

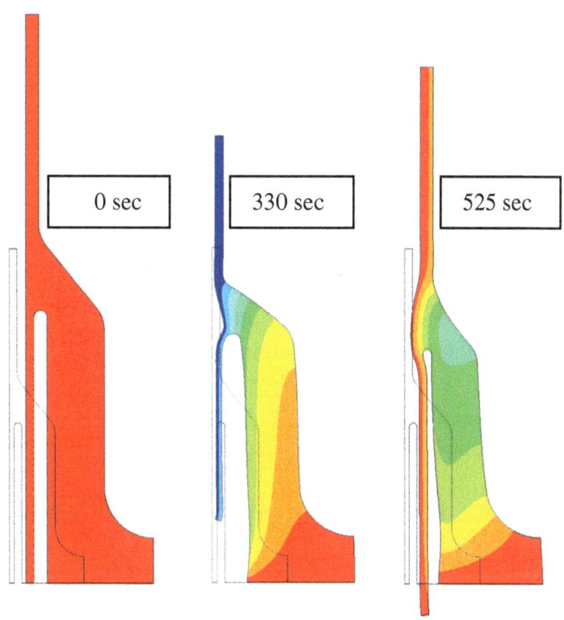

Fig. 7.16 Structural displacements at different transient times; the colors correspond to the temperatures in Fig. 7.15

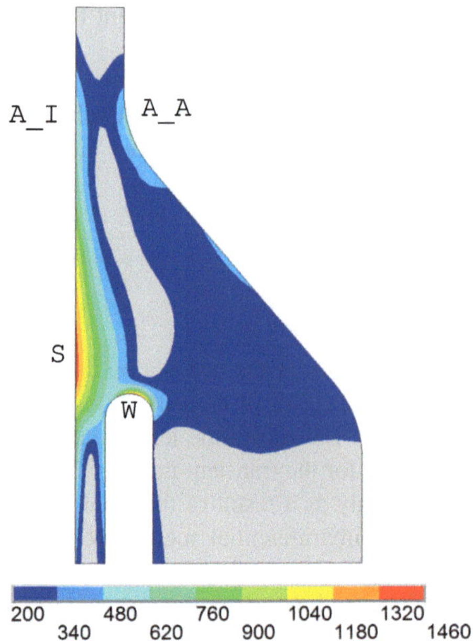

Fig. 7.17 Fictitious elastic calculated equivalent stress range (for clarity limited to values > 200 N/mm^2)

The maximum equivalent stress ranges can be found in the nozzle area for the times shortly after the end of the respective down- and up-transient. Figure 7.17 shows the distribution of the fictitious elastic calculated equivalent stress range formed from the times 330 s and 525 s in the relevant nozzle area. Their maximum value is 1,435 N/mm^2 and is located on the inner surface at about the level of the notch imposed by the sleeve connection at the point S. However, also in the notch itself, point W, and at the junction of the nozzle to the pipe, points A_i and A_a, stress concentrations are recognized.

The extreme stresses occur at different locations at different times. In addition, the stress components at a fixed location do not evolve synchronized to each other, so that no radial loading is present. The development of the fictitious elastic calculated stresses in the deviatoric stress space is shown in Fig. 7.18 for the points W, A_i, and S for a full load cycle. At these locations only direct stresses exist, no shear stresses. Because a full temperature equalization to 350 °C was enforced after 2,000 s, the trajectory ends back at the coordinate origin.

7.2.4.4 Incremental Elastic-Plastic Analysis (Transient)

The incremental analysis is performed with the Besseling model for a trilinear stress–strain diagram with temperature-dependent yield stresses. The material data are shown in Fig. 6.7. The trilinearization also shown there leads to the material parameters of Table 7.7.

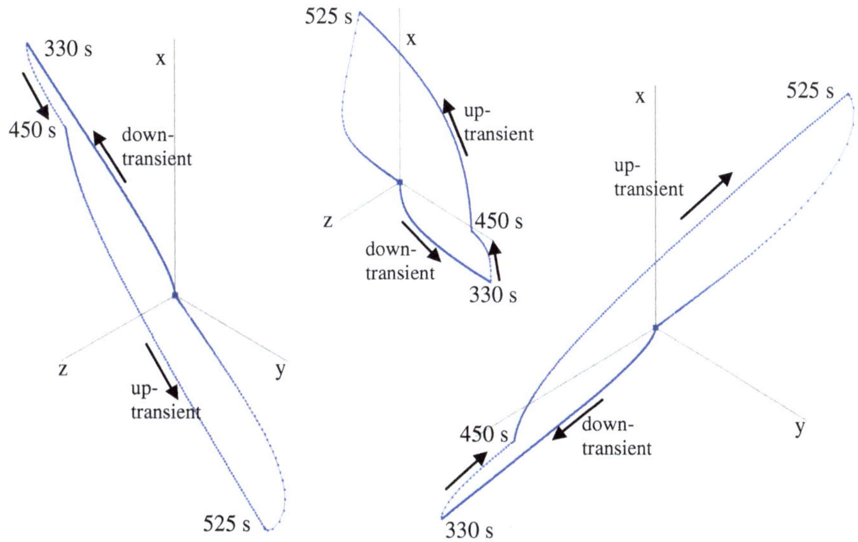

Fig. 7.18 Trajectories of fictitious elastic calculated stresses in the deviatoric stress space at the points W, A_i, and S

Table 7.7 Material parameters of the trilinear material model

Material parameters					
E (N/mm^2)	180,750				
ν	0.3				
α_T (K^{-1})	2.5×10^{-5}				
E_{t1} (N/mm^2)	21,168				
E_{t2} (N/mm^2)	7,551				
	20 °C	100 °C	200 °C	300 °C	400 °C
f_{y1} (N/mm^2)	156.40	131.10	112.55	93.81	83.65
f_{y2} (N/mm^2)	215.61	181.27	156.73	134.18	121.72

If a transient analysis is performed, in which the temperature transient is passed through and examined in detail with several hundred load steps per cycle, then one obtains the trajectories of the stress paths in the deviatoric stress space for the points W, A_i, and S shown in Fig. 7.19. Although, as in the fictitious elastic analysis, at each end of the cycle, i.e., every 2,000 s, a complete temperature equalization to 350 °C is enforced, the trajectory does not return back in the coordinate origin, because residual stresses are incurred. Because ratcheting does not matter, just a few cycles suffice until the trajectories are closed (represented are two cycles). Figure 7.19 also shows the yield surfaces (radius f_{y1}) belonging to the lower of the

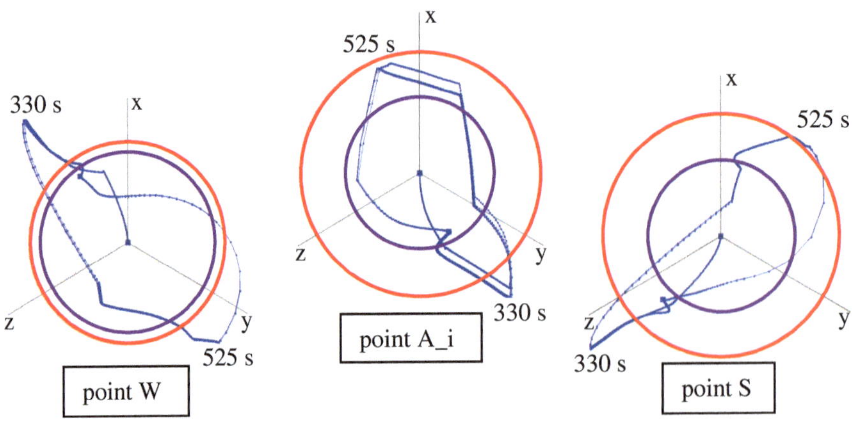

Fig. 7.19 Trajectories of incremental elastic-plastic calculated stresses in the deviatoric stress space at the points W, A_i, and S for complete temperature transient

two yield stresses at the temperatures reached at times 330 s and 525 s of each point. It can be seen that the differences between the two yield surfaces at the points that are located on the inner surface are substantially greater than at the point W, which undergoes the temperature transient only damped.

7.2.4.5 Incremental Elastic-Plastic Analysis (Load Pair)

As a basis for comparison with the STPZ, another incremental analysis is performed in which a load cycle does not consist of the entire temperature transient passed through step-by-step, but only of the pair of loading conditions formed by the temperature fields at the transient times 330 s and 525 s as determined in Sect. 7.2.4.2, i.e., only at the end of the down- and the up-transients.

The associated trajectories of the deviatoric stresses are illustrated in Fig. 7.20, again for comparison purposes with the yield surfaces corresponding to the lower of the two yield surfaces (radius f_{y1}) at the temperatures present at the respective points at the times 330 s and 525 s. Compared to Fig. 7.19, the trajectories are less filled.

However, the stresses and strains in the load reversal points hardly differ. By way of example, the strain components at transient time 525 s in the tenth cycle are compared in Table 7.8.

For the relevant area of the nozzle the direct stress components are shown in Fig. 7.21 and the strain components in Fig. 7.22 as an example for the time 330 s in the tenth cycle.

The development of the TIV cannot be examined for the total state with trilinear material behavior, but only individually for each layer. The associated material

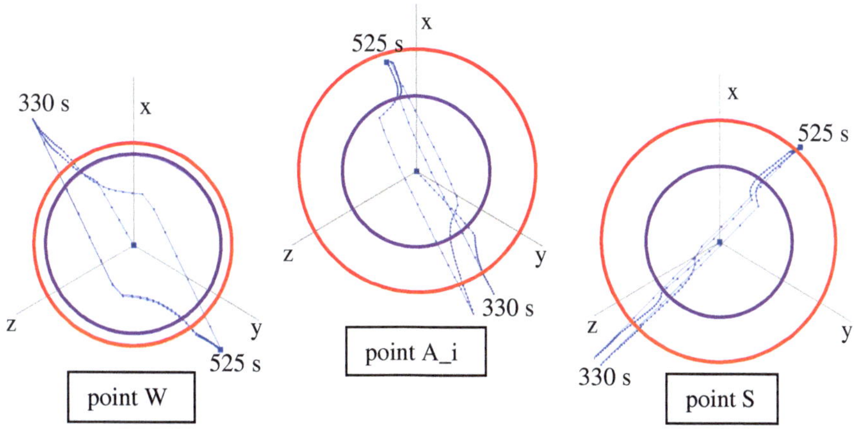

Fig. 7.20 Trajectories of incremental elastic-plastic calculated stresses in the deviatoric stress space at the points W, A_i, and S (for load pair 330 and 525 s)

Table 7.8 Strain components in incremental elastic-plastic analysis: calculation of the entire transient or only the pair 330 and 525 s

		ε_x (%)	ε_y (%)	ε_z (%)
Point W	Entire transient	−0.370	+0.386	−0.087
	Load pair	−0.368	+0.383	−0.084
Point A_i	Entire transient	+0.311	-0.235	−0.132
	Load pair	+0.312	-0.243	−0.128
Point S	Entire transient	+0.427	+0.048	−0.523
	Load pair	+0.413	+0.056	−0.520

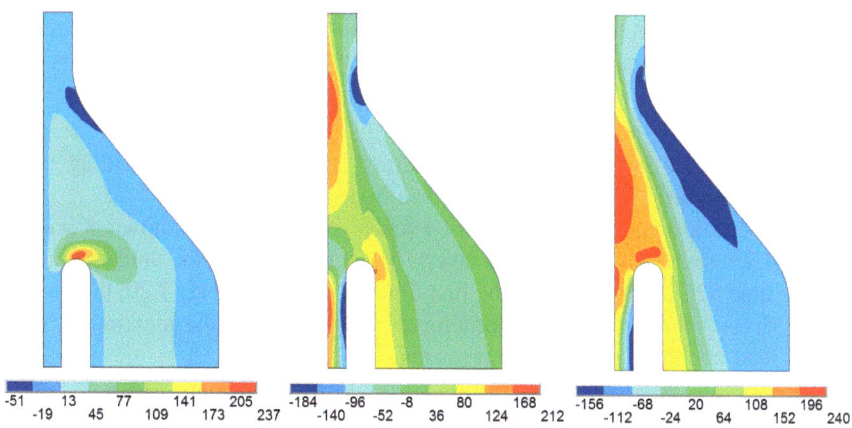

Fig. 7.21 Stress components σ_x, σ_y, σ_z in a part of the nozzle from an incremental analysis after 10 cycles of the load pair 330 and 525 s (for time 330 s)

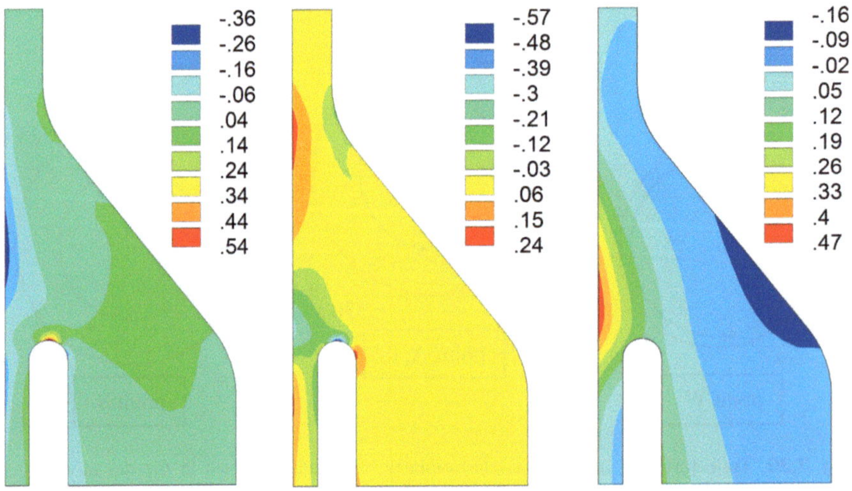

Fig. 7.22 Strain components ε_x, ε_y, ε_z (%) in a part of the nozzle from an incremental analysis after 10 cycles of the load pair 330 and 525 s (for time 330 s)

Table 7.9 Material parameters of the two layers of the trilinear material model

Material parameters					
E_α (N/mm²)	168,239				
E_β (N/mm²)	12,511				
ν	0.3				
α_T (K^{-1})	2.5×10^{-5}				
$E_{t\alpha}$ (N/mm²)	7,028.3				
$E_{t\beta}$ (N/mm²)	522.7				
	20 °C	100 °C	200 °C	300 °C	400 °C
$f_{y\alpha}$ (N/mm²)	145.57	122.03	104.76	87.32	77.86
$f_{y\beta}$ (N/mm²)	50.58	42.76	37.45	33.59	31.35

parameters of the individual layers, identified from Table 7.7 according to the relations presented in Sect. 6.3.6, are given in Table 7.9.

First the trajectories of the deviatoric stresses are exemplarily shown at point W in Fig. 7.23, separately for the two layers, but at different scales, until the loading state at time 525 s is attained for the first time. In contrast to Fig. 7.20, the position of the Mises yield surfaces at the minimum (330 s, red) and the maximum (525 s, violet) load condition can also be provided.

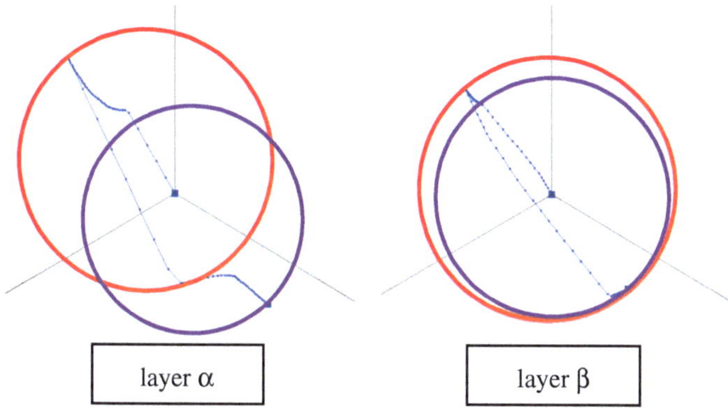

Fig. 7.23 Trajectories of incremental elastic-plastic calculated stresses of the two layers in the deviatoric stress space at the point W (for load pair 330 and 525 s)

For comparison with Fig. 7.21, the distribution of the direct stress components in the two layers at time 330 s are shown in Fig. 7.24. As can be seen on basis of the radial and axial stress components, the extreme stresses of the two layers partially show up at different locations. In view of the radial stresses σ_x, it is apparent that the sum of the two layers only vanishes on the free surface, but not in the individual layers.

Using the example of point W in Fig. 7.25, one also recognizes that the development of the TIV in the two layers, although not shown in the same scale, is not synchronous, but in some cases may even be in opposite direction. Thus the extreme stresses and strains in the two layers may occur not only at different locations but also at different times. This is also because stresses normal to a free surface do not necessarily disappear in the two layers, but simply must have opposite signs.

7.2.4.6 STPZ

As an example for the results of the STPZ, the elastic-plastic strains at loading time 330 s are shown in Fig. 7.26 after the fifth MEA (the jagged black lines indicate the border of the plastic zone). Visually they are very close to the results shown in Fig. 7.22 determined by incremental calculation.

Differences appear only upon more detailed examination, such as when the stresses and strains calculated with the STPZ in the load reversal points are compared with the stress–strain hysteresis of the incremental analysis, as shown in Fig. 7.27 for point W. As this is located in the notch root with horizontal tangent, the direct stress component in the y-direction as well as the shear stress vanish.

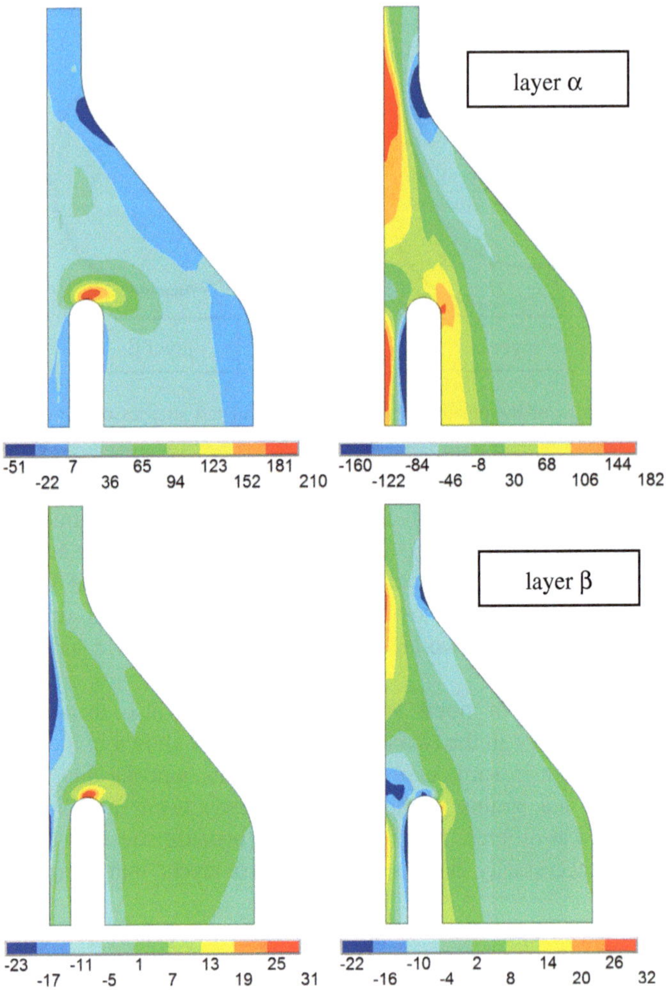

Fig. 7.24 Stress components σ_x, σ_y in the two layers in a part of the nozzle from an incremental analysis after 10 cycles of the load pair at 330 and 525 s (for time 330 s)

Table 7.10 shows how the extreme strains at load time 330 s and the extreme strain ranges of the structure develop with the number of MEAs.

After the first four MEAs have been incurred to calculate the range quantities, there are only small changes from the fourth additional MEA to calculate the strain states in the load reversal points. At the latest with the second MEA the correct location of the extreme stresses and strains is also identified by the STPZ within one

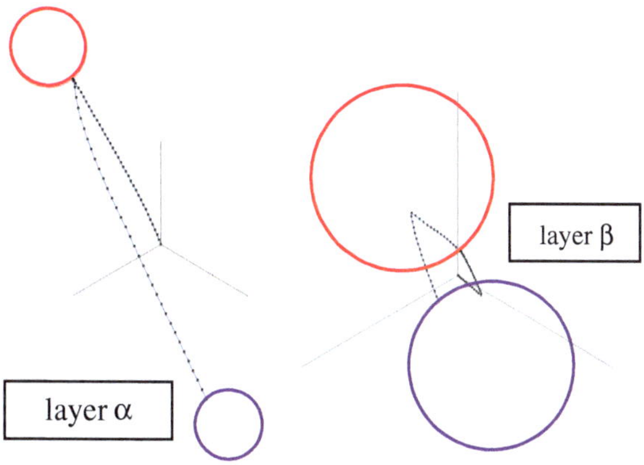

Fig. 7.25 Trajectories of incremental elastic-plastic calculated TIV in the TIV-space of the two layers at point W (for load pair 330 and 525 s)

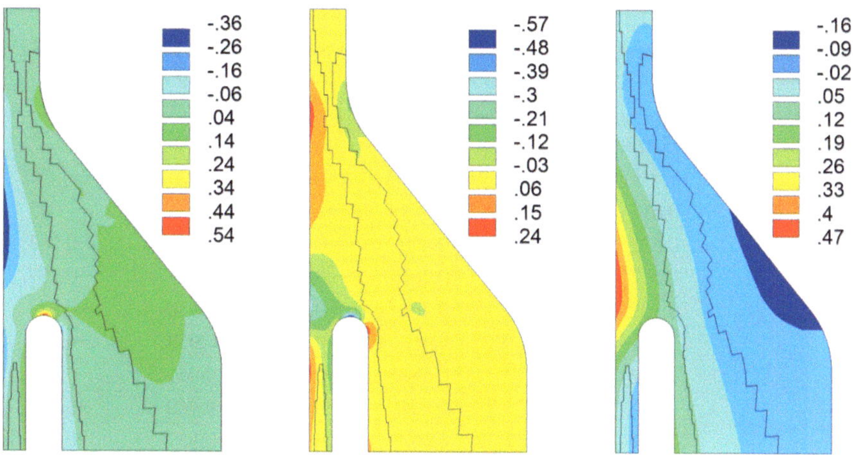

Fig. 7.26 Strain components ε_x, ε_y, ε_z, (%) for a part of the nozzle from the STPZ after five MEAs (for time 330 s)

element width around points W or S. The strain ranges are in good approximation to the incremental analysis, and the circumferential strains at loading time 330 s as well. Only the radial and axial strains at load time 330 s fall off somewhat in their accuracy.

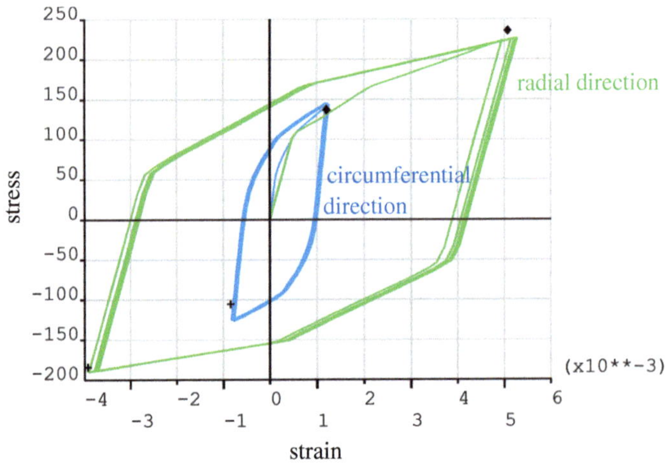

Fig. 7.27 Stress–strain hysteresis of the incremental analysis for load pair 330 and 525 s at point W, and stress–strain couples for load reversal points after five MEAs of the STPZ

Table 7.10 Development of the extreme strains with the number of MEAs of the STPZ compared with the tenth cycle of an incremental analysis (for load pair 330 and 525 s)

STPZ, MEA	Min. loading condition (330 s)			Range 525–330 s		
	ε_x (%)	ε_y (%)	ε_z (%)	$\Delta\varepsilon_x$ (%)	$\Delta\varepsilon_y$ (%)	$\Delta\varepsilon_z$ (%)
1	0.617	−0.705	0.544	−1.117	1.259	−1.087
2	0.522	−0.561	0.468	−0.942	1.009	−0.983
3	0.502	−0.541	0.467	−0.895	0.957	−0.977
4	0.505	−0.543	0.468	−0.897	0.956	−0.979
5	0.511	−0.545	0.468	−0.905	0.958	−0.978
6	0.512	−0.547	0.468	−0.905	0.961	−0.978
7	0.512	−0.546	0.468	−0.906	0.960	−0.978
8	0.513	−0.547	0.468	−0.906	0.959	−0.978
Incremental	0.532	−0.564	0.463	−0.899	0.942	−0.977

References

1. Hübel, H.: Simplified theory of plastic zones for cyclic loading and multilinear hardening. Int. J. Press. Vessels Pip. **129–130**, 19–31 (2015). doi:10.1016/j.ijpvp.2015.03.002
2. Zarka, J.: Direct analysis of elastic-plastic structures with 'overlay' materials during cyclic loading. Int. J. Numer. Methods Eng. **15**, 225–235 (1980)
3. Tribout, J., Inglebert, G., Casier, J.: A simplified method for the inelastic analysis of structures under cyclic loading, transactions of the ASME. J. Press. Vessel Technol. **105**, 222–226 (1983)
4. Hübel, H., et al.: Performance study of the simplified theory of plastic zones and the Twice-Yield method for the fatigue check. Int. J. Press. Vessels Pip. **116**, 10–19 (2014). doi:10.1016/j.ijpvp.2014.01.003

8 Plastic Limit Load

Many of the calculation methods for shakedown and ratcheting analyses mentioned in Sect. 2.9.4 can also be used to determine the plastic limit load, such as the Generalized Local Stress–Strain (GLOSS) Method [1], the Elastic Compensation Method (ECM), and the Linear Matching Method (LMM). This also applies to Zarka‘s method (see, for example, the application on piping elbows in [2]) and thus also to the STPZ.

The potential for using the simplified theory of plastic zones (STPZ) is seen mainly in the simplified determination of strain ranges and accumulated strains in the frame of service life analyses, which was the main subject considered in the previous chapters. Below is an investigation of how far the STPZ can also be used to determine the plastic limit load. It is considered with preparatory work of Zarka [3], Maier et al. [4], and Borhani [5].

8.1 Calculation Steps

The purpose of a plastic limit load analysis is to determine up to what level a load may be increased monotonically until the structure shows a certain reaction, namely becomes kinematic. This task is an inversion of the work in Chap. 3, where the response of the structure was calculated for a monotonic increase of load up to a given load level.

Although the failure of a structure is often accompanied by large displacements and rotations forming a kinematic chain, small displacements and small rotations are assumed in the following, frequently referred to as first-order theory. As is clear from the preceding chapters, the STPZ requires multilinear stress–strain behavior with kinematic hardening. However, a structure with unlimited hardening cannot develop a kinematic chain, so in regard to determination of the plastic limit load, the slope in the last segment of a multilinear stress–strain diagram must equal zero. In the following only linear elastic–linear kinematic hardening behavior is assumed, resulting in linear elastic-perfectly plastic behavior for the limit value $E_t \rightarrow 0$. The

H. Hübel, *Simplified Theory of Plastic Zones*,
DOI 10.1007/978-3-319-29875-7_8

yield stress f_y may vary locally, e.g., because of a temperature distribution present in the state of the limit load.

A limit analysis with the STPZ is done in several steps, which are set out below. First, an asymptotic state is determined (Sect. 8.2), which provides an estimate of the direction of the stress increments at an infinitely high load level. This allows an estimation of the TIV, used for a first estimation of the plastic limit load, assuming that the entire volume of the structure is plasticizing (Sect. 8.4), which can subsequently then be gradually improved (Sect. 8.6).

The STPZ is also implemented in ANSYS [6] for plastic limit load analyses as a user subroutine.

8.2 Asymptotic State

Zarka already described in [3] how an estimate of the plastic limit load can be obtained. The first step for this is to determine the maximum possible directional stress redistribution produced by plasticizing. For this purpose it is assumed that the stress level is infinitely high. Based on the assumption of unlimited kinematic hardening, this means that each location of the structure is plasticized, that is, the plastic zone V_p is identical to the total volume V of the structure:

$$V_p = \{\underline{x} | \sigma_v \geq f_y\} \quad ; \quad \sigma_v \to \infty \quad \Rightarrow \quad V_p = V \quad ; \quad V_e \not\subset V. \tag{8.1}$$

The stress and strain rates or increments involved can be determined by considering the asymptotic state. From the representation in the space of the transformed internal variables (TIV) in Fig. 8.1 it is evident that in asymptotic approximation to an infinitely high load level, the direction and magnitude of the TIV Y_i are known:

$$Y_i^{\infty} \to \sigma_i^{\prime \text{fel},\infty}. \tag{8.2}$$

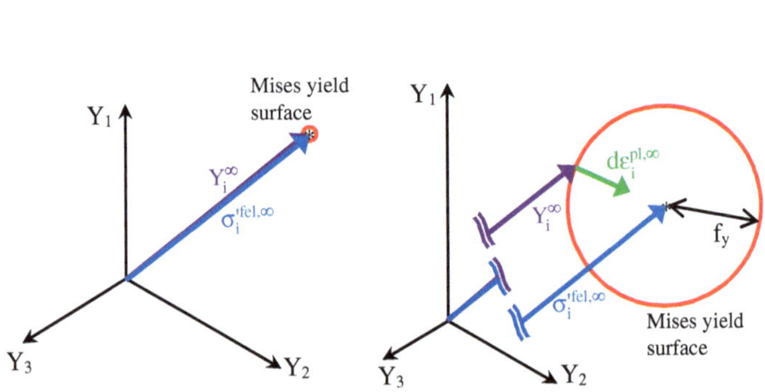

Fig. 8.1 Asymptotic state at a location of the structure at very high load level in the TIV-space

The direction of the plastic strain increment $\mathrm{d}\varepsilon_i^{\mathrm{pl},\infty}$, i.e., the inner normal to the Mises circle at the point Y_i^∞, initially remains unknown. However, it can be determined according to the usual procedure of STPZ as described in the previous chapters, if the increments of stresses, strains, etc., are considered at asymptotic approximation to an infinitely high load level instead of the stresses, strains, etc. Instead of Eqs. (3.66) and (3.107), the following applies for this quasi-steady state:

$$\mathrm{d}Y_i^\infty = \mathrm{d}\sigma_i^{\prime\mathrm{fel},\infty}. \tag{8.3}$$

The incremental initial strains thus become, in place of Eq. (3.14):

$$\mathrm{d}\varepsilon_{i,0}^\infty = \frac{3}{2C}\mathrm{d}Y_i^\infty = \frac{3}{2C}\mathrm{d}\sigma_i^{\prime\mathrm{fel},\infty} \tag{8.4}$$

and the incremental initial stresses, instead of Eq. (3.23):

$$\mathrm{d}\sigma_{i,0}^\infty = -E_{ij}^*\,\mathrm{d}\varepsilon_{j,0}^\infty = -\frac{3}{2}\frac{1-\frac{E_t}{E}}{1+\nu^*}\mathrm{d}\sigma_i^{\prime\mathrm{fel},\infty}, \tag{8.5}$$

with the modified elastic material parameters E^* and ν^* in the modified elasticity matrix E_{ij}^* being given by Eqs. (3.15) and (3.16).

With these initial strains or initial stresses serving as modified loading, and the modified elastic material parameters, a modified elastic analysis (MEA) is performed which delivers the residual stress increments $\mathrm{d}\rho_i^\infty$. From the defining Eqs. (2.25) and (2.66), the additivity of elastic and plastic deformations, Eq. (1.15), and the elastic and the modified elastic material law, Eqs. (1.3) and (3.12), each with restatement for the increments, results in

$$\mathrm{d}\varepsilon_i^{\mathrm{pl},\infty} = \frac{3}{2C}\mathrm{d}\sigma_i^{\prime\infty}. \tag{8.6}$$

The plastic strain increment has therefore, in the asymptotic state of infinitely high load level, the same direction as the elastic-plastic deviatoric stress increment, whereas at finite load level the plastic strain increment is directed in the direction of the reduced deviatoric stress because of the normality rule, Eq. (1.21). This is immediately plausible because, with infinitely high load, the increment of the backstress $\mathrm{d}\xi_i^\infty$ is proportional to the increment of the deviatoric stress $\mathrm{d}\sigma_i^{\prime\infty}$.

More MEAs for iterative improvement of this result are for now not possible, because on one hand the plastic zone is unchangeable, extending over the entire structure as with Eq. (8.1) ($V_p = V$), and on the other hand the incremental TIV $\mathrm{d}Y_i^\infty$ cannot be improved by projecting the incremental backstresses on the Mises yield surface, as opposed to the consideration of finite load levels. This is because the yield surface has shrunk to a point in the quasi-steady state of an asymptotic approach to an infinitely high load, which is equivalent to the transition $f_y \to 0$.

As is seen later, the signs of the incremental deviatoric stress components $d\sigma_i^{\prime\infty}$ play a very important role, but cannot be reliably identified correctly with the described procedure. For uniaxial stress state it arises, for example, that the plastic strain increments in the fully plastic state have the same signs as the fictitious elastic stresses, but this is not always true. Because even if in fact any location of a structure is plasticized, tension stresses can occur because of local stress redistribution at locations which experience compression stresses in the fictitious elastic state, and vice versa. This applies, for example, to a beam cross-section under bending stress, whose neutral axis is twisted or shifted because of plasticizing, as shown in Fig. 2.1 for a T-shaped beam cross-section.

At this point, it should be pointed out that the MEAs sometimes lead to vanishing residual stress increments. A justification for this is given in [7], where Schimmöller proves that residual stresses vanish when eigenstates exist which satisfy the compatibility conditions

$$\left.\begin{aligned} \frac{\partial^2 \varepsilon_x}{\partial y^2} + \frac{\partial^2 \varepsilon_y}{\partial x^2} - \frac{\partial^2 \gamma_{xy}}{\partial x \partial y} = 0 \\ \frac{\partial^2 \varepsilon_y}{\partial z^2} + \frac{\partial^2 \varepsilon_z}{\partial y^2} - \frac{\partial^2 \gamma_{yz}}{\partial y \partial z} = 0 \\ \frac{\partial^2 \varepsilon_z}{\partial x^2} + \frac{\partial^2 \varepsilon_x}{\partial z^2} - \frac{\partial^2 \gamma_{xz}}{\partial x \partial z} = 0 \\ \frac{\partial^2 \varepsilon_x}{\partial y \partial z} - \frac{\partial}{\partial x}\left(-\frac{\partial \gamma_{yz}}{\partial x} + \frac{\partial \gamma_{xz}}{\partial y} + \frac{\partial \gamma_{xy}}{\partial z}\right) = 0 \\ \frac{\partial^2 \varepsilon_y}{\partial x \partial z} - \frac{\partial}{\partial y}\left(-\frac{\partial \gamma_{xz}}{\partial y} + \frac{\partial \gamma_{xy}}{\partial z} + \frac{\partial \gamma_{yz}}{\partial x}\right) = 0 \\ \frac{\partial^2 \varepsilon_z}{\partial x \partial y} - \frac{\partial}{\partial z}\left(-\frac{\partial \gamma_{xy}}{\partial z} + \frac{\partial \gamma_{yz}}{\partial x} + \frac{\partial \gamma_{xz}}{\partial y}\right) = 0 \end{aligned}\right\} . \tag{8.7}$$

If each location of the structure is assumed to be plasticized (Eq. (8.1), $V_p = V$), the incremental initial strains according to Eq. (8.4) always meet the compatibility conditions. The proof in [7], however, only applies to the case that there are only "natural" boundary conditions, so no displacement boundary conditions. If there are also "essential" boundary conditions, the residual stresses from compatible eigenstates vanish only when the associated reaction forces are zero.

8.3 Examples of Asymptotic States

8.3.1 One Element with Biaxial Stress State

A single material element, which can be viewed as finite element with four nodes, is subjected to a homogeneous biaxial stress state. A tension stress is applied in the x-direction. A stress in the y-direction may also arise as a result of prevented lateral deformation (Fig. 8.2).

For a Poisson's ratio $\nu = 0$, however, no transverse stresses occur at elastic or fictitious elastic behavior:

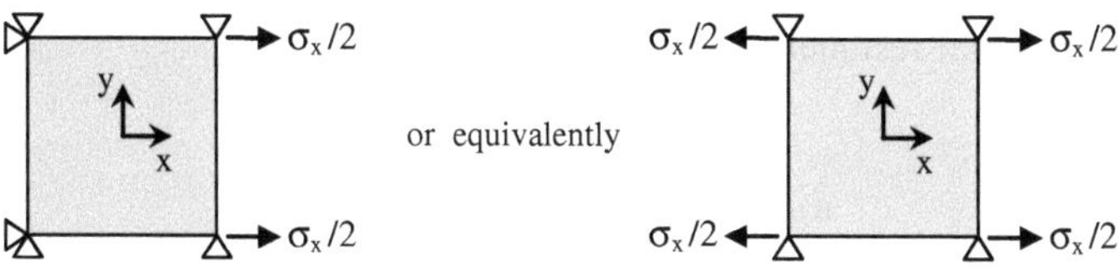

Fig. 8.2 One element under plane stress condition

$$\sigma_i^{\mathrm{fel}} = \begin{pmatrix} \sigma_x \\ 0 \\ 0 \end{pmatrix} \quad ; \quad \sigma_i^{\prime\mathrm{fel}} = \begin{pmatrix} \frac{2}{3}\sigma_x \\ -\frac{1}{3}\sigma_x \\ -\frac{1}{3}\sigma_x \end{pmatrix} \tag{8.8}$$

or in the asymptotic state

$$\mathrm{d}\sigma_i^{\mathrm{fel},\infty} = \begin{pmatrix} \mathrm{d}\sigma_x^\infty \\ 0 \\ 0 \end{pmatrix} \quad ; \quad \mathrm{d}\sigma_i^{\prime\mathrm{fel},\infty} = \begin{pmatrix} \frac{2}{3}\mathrm{d}\sigma_x^\infty \\ -\frac{1}{3}\mathrm{d}\sigma_x^\infty \\ -\frac{1}{3}\mathrm{d}\sigma_x^\infty \end{pmatrix}. \tag{8.9}$$

Because of

$$\nu^* = \frac{1}{2}\left(1 - \frac{E_t}{E}\right) \tag{8.10}$$

the incremental initial stresses arising from Eq. (8.5) are

$$\mathrm{d}\sigma_{0,i}^\infty = -3\frac{1 - \frac{E_t}{E}}{3 - \frac{E_t}{E}}\mathrm{d}\sigma_i^{\prime\mathrm{fel},\infty}. \tag{8.11}$$

An MEA performed with the "natural" boundary conditions in the x- and z-directions

$$\mathrm{d}\rho_x^\infty = \mathrm{d}\rho_z^\infty = 0 \tag{8.12}$$

and the "essential" boundary condition in the y-direction

$$\mathrm{d}\varepsilon_y^{*,\infty} = 0 \tag{8.13}$$

provides the incremental residual strains

$$\mathrm{d}\varepsilon_i^{*,\infty} = \frac{1}{4}\begin{pmatrix} 3 - 2\frac{E_t}{E} - \left(\frac{E_t}{E}\right)^2 \\ 0 \\ -3 + 4\frac{E_t}{E} - \left(\frac{E_t}{E}\right)^2 \end{pmatrix}\frac{\mathrm{d}\sigma_x^\infty}{E_t} \tag{8.14}$$

and the incremental residual stresses

$$\mathrm{d}\rho_i^\infty = \frac{1}{2}\left(1 - \frac{E_t}{E}\right)\begin{pmatrix} 0 \\ 1 \\ 0 \end{pmatrix} \mathrm{d}\sigma_x^\infty. \tag{8.15}$$

These are not zero, although the initial strains are distributed homogeneously, thus satisfying the compatibility conditions Eq. (8.7), but the reaction forces in the y-direction cannot be determined solely by equilibrium conditions and therefore do not vanish in the residual stress state. This is immediately obvious when the initial strains are imagined as free thermal expansion strains.

For the combined elastic-plastic state we get

$$\mathrm{d}\varepsilon_i^{\mathrm{el-pl},\infty} = \frac{1}{4}\begin{pmatrix} 3 + 2\frac{E_t}{E} - \left(\frac{E_t}{E}\right)^2 \\ 0 \\ -3 + 4\frac{E_t}{E} - \left(\frac{E_t}{E}\right)^2 \end{pmatrix} \frac{\mathrm{d}\sigma_x^\infty}{E_t} \tag{8.16}$$

$$\mathrm{d}\sigma_i^\infty = \begin{pmatrix} 1 \\ \frac{1}{2}\left(1 - \frac{E_t}{E}\right) \\ 0 \end{pmatrix} \mathrm{d}\sigma_x^\infty \quad ; \quad \mathrm{d}\sigma_i'^\infty = \frac{1}{6}\begin{pmatrix} 3 + \frac{E_t}{E} \\ -2\frac{E_t}{E} \\ -3 + \frac{E_t}{E} \end{pmatrix} \mathrm{d}\sigma_x^\infty. \tag{8.17}$$

The plastic strain increment in the asymptotic state acts according to Eq. (8.6) in the direction of the incremental deviatoric stress vector:

$$\mathrm{d}\varepsilon_i^{\mathrm{pl},\infty} = \frac{1}{4}\left(1 - \frac{E_t}{E}\right)\begin{pmatrix} +3 + \frac{E_t}{E} \\ -2\frac{E_t}{E} \\ -3 + \frac{E_t}{E} \end{pmatrix} \frac{\mathrm{d}\sigma_x^\infty}{E_t}, \tag{8.18}$$

so in the case of the transition $E_t \rightarrow 0$,

$$\mathrm{d}\varepsilon_i^{\mathrm{pl},\infty} \propto \begin{pmatrix} +1 \\ 0 \\ -1 \end{pmatrix}, \tag{8.19}$$

which immediately seems plausible because of vanishing elastic strain increments in the asymptotic state because of the incompressibility of plastic deformations. This applies irrespective of the choice of Poisson's ratio ($\nu = 0$) made above, so the plastic strain increments always have the same direction for $E_t \rightarrow 0$ regardless of ν.

In this example, no local redistribution of the stresses occurs because of plasticizing but instead because of directional redistribution, because the ratio of the strain components to each other or the ratio of the stress components to each other is changed compared to purely elastic behavior.

8.3.2 Bending Beams and Truss Members

Directional stress redistribution is ruled out in bending beams and truss rods because of the uniaxial stress states. However local stress redistributions occur under certain circumstances.

As in the analysis of the asymptotic state, the entire structure is considered as plasticized and the stiffness distribution in the modified elastic analysis is the same as in the fictitious elastic analysis, at least as long as E and E_t do not vary spatially. Because of the uniaxial stress state, and because of the proportionality between the incremental initial stresses $\mathrm{d}\sigma_{i,0}^{\infty}$ and the fictitious elastic stresses, non-vanishing incremental residual strains $\mathrm{d}\varepsilon_i^{*,\infty}$ are obtained by the MEA, which are identical with the incremental initial strains, but the incremental residual stresses $\mathrm{d}\rho_i^{\infty}$ are zero, because the incremental initial strains can freely translate into displacements without restraints. The incremental stress vector $\mathrm{d}\sigma_i^{\infty}$ is therefore identical with the respective incremental fictitious elastic stress vector $\mathrm{d}\sigma_i^{\mathrm{fel},\infty}$ at each location of the structure. If the axis of the beam or truss extends in the x-direction, the direction of the plastic strain increment is given by the vector

$$\mathrm{d}\varepsilon_{i(x,y,z)}^{\mathrm{pl},\infty} \propto \mathrm{d}\sigma_{i(x,y,z)}^{\prime\infty} \left[= \mathrm{d}\sigma_{i(x,y,z)}^{\prime\mathrm{fel},\infty} \right] \propto \sigma_{x(x,y,z)}^{\mathrm{fel}} \begin{pmatrix} +1 \\ -0.5 \\ -0.5 \end{pmatrix}. \tag{8.20}$$

As already mentioned in Sect. 8.2, the proportionality factor between $\mathrm{d}\sigma_{x(x,y,z)}^{\infty}$ and $\sigma_{x(x,y,z)}^{\mathrm{fel}}$ is always positive according to the procedure discussed here, but may be negative in reality, so Eq. (8.20) represents only an initial estimate.

8.3.3 Thick-Walled Cylinder Under Internal Pressure

For a thick-walled cylinder under internal pressure, as already considered in Sect. 3.6.4, the fictitious elastic stress increments in the asymptotic state are

$$\mathrm{d}\sigma_i^{\mathrm{fel},\infty} = \mathrm{d}p \frac{r_i^2}{r_a^2 - r_i^2} \begin{pmatrix} 1 - \left(\frac{r_a}{x}\right)^2 \\ 1 \\ 1 + \left(\frac{r_a}{x}\right)^2 \end{pmatrix} \quad ; \quad \mathrm{d}\sigma_i^{\prime\mathrm{fel},\infty} = \mathrm{d}p \frac{r_i^2 r_a^2}{\left(r_a^2 - r_i^2\right) x^2} \begin{pmatrix} -1 \\ 0 \\ +1 \end{pmatrix}. \tag{8.21}$$

Because these are independent of the Poisson's ratio, and as no stiffness differences are present in the MEA, because the same modified Young's modulus $E^* = E_t$ is attributed to any location in the whole structure, the increments of the initial strains determined according to Eq. (8.4) can adjust freely. As result of the MEA, the residual stress increments are zero, because the incremental initial strain field arising from Eq. (8.4) satisfies the compatibility conditions Eq. (8.7) and there exist only "natural" boundary conditions.

Similar to the uniaxial stress state, the plastic deformation increments in the asymptotic state also act here at any location x of the structure in the direction of the respective deviatoric fictitious elastic stresses, regardless of the hardening modulus E_t:

$$d\varepsilon_{i(x)}^{\mathrm{pl},\infty} \propto d\sigma_{i(x)}^{\prime\infty} \propto dp\frac{r_i^2 r_a^2}{\left(r_a^2 - r_i^2\right)x^2}\begin{pmatrix}-1\\0\\+1\end{pmatrix}. \tag{8.22}$$

8.3.4 Disc with a Hole

For the disc already presented in Sect. 3.6.3, two variants are examined, which differ in terms of the boundary conditions along the longitudinal edges.

8.3.4.1 Free Longitudinal Edges

According to Eq. (8.5), the incremental initial stresses are proportional to the fictitious elastic calculated deviatoric stresses and are shown in Fig. 8.3 for the structural system given in Fig. 3.16 for the increment of the distributed tensile load $dp = 100$.

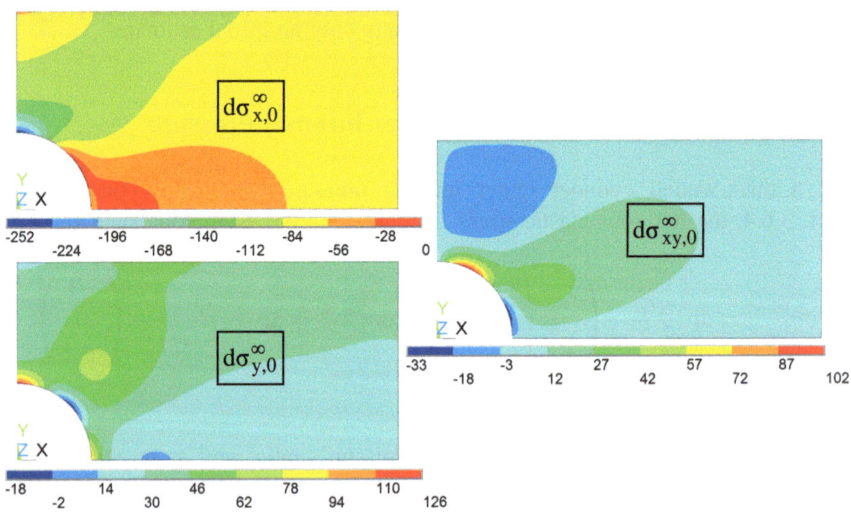

Fig. 8.3 Increments of the initial stresses in the asymptotic state for the load increment $dp = 100$, $E_t/E = 10^{-6}$, $\nu = 0.3$

Residual stresses are not induced in the MEA of the asymptotic state, because the associated initial strain field satisfies the compatibility conditions Eq. (8.7) and only "natural" boundary conditions exist.

Thus, the deviatoric stress increments and the plastic deformation increments in the asymptotic state have the same direction as the fictitious elastic deviatoric stresses.

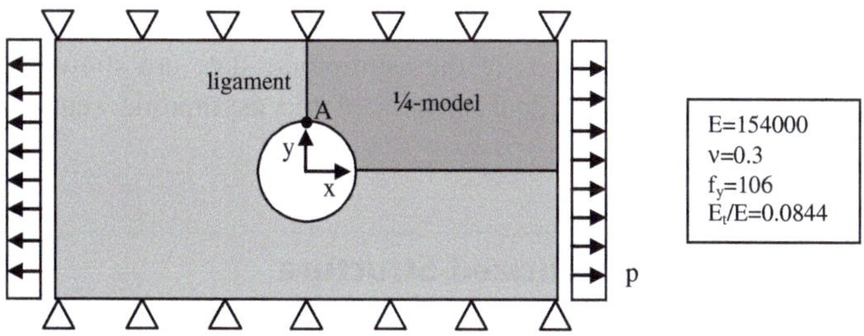

Fig. 8.4 Disc with a hole and supported longitudinal edges

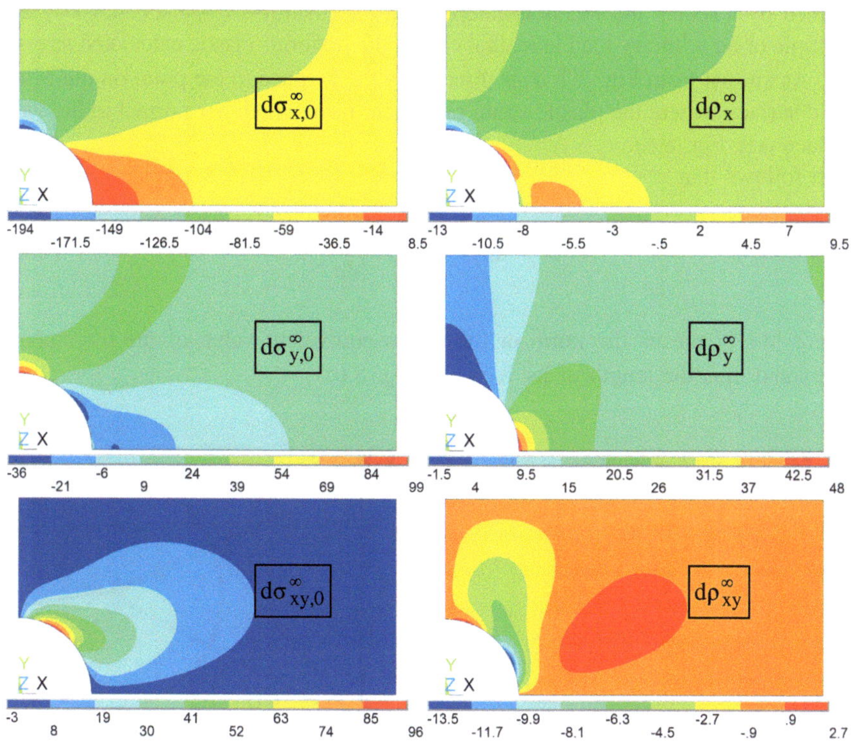

Fig. 8.5 Increments of the initial stresses in the asymptotic state for the load increment $dp = 100$, $E_t/E = 10^{-6}$, $\nu = 0.3$; as well as the incremental residual stresses in the asymptotic state

8.3.4.2 Supported Longitudinal Edges

Even if the edge y = const. can move only parallel because of a stiffening of the edge, such as in an I-profile with hole in the web and rigid flanges, only “natural” boundary conditions are present, so no constraints occur as a result of the MEA of the asymptotic state.

However, this does not apply if a shift of the longitudinal edges in the transverse direction is prevented, as shown in Fig. 8.4, because in such a case the reaction forces corresponding to the “essential” boundary conditions are not null and so constraints in the MEA of the asymptotic state arise.

The incremental initial stresses in the asymptotic state are shown in Fig. 8.5, along with the incremental residual stresses of the asymptotic state called forth through them.

8.4 Completely Plasticized Structure

The extreme direction of the plastic strain increment $(\mathrm{d}\varepsilon_i^{\mathrm{pl},\infty} \propto \mathrm{d}\sigma_i'^{\infty})$ determined in accordance with Sect. 8.2 for an infinitely high load level is used to estimate the TIV Y_i for a finite level of loading. The load level is initially chosen arbitrarily and defined by a scalar and location-independent load factor α $(-\infty < \alpha < +\infty)$ as a multiple of an arbitrary load level that causes the fictitious elastic calculated stresses $\bar{\sigma}_i^{\mathrm{fel}}$. As known from Fig. 3.1, Y_i is directed in the Y-space to the point on the Mises yield surface, where the plastic strain increment forms the inner normal to the yield surface (see Fig. 8.6).

It follows that

$$Y_i = \alpha \bar{\sigma}_i'^{\mathrm{fel}} - \mathrm{d}\sigma_i'^{\infty} \frac{f_y}{\left|\mathrm{d}\sigma_i'^{\infty}\right|}, \tag{8.23}$$

where $\left|\mathrm{d}\sigma_i'^{\infty}\right|$ means the norm or the Mises equivalent value of the incremental stress and thus the length of its vector in Fig. 8.6:

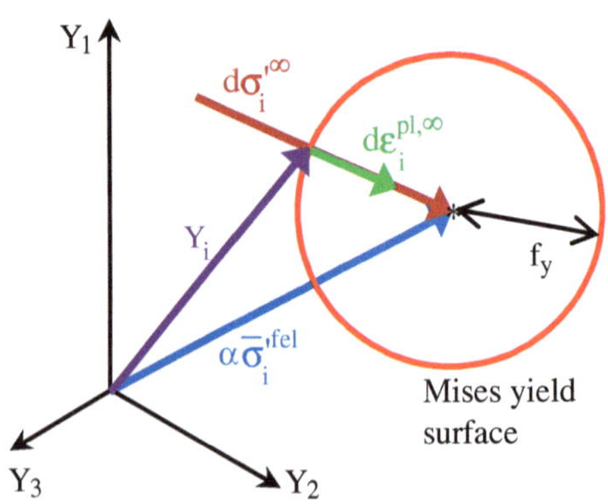

Fig. 8.6 TIV-space at finite load level (load factor α)

$$\left|\mathrm{d}\sigma_i^{\prime\infty}\right| = \sqrt{\frac{3}{2}}\sqrt{\left(\mathrm{d}\sigma_x^{\prime\infty}\right)^2 + \left(\mathrm{d}\sigma_y^{\prime\infty}\right)^2 + \left(\mathrm{d}\sigma_z^{\prime\infty}\right)^2 + 2\left[\left(\mathrm{d}\sigma_{xy}^{\infty}\right)^2 + \left(\mathrm{d}\sigma_{xz}^{\infty}\right)^2 + \left(\mathrm{d}\sigma_{yz}^{\infty}\right)^2\right]}. \tag{8.24}$$

In the special case of a uniaxial stress state $\mathrm{d}\sigma^{\infty}(=\frac{3}{2}\mathrm{d}\sigma_x^{\prime\infty})$ this becomes

$$\left|\mathrm{d}\sigma_i^{\prime\infty}\right| = \mathrm{abs}(\mathrm{d}\sigma^{\infty}), \tag{8.25}$$

so that Eq. (8.23) can be written after making use of $Y=\frac{3}{2}Y_x$ as

$$Y = \alpha\bar{\sigma}^{\mathrm{fel}} - f_y \mathrm{sgn}(\mathrm{d}\sigma^{\infty}). \tag{8.26}$$

Assuming here again that the entire structure is completely plasticized (see Eq. (8.1)), the initial strains can be determined for any location in the structure according to Eq. (3.14) or the initial stresses according to Eq. (3.23) and an MEA can be performed as described in Sect. 3.1.3.

Because of the assumption $V_p = V$, the structural stiffness is distributed in the modified elastic analysis in the same way as in the fictitious elastic analysis. The modified load $\sigma_{i,0}$ is proportional to Y_i and consists of two additive portions. One of them is independent of the actual load level, whereas the other depends linearly on the load factor α. Therefore, the result of the MEA, in the form of residual stresses and residual strains and displacements etc., also consists of a load-independent term and a term proportional to the actual load level. So ultimately there is a linear relationship between residual stresses and the load factor. The strains, the plastic strains, the stresses, etc., also depend linearly on the load factor as a result from the linear superposition according to the defining equations already known from Eqs. (3.24) and (3.25):

$$\sigma_i = \sigma_i^{\mathrm{fel}} + \rho_i = \alpha\bar{\sigma}_i^{\mathrm{fel}} + \rho_i \tag{8.27}$$

$$\varepsilon_i^{\mathrm{el-pl}} = \varepsilon_i^{\mathrm{fel}} + \varepsilon_i^{*}. \tag{8.28}$$

This is exemplarily shown for the plastic strain share at any point of a structure in Fig. 8.7.

Both the ordinate at $\varepsilon^{\mathrm{pl}} = 0$, and the slope of the straight line represented in Fig. 8.7 depend on the hardening parameter E_t. In the limit of $E_t \to 0$ the slope must disappear so that the load factor $\alpha = \alpha_L$ at $\varepsilon^{\mathrm{pl}} = 0$ is the limit load factor or at least a first estimate of the load factor, as is discussed later. Thus, the requirement to determine the limit load factor is[1]

[1]Zarka et al. used a different criterion in [3], which leads to less good results according to the experience of the present author.

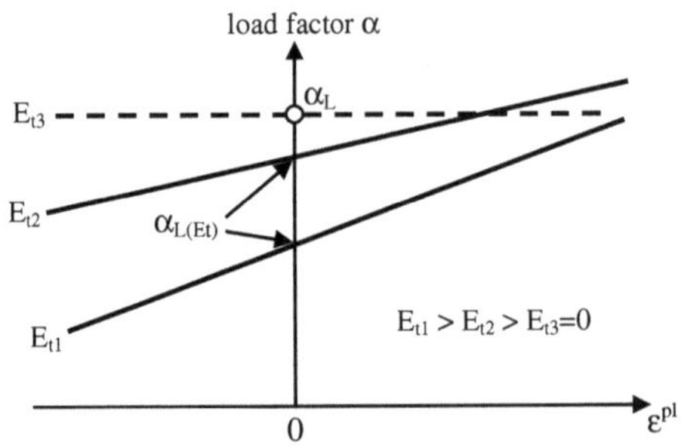

Fig. 8.7 Linear dependence of the structural response of the load level (load factor α) for different hardening modules

$$\varepsilon^{pl}_{(E_t=0)} \overset{!}{=} 0 \quad \rightarrow \quad \alpha_L. \tag{8.29}$$

In the case where $E_t/E = 0$ this requirement is equivalent to the requirements

$$\varepsilon^{*}_{(E_t=0)} \overset{!}{=} 0 \quad \text{or} \quad \varepsilon^{el-pl}_{(E_t=0)} \overset{!}{=} 0. \tag{8.30}$$

The fact that the load factor depends on E_t and thus on the plastic material behavior even in the case of $\varepsilon^{pl} = 0$ may be surprising. However, the condition $\varepsilon^{pl} = 0$ cannot be associated in the context described here simply with purely elastic behavior, because plastic straining was assumed at each location of the structure ($V_p = V$). Instead, the yield stress f_y enters only in the determination of initial stresses. Accordingly, plastic strains, stresses, and residual stresses are obtained formally even for the load factor $\alpha = 0$, i.e., for the unloaded structure!

If at any value of E_t/E an MEA is performed for each of two different arbitrarily chosen load factors, $\alpha_{L(Et)}$ can be determined by linear inter- or extrapolation to $\varepsilon^{pl} = 0$ (if there occur displacements of astronomical scale as a result of the freely chosen load factors, this does not matter). $\alpha_{L(Et)}$ can be greater or less than α_L.

However, it is best if a value of E_t/E is chosen that is only slightly greater than zero. E_t/E cannot be chosen as equal to zero because the structure then has no stiffness in the MEA. A too small value of E_t, many orders of magnitude less than E, e.g., $E_t/E = 10^{-8}$, however, can cause problems in an application in a finite element environment, although there are no stiffness differences in the structure, which could lead to an ill-conditioned system of equations. In that regard, a small value of the modified Young's modulus $E^* = E_t$ is unproblematic. However, the modified Poisson's ratio ν^* is then very close to 0.5, which corresponds to an incompressible elastic material, and requires special element formulations. For this purpose, mixed u-P elements are available, with the degrees of freedom not only consisting of displacements, but also of the hydrostatic pressure.

For practical applications, $E_t/E = 10^{-3}$–10^{-5} (instead of 0) usually suffices for a good approximation:

$$\alpha_L \approx \alpha_{L(Et)}. \tag{8.31}$$

Thus, α_L can be determined by inter- or extrapolation after two MEAs performed with different load factors α_1 and α_2, in addition to the one MEA for determining the asymptotic state:

$$\alpha_L = \frac{\alpha_1 \varepsilon^{\text{pl}}_{(\alpha_2)} - \alpha_2 \varepsilon^{\text{pl}}_{(\alpha_1)}}{\varepsilon^{\text{pl}}_{(\alpha_2)} - \varepsilon^{\text{pl}}_{(\alpha_1)}}, \tag{8.32}$$

or after reformulation

$$\alpha_L = \frac{\alpha_1 \rho'_{(\alpha_2)} - \alpha_2 \rho'_{(\alpha_1)} - (\alpha_1 - \alpha_2) f_y \frac{\mathrm{d}\sigma'^{\infty}}{\mathrm{d}\sigma_v^{\infty}}}{\rho'_{(\alpha_2)} - \rho'_{(\alpha_1)} - (\alpha_1 - \alpha_2)\bar{\sigma}'^{\text{fel}}}. \tag{8.33}$$

Equations (8.32) and (8.33) apply in principle to each component of the plastic strain and residual stress vector, and Eq. (8.32) also for the equivalent plastic strain. However, a different result may be obtained in general for each component. The component leading to a kinematic chain arising in the limit load condition prevails.

In cases where the incremental initial strains Eq. (8.4) or incremental initial stresses Eq. (8.5), which are based on the TIV-increments $\mathrm{d}Y_i^{\infty}$ according to Eq. (8.3), do not cause residual stress increments $\mathrm{d}\rho_i^{\infty}$ (as described at the end of Sect. 8.2), the fraction $\alpha\bar{\sigma}_i'^{\text{fel}}$ in Eq. (8.23) also does not cause residual stresses in the structure assumed as completely plasticized ($V_p = V$). The residual stresses depend then only on the second term in Eq. (8.23) and are thus independent of the load factor α. In these cases, the calculation of residual stresses with only one load factor, which may be chosen arbitrarily and thus may also be zero, is sufficient for determining the plastic limit load. Equation (8.33) is then reduced to

$$\alpha_L = \frac{f_y \frac{\mathrm{d}\sigma'^{\infty}}{\mathrm{d}\sigma_v^{\infty}} - \rho'}{\bar{\sigma}'^{\text{fel}}}. \tag{8.34}$$

If, in addition, the stress state is uniaxial, we get

$$\alpha_L = \frac{f_y \operatorname{sgn}(\mathrm{d}\sigma^{\infty}) - \rho}{\bar{\sigma}^{\text{fel}}}. \tag{8.35}$$

If necessary, an extrapolation to $E_t = 0$ can be made on the basis of two different $\alpha_{L(Et)}$-values obtained for two different small non-zero values of E_t, because $\alpha_{L(Et)}$ always depends approximately linearly on E_t in the vicinity of $E_t \approx 0$. However, this doubles the computational effort because now a total of six or four instead of three or two MEAs are required.

8.5 Examples of Plastic Limit Load for Completely Plasticized Structures

8.5.1 One Element with Biaxial Stress State

For the asymptotic state treated in Sect. 8.3.1 (with $\nu = 0$) the norm of the incremental deviatoric stress calculated according to Eq. (8.24) results in

$$\left|\mathrm{d}\sigma_i'^{\infty}\right| = \frac{1}{2}\sqrt{3+\left(\frac{E_t}{E}\right)^2}\,\mathrm{d}\sigma_x \tag{8.36}$$

and thus the TIV for a finite load level according to Eq. (8.23) in

$$Y_i = \alpha\bar{\sigma}_x\begin{pmatrix}\frac{2}{3}\\ -\frac{1}{3}\\ -\frac{1}{3}\end{pmatrix} - \frac{f_y}{\sqrt{3+\left(\frac{E_t}{E}\right)^2}}\begin{pmatrix}1+\frac{1}{3}\frac{E_t}{E}\\ -\frac{2}{3}\frac{E_t}{E}\\ -1+\frac{1}{3}\frac{E_t}{E}\end{pmatrix}. \tag{8.37}$$

An MEA for the initial stresses determined according to Eq. (3.23)

$$\sigma_{i,0} = \frac{1-\frac{E_t}{E}}{1+\nu^*}\left[-\alpha\bar{\sigma}_x\begin{pmatrix}1\\ -\frac{1}{2}\\ -\frac{1}{2}\end{pmatrix} + \frac{f_y}{\sqrt{3+\left(\frac{E_t}{E}\right)^2}}\begin{pmatrix}\frac{3}{2}+\frac{1}{2}\frac{E_t}{E}\\ -\frac{E_t}{E}\\ -\frac{3}{2}+\frac{1}{2}\frac{E_t}{E}\end{pmatrix}\right] \tag{8.38}$$

with the modified elastic material parameters E^* and ν^* according to Eqs. (3.15) and (3.16) and with the boundary conditions

$$\rho_x = \rho_z = 0 \quad ; \quad \varepsilon_y^* = 0 \tag{8.39}$$

provides the residual strain in the x-direction, which is identical to the plastic strain component in the x-direction here, because the elastic strain share of the total elastic-plastic strain is identical to the fictitious elastic strain in the x-direction because of the stress-controlled load in the x-direction:

$$\varepsilon_x^* = \varepsilon_x^{\mathrm{pl}} = \frac{1-\frac{E_t}{E}}{E_t}\left[\frac{1}{4}\alpha\bar{\sigma}_x\left(3+\frac{E_t}{E}\right) - \frac{1}{2}f_y\sqrt{3+\left(\frac{E_t}{E}\right)^2}\right] \tag{8.40}$$

or the elastic-plastic strain

$$\varepsilon_x^{\mathrm{el-pl}} = \varepsilon_x^{\mathrm{pl}} + \varepsilon_x^{\mathrm{el}} = \frac{1}{4}\frac{\alpha\bar{\sigma}_x}{E_t}\left[3+2\frac{E_t}{E}-\left(\frac{E_t}{E}\right)^2\right] - \frac{1}{2}\left(1-\frac{E_t}{E}\right)\frac{f_y}{E_t}\sqrt{3+\left(\frac{E_t}{E}\right)^2}, \tag{8.41}$$

and the residual stress in the y-direction

$$\rho_y = \left(1 - \frac{E_t}{E}\right)\left[\frac{1}{2}\alpha\bar{\sigma}_x - f_y \frac{\frac{E_t}{E}}{\sqrt{3 + \left(\frac{E_t}{E}\right)^2}}\right]. \tag{8.42}$$

Application of Eq. (8.32) or Eq. (8.33) for the x-component, or direct insertion of the requirement $\varepsilon_x^{\mathrm{pl}} = 0$ in Eq. (8.40), leads to

$$\alpha_{L(Et)}\bar{\sigma}_x = 2f_y \frac{\sqrt{3 + \left(\frac{E_t}{E}\right)^2}}{3 + \left(\frac{E_t}{E}\right)}. \tag{8.43}$$

If the elastic limit load is chosen as the reference level for the load, i.e.,

$$\bar{\sigma}_x = f_y, \tag{8.44}$$

then $\alpha_{L(Et)}$ specifies the load factor based on the elastic limit load:

$$\alpha_{L(Et)} = 2\frac{\sqrt{3 + \left(\frac{E_t}{E}\right)^2}}{3 + \left(\frac{E_t}{E}\right)}. \tag{8.45}$$

Figure 8.8 illustrates the dependency of this load factor on the hardening modulus.

With decreasing E_t/E, $\alpha_{L(Et)}$ asymptotically reaches the exact limit load factor for $E_t/E \rightarrow 0$:

$$\alpha_L = \frac{2}{\sqrt{3}} = 1.1547\ldots, \tag{8.46}$$

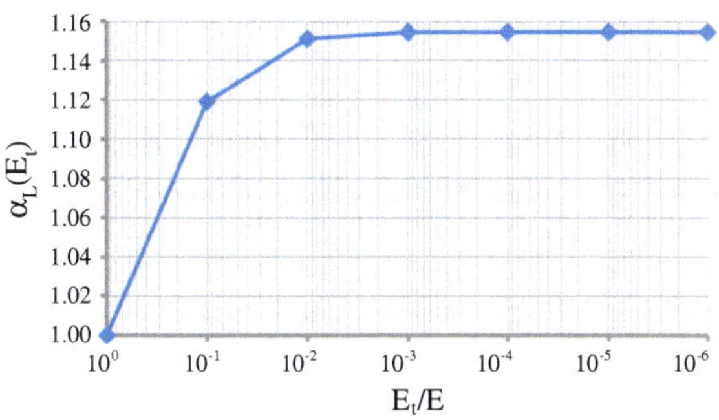

Fig. 8.8 Dependency of the load factor $\alpha_{L(Et)}$ on the hardening modulus E_t/E

already reached in good approximation ($\alpha_{L(Et)} = 1.1543\ldots$) at $E_t/E = 10^{-3}$, and achieved exactly within six digits at $E_t/E = 10^{-5}$.

The theoretically exact plastic limit load could be obtained in this example with relatively little computational effort to almost any degree of accuracy, namely with two linear analyses: one MEA for the asymptotic state and another MEA for a finite load level. This is because the structure is completely plasticized when reaching the plastic limit load, so that the assumption $V_p = V$ made in all MEAs is precisely met here.

8.5.2 Bending Beam

Because of the uniaxial stress state, the initial stress of a bending beam in the x-y-plane with bending about the z-axis is

$$\sigma_{0(x,y)} = -\left(1 - \frac{E_t}{E}\right)\left[\alpha\bar{\sigma}^{\text{fel}}_{(x,y)} - \text{sgn}\left(\bar{\sigma}^{\text{fel}}_{(x,y)}\right)f_y\right]. \tag{8.47}$$

Accordingly, the initial stress consists of two parts, the first of which is variable in terms of magnitude and sign along the y-axis and may also be variable along the x-axis, and the second part is at least sectional constant in longitudinal direction and distributed constantly each above and below the neutral fiber.

The initial stresses of Eq. (8.47) cause an initial bending moment in each section

$$M_{0(x)} = -\left(1 - \frac{E_t}{E}\right)\left[\alpha\bar{M}^{\text{fel}}_{(x)} - \text{sgn}\left(\bar{M}^{\text{fel}}_{(x)}\right)M_{\text{pl}}\right], \tag{8.48}$$

where M_{pl} is the fully plastic moment of the cross-section for a non-hardening material ($E_t = 0$), i.e., in the case of a rectangular cross-section

$$M_{\text{pl}} = \frac{1}{4}f_y bh^2. \tag{8.49}$$

The share of the initial stress or the initial bending moment which is proportional to $\bar{\sigma}^{\text{fel}}_{(x,y)}$ or $\bar{M}^{\text{fel}}_{(x)}$ does not generate constraints in the MEA, and each second share can generate bending strains without constraints only at statically determinate structures, but not at statically indeterminate ones. Moreover, the share of the initial stresses with the magnitude f_y distributed blockwise across the height of the beam from the sgn-function cannot ensure that a section remains plane, so that to fulfill the Bernoulli hypothesis even more residual stresses are produced.

Because the term $\alpha\bar{\sigma}^{\text{fel}}_{(x,y)}$ does not cause residual stresses, they are independent of the load factor α. The plastic limit load factor α_L based on the requirement $\varepsilon^{\text{pl}} = 0$ therefore reduces for uniaxial stress states to

$$\alpha_{L(x,y)} = \frac{\operatorname{sgn}\left(\bar{\sigma}^{\text{fel}}_{(x,y)}\right) f_y - \rho_{(x,y)}}{\bar{\sigma}^{\text{fel}}_{(x,y)}}. \tag{8.50}$$

In the case of $E_t = 0$, the numerator corresponds to the linearized form of the initial stresses distributed blockwise across the height of the beam (Fig. 8.9) in statically determinate structures. In statically indeterminate structures, the numerator also contains bending residual stresses as a result of constraints. Anyway, the numerator as well as the denominator are proportional to the coordinate y, making the limit load factor at $E_t = 0$ constant across the beam height h. If the limit load factor α_L had been determined on the basis of the requirement $\varepsilon^{\text{el-pl}} = 0$ instead of $\varepsilon^{\text{pl}} = 0$, then it would always be constant over the beam height regardless of E_t.

In the case of a line model, Eq. (8.50) can also be expressed by fictitious elastic and residual bending moments:

$$\alpha_{L(x)} = \frac{\operatorname{sgn}\left(\overline{M}^{\text{fel}}_{(x)}\right) M_{\text{pl}} - M^{\rho}_{(x)}}{\overline{M}^{\text{fel}}_{(x)}}. \tag{8.51}$$

Because α_L depends on E_t via the residual stresses, a numerically very small value of E_t should be used when determining the initial stresses σ_0 and the modified Young's modulus E^* for performing the MEA.

Thus, a total number of a single MEA with arbitrarily chosen load factor, for example, with $\alpha = 0$, is sufficient here to determine the plastic limit load, or at least its first estimate, because the asymptotic state could be determined without an MEA.

8.5.2.1 Example for Statically Determinate Bearing

A beam with solid rectangular cross-section is subjected to a uniformly distributed load q. The dimensions are chosen so that shear stresses can be neglected ($L >> b, h$) (Fig. 8.10).

As the initial bending moment can freely transform to bending strains without constraints because of the statically determinate bearing, the residual stresses arise only from the nonlinear share of the initial stress distribution across the height of the beam:

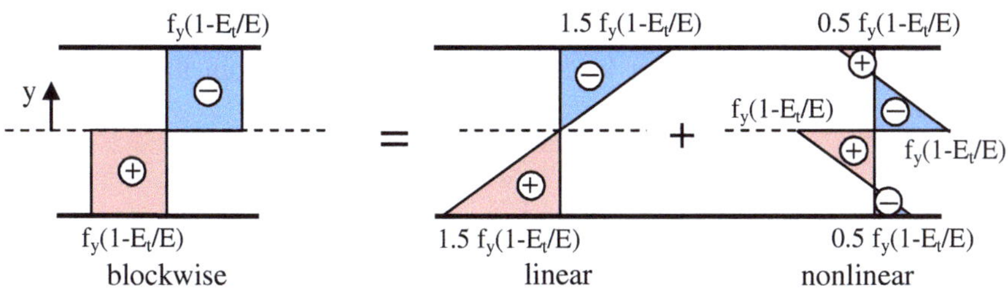

Fig. 8.9 Decomposition of initial stresses distributed blockwise across the beam height

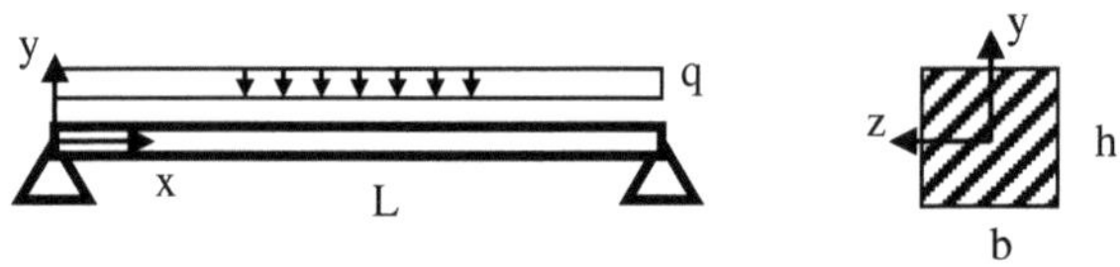

Fig. 8.10 Statically determinate bending beam with solid rectangular cross-section

$$\rho_{(y)} = \left(1 - \frac{E_t}{E}\right) f_y \left[\frac{3}{2}\frac{y}{h/2}\,\mathrm{sgn}\left(\bar{M}^{\mathrm{fel}}_{(x)}\right) + \mathrm{sgn}\left(\bar{\sigma}^{\mathrm{fel}}_{(x,y)}\right)\right]. \tag{8.52}$$

The residual stresses are thus independent of the load factor α, i.e., they exist even in the unloaded condition, $\alpha = 0$, and are constant along the longitudinal direction, because the fictitious elastic bending moment everywhere has the same sign, and the sign of $\bar{\sigma}^{\mathrm{fel}}_{(x,y)}$ depends on y but not on x.

The residual strains result from the modified elastic material law for a uniaxial stress state

$$\varepsilon^*_{(x,y)} = \frac{\rho_{(y)}}{E_t} + \varepsilon_{0(x,y)} = \frac{\rho_{(y)}}{E_t} - \frac{\sigma_{0(x,y)}}{E_t} \tag{8.53}$$

and evaluate with Eqs. (8.47) and (8.52) to

$$\varepsilon^*_{(x,y)} = \frac{1 - \frac{E_t}{E}}{E_t}\left[\frac{3}{2}\frac{y}{h/2} f_y + \alpha\bar{\sigma}^{\mathrm{fel}}_{(x,y)}\right]. \tag{8.54}$$

From these, we can determine the elastic-plastic or the plastic strains:

$$\varepsilon^{\mathrm{el-pl}}_{(x,y)} = \varepsilon^*_{(x,y)} + \alpha\bar{\varepsilon}^{\mathrm{fel}}_{(x,y)} = \frac{1}{E_t}\left[\left(1 - \frac{E_t}{E}\right)\frac{3}{2}\frac{y}{h/2} f_y + \alpha\bar{\sigma}^{\mathrm{fel}}_{(x,y)}\right] \tag{8.55}$$

$$\varepsilon^{\mathrm{pl}}_{(x,y)} = \varepsilon^{\mathrm{el-pl}}_{(x,y)} - \varepsilon^{\mathrm{el}}_{(x,y)} = \varepsilon^*_{(x,y)} - \frac{\rho_{(y)}}{E} \tag{8.56}$$

$$= \frac{1 - \frac{E_t}{E}}{E_t}\left\{\left[\left(1 - \frac{E_t}{E}\right)\frac{3}{2}\frac{y}{h/2} + \frac{E_t}{E}\,\mathrm{sgn}\left(\bar{\sigma}^{\mathrm{fel}}_{(x,y)}\right)\right] f_y + \alpha\bar{\sigma}^{\mathrm{fel}}_{(x,y)}\right\}. \tag{8.57}$$

The limit load or the limit load factor $\alpha_{L(Et)}$ resulting from the requirement $\varepsilon^{\mathrm{pl}} = 0$ is

$$\alpha_{L(Et)} = -\left[\left(1 - \frac{E_t}{E}\right)\frac{3}{2}\frac{y}{h/2} - \frac{E_t}{E}\,\mathrm{sgn}\left(\bar{\sigma}^{\mathrm{fel}}_{(x,y)}\right)\right]\frac{f_y}{\bar{\sigma}^{\mathrm{fel}}_{(x,y)}}. \tag{8.58}$$

After transition $E_t \rightarrow 0$ (as in Eq. (8.29)), α_L only depends on the x-coordinate:

$$\alpha_L = -\frac{3}{2}\frac{f_y}{\bar{\sigma}^{\text{fel}}_{(x,y=+h/2)}}. \tag{8.59}$$

The same result is also obtained for the criterion Eq. (8.30).

In a cross-section x = const., in which the elastic limit state is reached, i.e., the fictitious elastic stress at the top of the beam takes on the value of the negative yield stress, the limit load factor thus turns out to meet the plastic support factor for a rectangular cross-section under bending:

$$\min \alpha_L = \frac{3}{2} \tag{8.60}$$

and thus provides the exact limit load factor for the structure. For all other sections of this beam, the α_L-values are larger, so each represents an upper bound. Thus, the smallest value of all upper bounds here is the actual plastic limit load.

Now, it is already obvious by intuition in this example where the location of the most highly stressed cross-section is and that, upon reaching α_L in this cross-section, the plastic limit load of the entire system has already been reached. However, this is not readily detectable from the equations presented here, because, for an arbitrary load factor α, the largest stresses and the largest strains do not necessarily occur at midspan!

As an example, initial stresses and elastic-plastic stresses of a statically determinate beam with solid rectangular cross-section are shown in Fig. 8.11

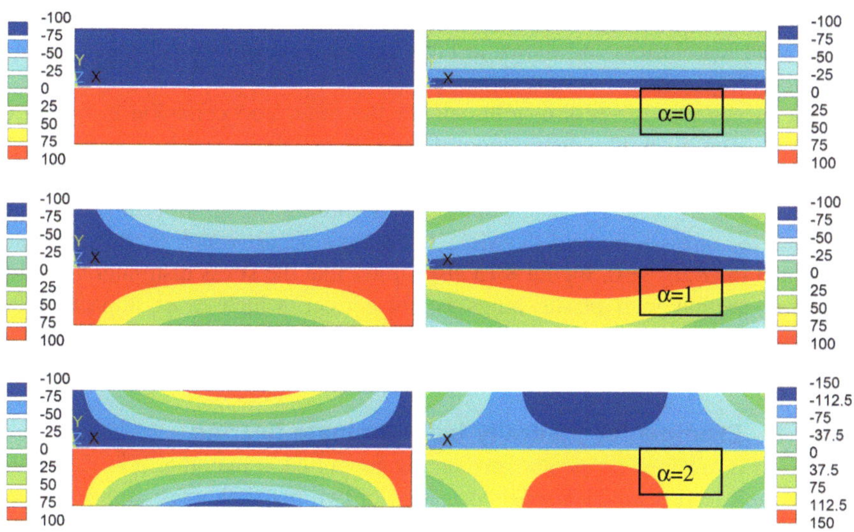

Fig. 8.11 Initial stresses (*left*) and elastic-plastic stresses (*right*) for various load factors α, $E_t/E \rightarrow 0$, f_y = 100

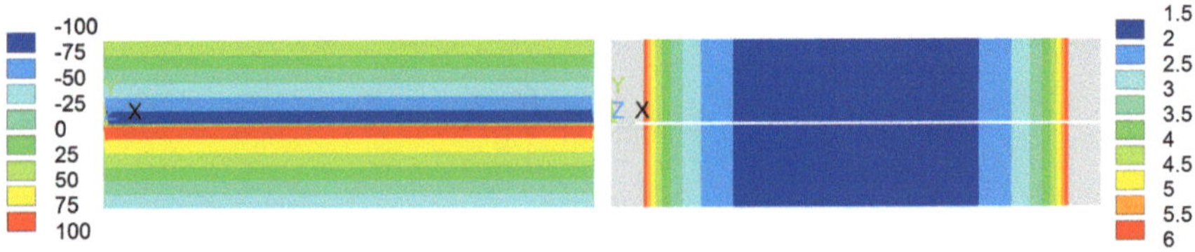

Fig. 8.12 Residual stresses (*left*) (at $E_t/E \to 0, f_y = 100$) and distribution of the limit load factors (*right*) in a simply supported beam

Fig. 8.13 Elastic-plastic stress distribution of the simply supported beam for $E_t/E \to 0, f_y = 100$, $\alpha = \min \alpha_L = 1.5$

(exaggerated aspect ratio) for the load factors $\alpha = 0/1/2$ at $E_t \to 0$ ($E_t/E = 10^{-7}$). Basic, and thus defining the load factor $\alpha = 1$, is a load level corresponding to the elastic limit load. In the vicinity of the neutral fiber the representation is omitted because of numerical inaccuracies at very small stress levels caused by the finite element model in the neutral fiber and at the bearings.

The residual stresses and the distribution of the limit load factor are each the same at all load factors and shown in Fig. 8.12. Limit load factors above 6 are not depicted (gray color).

The elastic-plastic stress distribution for the limit load condition is shown in Fig. 8.13. The block-like distribution of the stresses above and below the neutral fiber in the midspan with the absolute value of $f_y = 100$ associated with this particular load level is clearly visible, although, according to Fig. 8.11, at load factors $\alpha > \min \alpha_L$, stresses beyond the yield stress may occur, which is, however, not permitted for $E_t/E \to 0$. Figure 8.13 also shows that the stress distribution outside the cross-section with plastic hinge is not correct, which is an impact of the assumption $V_p = V$, which is not true, although still the exact plastic limit load was found in this case.

8.5.2.2 Example for Statically Indeterminate Bearing

If a beam is clamped at one end and simply supported at the other (Fig. 8.14), the plastic limit load is achieved after formation of two plastic hinges: at the clamped end and in the span. However, the position of the plastic hinge in the span is not known a priori. Because of local stress redistribution, it is in any case a different

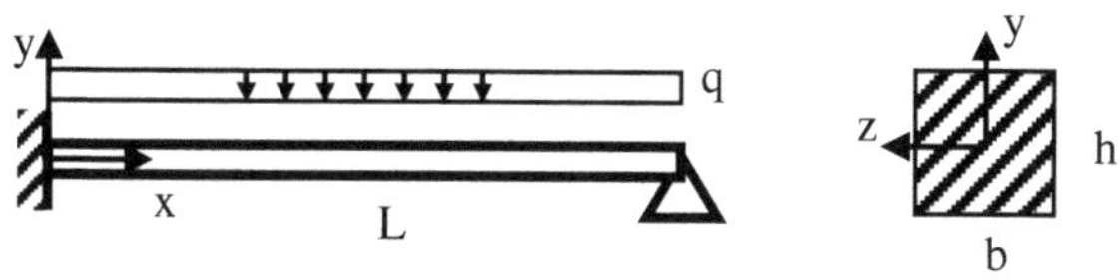

Fig. 8.14 Bending beam clamped at one end and simply supported at the other end

location than the one that undergoes the maximum stress in the span from the fictitious elastic analysis.

Although the initial curvature associated with the initial moment of Eq. (8.48) causes displacements without constraints in a statically determinate beam, section forces are induced by the share $\left(1-\frac{E_t}{E}\right)M_{\rm pl}$ in a statically indeterminate structure. The share of the initial curvature w_0'' responsible for these section forces amounts to

$$w''_{0(x)} = -\mathrm{sgn}\left(\bar{M}^{\rm fel}_{(x)}\right)\left(1 - \frac{E_t}{E}\right)\frac{3f_y}{E_t h} \tag{8.61}$$

in a solid rectangular cross-section. It is therefore a sectional constant in the longitudinal direction. The change of sign is located at the zero position $x = x_N$ of the fictitious elastic moment line

$$M^{\rm fel}_{(x)} = -\frac{qL^2}{8}\left[1 - 5\frac{x}{L} + 4\left(\frac{x}{L}\right)^2\right], \tag{8.62}$$

i.e., at

$$\frac{x_N}{L} = \frac{1}{4}. \tag{8.63}$$

The residual bending moment caused as a reaction to the initial curvature because of the statically indeterminate bearings is distributed linearly for equilibrium reasons. It can be determined by means of the force method and at the clamped end is

$$X = \frac{3}{64}\left(1 - \frac{E_t}{E}\right)f_y bh^2. \tag{8.64}$$

The residual stresses arising from this moment, acting in addition to those given by Eq. (8.52) for the statically determinate structure, are distributed linearly across the beam height and also linearly variable along the beam axis:

$$\rho_{X(x,y)} = -\frac{X\left(1 - \frac{x}{L}\right)y}{\frac{bh^3}{12}}. \tag{8.65}$$

The total of the residual stresses is thus given by

$$\rho_{(x,y)} = \left(1 - \frac{E_t}{E}\right) f_y \left\{ \frac{3}{2} \frac{y}{h/2} \left[\operatorname{sgn}\left(\bar{M}^{\text{fel}}_{(x)}\right) - \frac{3}{16}\left(1 - \frac{x}{L}\right) \right] + \operatorname{sgn}\left(\bar{\sigma}^{\text{fel}}_{(x,y)}\right) \right\}. \quad (8.66)$$

Accordingly, the limit load factor determined by Eq. (8.51) is

$$\alpha_{L(Et)} = -\left\{ \left(1 - \frac{E_t}{E}\right) \frac{3}{2} \frac{y}{h/2} \left[\operatorname{sgn}\left(\bar{M}^{\text{fel}}_{(x)}\right) - \frac{3}{16}\left(1 - \frac{x}{L}\right) \right] - \frac{E_t}{E} \operatorname{sgn}\left(\bar{\sigma}^{\text{fel}}_{(x,y)}\right) \right\} \frac{f_y}{\bar{\sigma}^{\text{fel}}_{(x,y)}}. \quad (8.67)$$

For $E_t/E \to 0$, this limit load factor becomes constant across the beam height:

$$\alpha_L = -\frac{3}{2} \left[\operatorname{sgn}\left(\bar{M}^{\text{fel}}_{(x)}\right) - \frac{3}{16}\left(1 - \frac{x}{L}\right) \right] \frac{f_y}{\bar{\sigma}^{\text{fel}}_{(x,y=+h/2)}}. \quad (8.68)$$

The smallest value is found at the clamped end ($x = 0$). Based on the elastic limit load ($\bar{\sigma}^{\text{fel}}_{(x=0,y=h/2)} = \mathrm{f}_y$) we obtain

$$\min \alpha_L = \frac{57}{32} \approx 1.781, \quad (8.69)$$

and as a second minimum in the span

$$\min \alpha_{L,\text{span}} = \frac{27}{32} \frac{1}{119 - 16\sqrt{55}} \approx 2.476, \quad (8.70)$$

located at

$$\frac{x_m}{L} = \frac{1}{3}\left(\sqrt{220} - 13\right) \approx 0.6108. \quad (8.71)$$

Thus, the position of maximum stress in the span shifted slightly compared to the fictitious elastic calculation ($x_m/L = 0.625$).

This result was achieved with only one MEA, because the asymptotic solution is known from the outset because of the uniaxial stress state (see Eq. (8.20)), and because the residual stresses are independent of the load factor α according to Eq. (8.66).

With the load factor of Eq. (8.69), only the section at the clamped end is completely plasticized, and no plastic hinge has developed yet in the span because of Eq. (8.70). Thus no kinematic chain is present and the plastic limit load is not yet reached. The actual limit load factor is therefore between 1.781 and 2.476. Later this range is reduced.

For comparison, the actual limit load factor determined with other methods is provided here:

$$\alpha_{L,\text{actual}} = \frac{3}{8}\left(3 + 2\sqrt{2}\right) \approx 2.184. \tag{8.72}$$

The corresponding plastic hinge in the span is located at

$$\left(\frac{x_m}{L}\right)_{\text{actual}} = \frac{2+\sqrt{2}}{3+2\sqrt{2}} \approx 0.5858. \tag{8.73}$$

Figure 8.15 shows the initial stresses and the elastic-plastic stresses at three different load factors and Fig. 8.16 the residual stresses and the distribution of the limit load factors, both of which are the same at all load factors, each obtained for $E_t/E \rightarrow 0$. The aspect ratio is exaggerated and limit load factors above 6 are grayed out.

It is evident from Fig. 8.15 that, at all load factors, the axial distribution of the stresses has a discontinuity at the same location, namely at the zero crossing of the fictitious elastic calculated moment line which erroneously leads to a discontinuity in the elastic-plastic moment line. This is attributable to the fact that the entire structure was considered to be plastic ($V_p = V$) and that therefore initial stresses were applied in the MEA as modified loading also in lower stressed locations. This is the reason that the actual plastic limit load is not found so far by the STPZ.

In addition, it can be seen that at all load factors considered in Fig. 8.15, the elastic-plastic stresses do nowhere exhibit the block-like distribution actually expected across the beam height in the case of $E_t/E = 0$. However, this is the case at the clamped end with the load factor of Eq. (8.69), but not in the span (Fig. 8.17). Conversely, a block-like distribution of stress is obtained with the load factor of

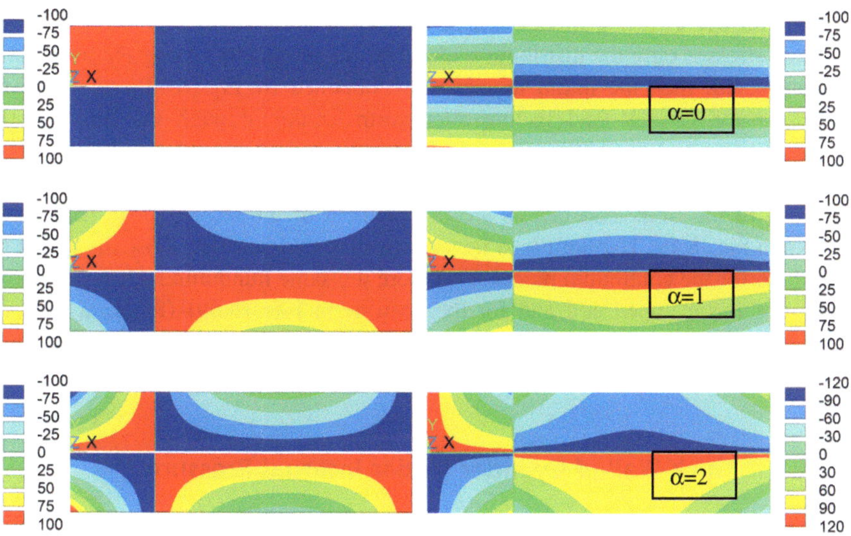

Fig. 8.15 Initial stresses (*left*) and elastic-plastic stresses (*right*) for various load factors α, $E_t/E \rightarrow 0$, $f_y = 100$

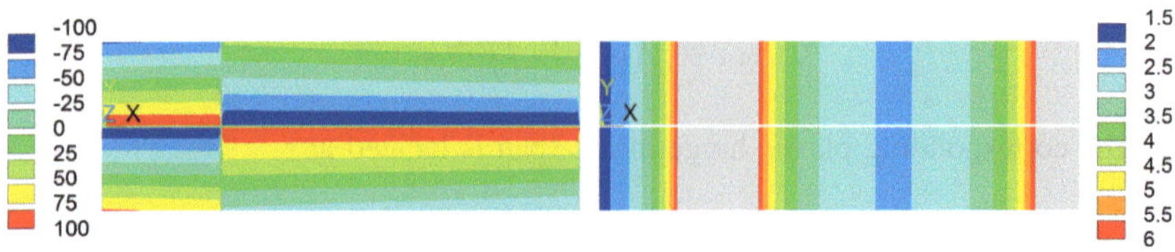

Fig. 8.16 Residual stresses (*left*) (at $E_t/E \to 0, f_y = 100$) and distribution of the first estimate of the limit load factors (*right*) in a statically indeterminate beam

Fig. 8.17 Elastic-plastic stress distribution of the statically indeterminate beam for the first estimate of the limit load factor $\alpha = \min \alpha_L = 57/32$ ($E_t/E \to 0, f_y = 100$)

Eq. (8.70) at the location specified by Eq. (8.71), but then not at the clamped end. It can also be seen that at load factors $\alpha > \min \alpha_L$, the yield stress ($f_y = 100$) is exceeded, which is, however, not permissible in a non-hardening material. These contradictions are caused by the fact that so far only a first estimate of the plastic limit load was carried out, which is improved later.

If the statically indeterminate beam is no longer considered as a two-dimensional continuum but as a line model, where instead of horizontally and vertically distributed stresses only the distribution of the bending moments along the beam axis is considered, then the initial moments are obtained by Eq. (8.48), as shown in Fig. 8.18 for various load factors after normalization through the fully plastic moment given by Eq. (8.49), as well as the corresponding normalized elastic-plastic moment lines. The resulting residual moments and the distribution of the first estimate of the limit load factors are independent of the selected load factor and shown in Fig. 8.19. The limit load factors are almost identical to those illustrated in Fig. 8.16 for the continuum model. Therefore the same minimum limit load factors apply for the clamped end and for the span as in Eqs. (8.69)–(8.71), at least in four-digit precision. Figure 8.20 shows the moment line, normalized by the fully plastic moment, obtained for the minimum plastic limit load factor.

8.5.3 Thick-Walled Cylinder Under Internal Pressure

The norm of the incremental stress in the asymptotic state according to Eq. (8.22) is, according to Eq. (8.24),

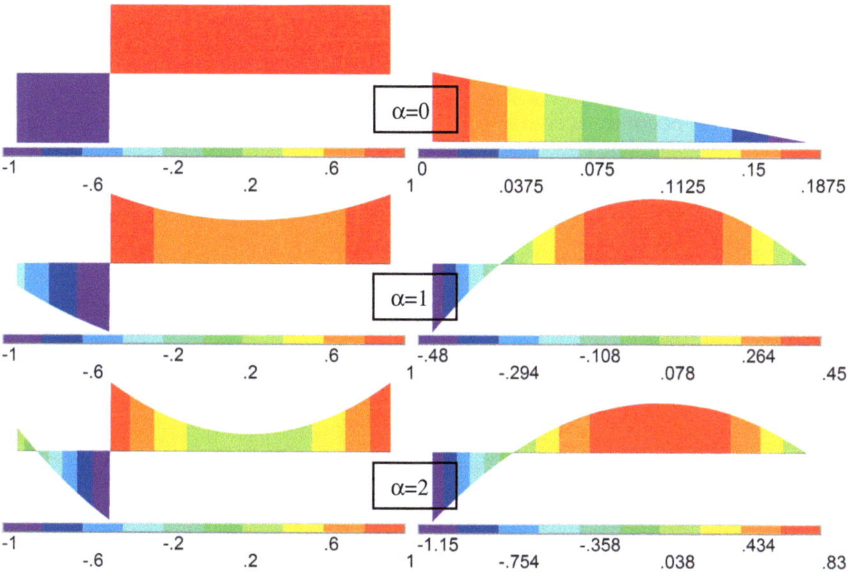

Fig. 8.18 Normalized initial moments (*left*) and elastic-plastic moments (*right*) for the line model of the statically indeterminate beam for various load factors α

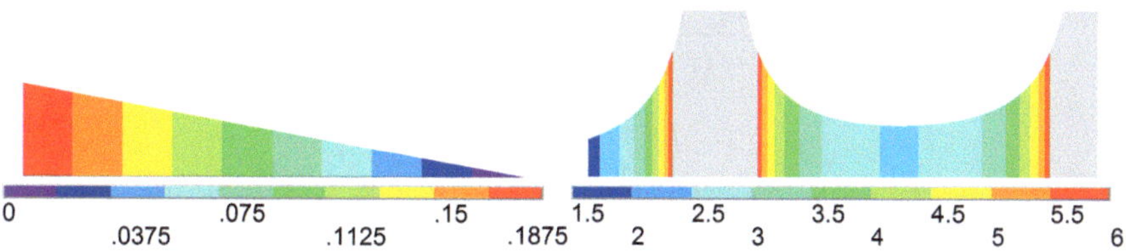

Fig. 8.19 Normalized residual moments and distribution of the first estimate of the limit load factor for the line model of the statically indeterminate beam

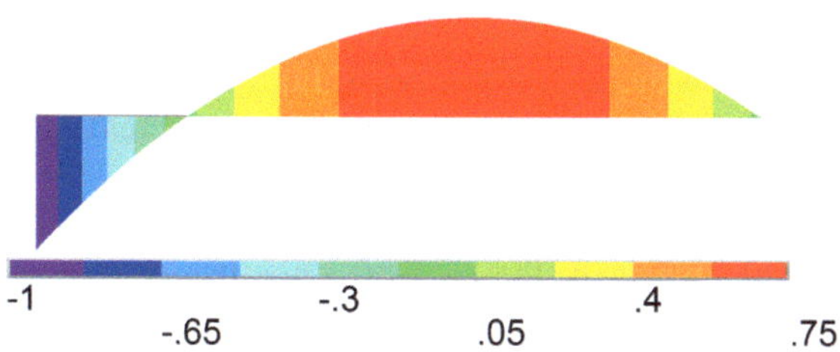

Fig. 8.20 Normalized elastic-plastic moment line for the line model of the statically indeterminate beam for the first estimate of the limit load factor, $\alpha = \min \alpha_L = 1.784$

$$\left|\mathrm{d}\sigma'^{\infty}_{i(x)}\right| \propto \mathrm{d}p \frac{r_i^2 r_a^2}{\left(r_a^2 - r_i^2\right)x^2}\sqrt{3}, \tag{8.74}$$

so that the TIV become, according to Eq. (8.23),

$$Y_{i(x)} = \alpha\bar{\sigma}'^{\mathrm{f.el}}_{i(x)} - \frac{f_y}{\sqrt{3}}\begin{pmatrix} -1 \\ 0 \\ +1 \end{pmatrix} \tag{8.75}$$

and the initial strains, according to Eq. (3.14), or the initial stresses, according to Eq. (3.23),

$$\varepsilon_{0,i(x)} = \frac{3}{2}\frac{1-\frac{E_t}{E}}{E_t}\left[\alpha\bar{p}\frac{r_i^2 r_a^2}{\left(r_a^2 - r_i^2\right)x^2} - \frac{f_y}{\sqrt{3}}\right]\begin{pmatrix} -1 \\ 0 \\ +1 \end{pmatrix} \tag{8.76}$$

$$\sigma_{0,i(x)} = -\frac{3}{2}\frac{1-\frac{E_t}{E}}{1+\nu^*}\left[\alpha\bar{p}\frac{r_i^2 r_a^2}{\left(r_a^2 - r_i^2\right)x^2} - \frac{f_y}{\sqrt{3}}\right]\begin{pmatrix} -1 \\ 0 \\ +1 \end{pmatrix}. \tag{8.77}$$

As a result of the MEA, we obtain after some calculation

$$\rho_x = \frac{\sqrt{3}}{2}f_y\frac{1-\frac{E_t}{E}}{1-(\nu^*)^2}\left[\ln\frac{x}{r_i} + \frac{r_a^2}{r_a^2 - r_i^2}\left(1 - \frac{r_i^2}{x^2}\right)\ln\frac{r_i}{r_a}\right] \tag{8.78}$$

$$\rho_y = \frac{\sqrt{3}}{2}f_y\frac{1-\frac{E_t}{E}}{1-(\nu^*)^2}\nu^*\left[1 + 2\ln\frac{x}{r_i} + 2\frac{r_a^2}{r_a^2 - r_i^2}\ln\frac{r_i}{r_a}\right] \tag{8.79}$$

$$\rho_z = \frac{\sqrt{3}}{2}f_y\frac{1-\frac{E_t}{E}}{1-(\nu^*)^2}\left[1 + \ln\frac{x}{r_i} + \frac{r_a^2}{r_a^2 - r_i^2}\left(1 + \frac{r_i^2}{x^2}\right)\ln\frac{r_i}{r_a}\right]. \tag{8.80}$$

Accordingly, the residual stresses are independent of the load level, i.e., independent of the load factor α, and thus arise only from the term of the initial strains on the right hand side in the square bracket of Eq. (8.76). This is after the context discussed in Sect. 8.4 in line with the fact that no incremental residual stress occurred in the asymptotic state (Sect. 8.3.3).

The material data (E_t/E, ν) enter the equations for the residual stresses only by a proportionality factor, which in the case of a non-hardening material ($E_t/E = 0$, $\nu^* = 0.5$) amounts to

$$\frac{\sqrt{3}}{2}f_y\frac{1-\frac{E_t}{E}}{1-(\nu^*)^2} = \frac{2}{\sqrt{3}}f_y. \tag{8.81}$$

The plastic limit load is formed according to Eq. (8.33). The residual stress component in the radial direction is chosen as reference. Because the residual stresses of Eqs. (8.78)–(8.80), however, do not depend on the load factor α, Eq. (8.33) can also be written as

$$\alpha_{L(Et)} = \frac{\rho_x - \frac{1}{2}\left(\rho_y + \rho_z\right) + \frac{\sqrt{3}}{2} f_y}{-\frac{3}{2}\bar{\sigma}_x^{\prime\mathrm{fel}}}, \tag{8.82}$$

where only Eqs. (8.78)–(8.80) must be inserted. Consequently, the limit load factor depends on the material parameters E_t/E and ν, and is variable in the radial direction. At low hardening ($E_t/E \approx 0$), ν^* tends to 0.5, and the residual stresses and hence the limit load factor according to Eq. (8.82) depend approximately linearly on E_t/E. With a non-hardening material ($E_t/E = 0$ and hence $\nu^* = 0.5$), the limit load factor turns out to be independent of Poisson's ratio and of the x-coordinate, so that it takes on a constant value for the entire structure:

$$\alpha_{\mathrm{L}} = \frac{2}{\sqrt{3}} \frac{f_y}{\bar{p}} \ln \frac{r_a}{r_i}. \tag{8.83}$$

When normalized to the elastic limit load

$$\bar{p} = p_{\mathrm{el\_limit}} = \frac{f_y}{\sqrt{3}}\left(1 - \frac{r_i^2}{r_a^2}\right), \tag{8.84}$$

Equation (8.83) becomes

$$\alpha_L = 2\frac{\ln \frac{r_a}{r_i}}{1 - \frac{r_i^2}{r_a^2}}. \tag{8.85}$$

With hardening, the limit load factor of Eq. (8.82) is location-dependent. It increases from the inside to the outside and is, in the relevant place, i.e., at the inner surface, smaller than given by Eq. (8.83) or Eq. (8.85) without hardening.

The result of Eq. (8.85) is the same as the plastic limit load derived in the literature. The assumption of a completely plasticized structure ($V_p = V$) has therefore led to the exact result, because it has proven to be actually true, which manifests itself in the location-independent limit load factor.

Although directional stress redistributions occur because of plastic straining, which are a function of location, a single MEA, in addition to the MEA for the asymptotic state, was sufficient to obtain the exact results, because the residual stresses do not depend on the load factor.

8.5.4 Disc with a Hole

8.5.4.1 Free Longitudinal Edges

In Sect. 8.3.4.1, the asymptotic state of a disc with a hole and free longitudinal edges was determined.

The initial stresses, calculated according to Eq. (3.14) with the TIV Y_i according to Eq. (8.23), are shown in Fig. 8.21 for an arbitrary load factor $\alpha = 1$, along with the residual stresses from the associated MEA.

The residual stresses are in this structure independent of the load factor (see Sect. 8.4), because the residual stress increments of the asymptotic state in Sect. 8.3.4.1 were already zero. So the distribution of the limit load factors can be determined after this MEA according to Eq. (8.34). Their representation in Fig. 8.22 is based on the component in the direction of the applied force, that is, the x-direction. The smallest and therefore decisive limit load factor amounts to

$$\min \alpha_L = 0.4807 \tag{8.86}$$

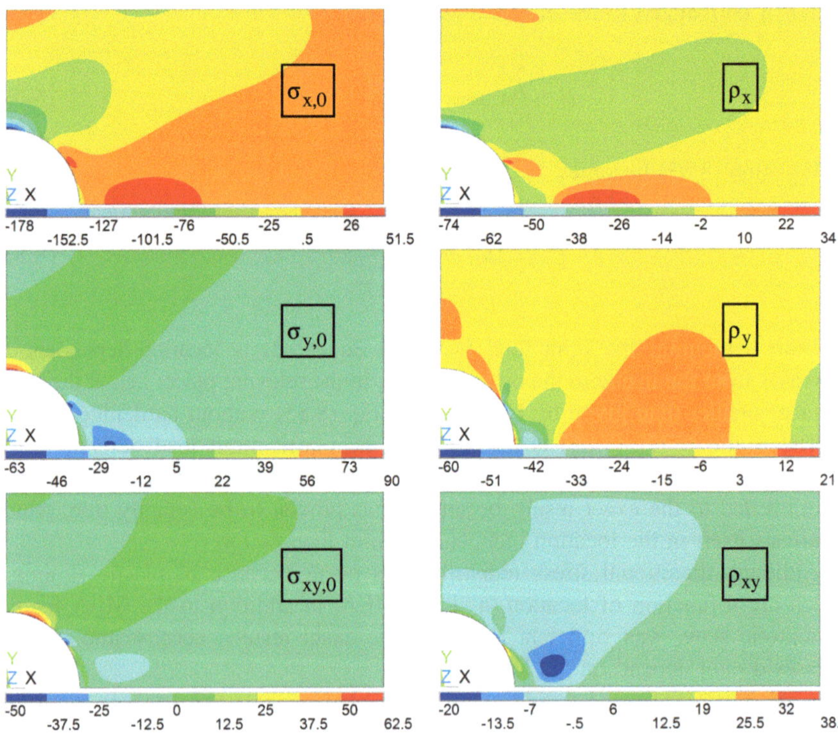

Fig. 8.21 Initial stresses for load factor $\alpha = 1$, as well as the residual stresses of the corresponding MEA ($E_t/E = 10^{-6}$, $\nu = 0.3$) for the disc with a hole and with free longitudinal edges

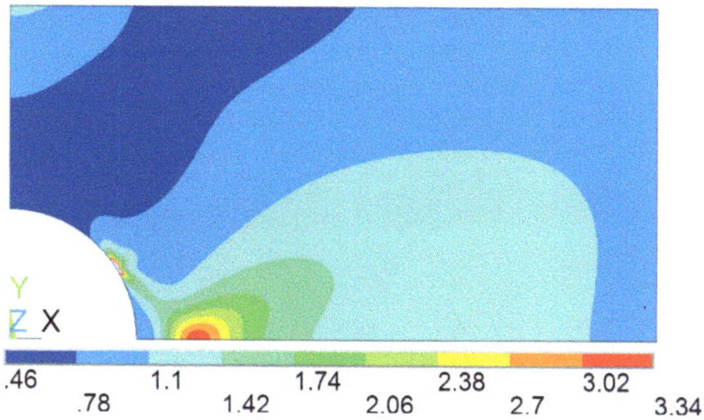

Fig. 8.22 First estimate of the distribution of limit load factors (in the x-direction, with respect to $p = 100$)

and is located in the ligament, a little above the hole edge, not at the fictitious elastic maximum stressed point, which is located right at the edge of the hole. Figure 8.23 shows the distribution of equivalent stresses for the stress level of Eq. (8.86).

A kinematic chain would be present if the smallest limit load factor of the system were distributed constantly along a yield line across the width of the disc in Fig. 8.22. However, this is not the case here, even if the dark blue colored area in Fig. 8.22 suggests that such an area is emerging. The actual limit load has therefore not yet been found. The reason for this is again the assumption $V_p = V$, which is not exactly true in this structure, as well as the fact that the TIV Y_i was not based on the respective limit load factor, but on the plastic strain increments at an infinitely high load level in Eq. (8.23). An improved estimate of the limit load is carried out later.

An incremental analysis performed with ANSYS suggests that the actual plastic limit load factor, based on the load $p = 100$, amounts to

$$\alpha_{L,\mathrm{actual}} = 0.6478. \tag{8.87}$$

The corresponding yield line can be seen in Fig. 8.24 (c.f. [8]).

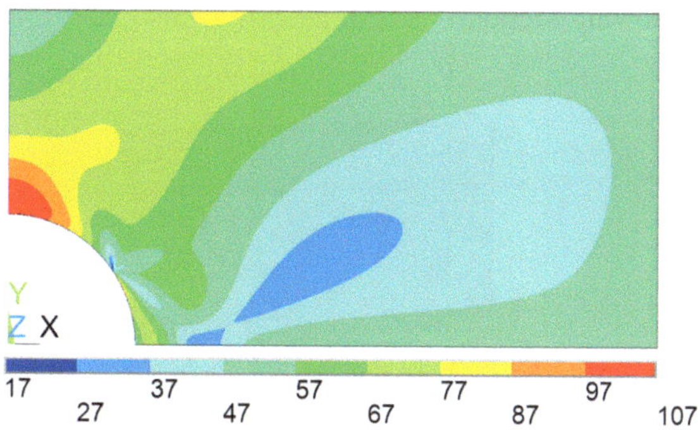

Fig. 8.23 Equivalent stresses for first estimate of the plastic limit load, $\alpha_L = 0.4807$

Fig. 8.24 Yield line associated with the actual plastic limit load according to an incremental analysis

8.5.4.2 Supported Longitudinal Edges

In Sect. 8.3.4.2, the asymptotic state of a disc with a hole, supported at the longitudinal edges against displacement in transverse direction, was determined.

For this system, two MEAs with different load factors are required for a first estimate of the plastic limit load, because the residual stresses depend on the load factor, because the incremental residual stresses in Sect. 8.3.4.2 were not zero. The load factors arbitrarily chosen here are $\alpha_1 = 1$ and $\alpha_2 = 2$. The representation of the corresponding initial stresses, determined according to Eq. (3.14) from the TIV Y_i according to Eq. (8.23) and the residual stresses from the two related MEAs are omitted here.

The distribution of the corresponding limit load factors after Eq. (8.33) based on the stress components in the x-direction is shown in Fig. 8.25. The smallest and therefore decisive limit load factor is located in the ligament (apart from a kind of a pole, as visible in Fig. 8.25, arising from computing dirt). Related to $p = 100$ it is

$$\min \alpha_L = 0.6518. \tag{8.88}$$

A kinematic chain has not formed yet, so the actual limit load is still not found with the STPZ, but only a first estimate. An improved estimate is made in the following sections.

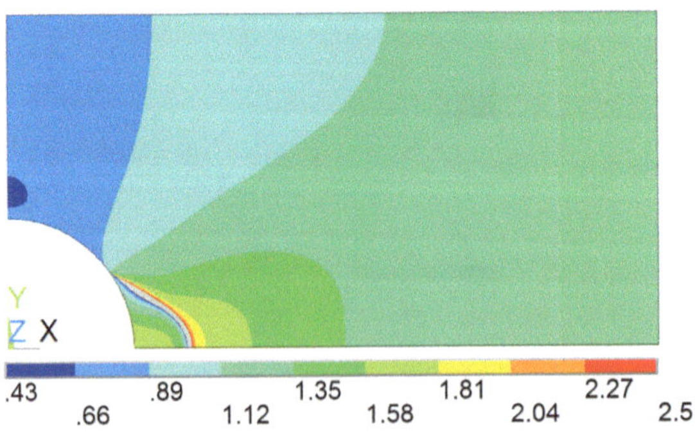

Fig. 8.25 First estimate of the distribution of limit load factors (in the x-direction, with respect to $p = 100$)

Fig. 8.26 Yield line associated with the actual plastic limit load according to an incremental analysis

According to an incremental analysis with ANSYS, the actual plastic limit load factor, based on the load $p = 100$, is

$$\alpha_{L,\text{actual}} = 0.7311. \tag{8.89}$$

The corresponding yield line can be seen in Fig. 8.26.

8.6 Structures not Completely Plasticized

In Sect. 8.4, the structures were considered to be completely plasticized ($V_p = V$) in the state of the plastic limit load, as already assumed for the asymptotic solution in Sect. 8.2. For a first estimate of the limit load factor, one or two other modified elastic analyses were then required beyond the asymptotic solution, making a total of two or three linear elastic analyses in addition to the obligatory fictitious elastic analysis. In some instances the correct solution has already been found, e.g., for the one-element model in Sect. 8.5.1 and the thick-walled cylinder in Sect. 8.5.3.

In cases where the structure is not completely plastic in the limit load state, this first estimation of the limit load factor turned out as a lower bound for the actual limit load factor, thus lying on the safe side (see Table 8.1). According to further examples, not presented here, this property of a lower bound is encountered at least very often. A proof that this is generally true, however, is not yet known.

This first estimate can now be iteratively improved by providing an improved estimate of the plastic zone V_p instead of the assumption $V_p = V$.

In each of these iteration steps, three MEAs are required, namely one MEA for the asymptotic state and two more MEAs for the determination of the residual stresses at the two different load factors α_1 and α_2 for determining the limit load factor from Eq. (8.33). The same plastic zone must be used in these three MEAs. Otherwise, the plastic strains would not depend linearly on the load factor as shown in Fig. 8.7, and thus a linear interpolation to find the limit load factor as in Eq. (8.32) based on the criterion of Eq. (8.29) would no longer be permitted.

Table 8.1 Overview of the examples considered so far (in Sects. 8.3 and 8.5)

Example	Stress state		Stress redistribution		$V_p = V$ correct?	Residual stress: $\rho_{\text{asympt}} = 0$, $\rho \neq \rho(\alpha)$	First estimate of plastic limit load
	Multiaxial	Homo-geneous	Local	Directional			
One element (Sect. 8.5.1)	Yes	Yes	No	Yes	Yes	No	Exact (after three MEAs)
Bending beam statically determinate (Sect. 8.5.2.1)	Uniaxial	No	Yes	No	No	Yes	Exact (after one MEA)
Bending beam statically indeterminate (Sect. 8.5.2.2)	Uniaxial	No	Yes	No	No	Yes	Conservative (after one MEA)
Cylinder (Sect. 8.5.3)	Yes	No	Yes	Yes	Yes	Yes	Exact (after two MEAs)
Disk with hole, free longitudinal edges (Sect. 8.5.4.1)	Yes	No	Yes	Yes	No	Yes	Conservative (after two MEAs)
Disc with hole, supported longitudinal edges (Sect. 8.5.4.2)	Yes	No	Yes	Yes	No	No	Conservative (after three MEAs)

After the structural volume V is partitioned in an elastic region V_e and the plastic zone V_p, the modified elastic material law Eq. (3.20) is used in all subsequent MEAs, as familiar from Chap. 3, according to which the modified elastic modulus E^* of Eq. (3.15) and the modified Poisson's ratio of Eq. (3.16) is used only in V_p, and the elastic material parameters E and ν remain unchanged in V_e. Correspondingly, no initial strains or initial stresses are applied in V_e, but only in V_p. Thus, the modified elastic material laws for the asymptotic state or for the residual states in order to estimate the limit load now read

$$\mathrm{d}\varepsilon_i^{*,\infty} = \begin{cases} \left(E_{ij}^*\right)^{-1} \mathrm{d}\rho_j^{\infty} + \mathrm{d}\varepsilon_{i,0}^{\infty} & \forall \underline{x} \in V_p \\ \left(E_{ij}\right)^{-1} \mathrm{d}\rho_j^{\infty} & \forall \underline{x} \in V_e \end{cases} \tag{8.90}$$

$$\varepsilon_i^* = \begin{cases} \left(E_{ij}^*\right)^{-1} \rho_j + \varepsilon_{i,0} & \forall \underline{x} \in V_p \\ \left(E_{ij}\right)^{-1} \rho_j & \forall \underline{x} \in V_e \end{cases} \tag{8.91}$$

with the initial strains, whereby, of course, the corresponding initial stresses could also be used alternatively,

$$\mathrm{d}\varepsilon_{i,0}^{\infty} = \begin{cases} \frac{3}{2}\frac{1-\frac{E_t}{E}}{E_t}\mathrm{d}\sigma_i^{\prime\mathrm{fel},\infty} & \forall \underline{x} \in V_p \\ 0 & \forall \underline{x} \in V_e \end{cases} \tag{8.92}$$

$$\varepsilon_{i,0} = \begin{cases} \frac{3}{2}\frac{1-\frac{E_t}{E}}{E_t} Y_i & \forall \underline{x} \in V_p \\ 0 & \forall \underline{x} \in V_e \end{cases}. \tag{8.93}$$

The TIV Y_i can be determined unchanged according to Eq. (8.23), but limited to the region V_p.

From Eqs. (8.92) and (8.93) it is immediately evident that, according to the partition of the structural volume in V_p and V_e, the initial strains can no longer meet the compatibility conditions at Eq. (8.7) in the entire structure because of the discontinuities at the boundary between these two regions. Thus, it is generally to be expected that the asymptotic residual stress increments no longer vanish and that the residual stresses depend on the finite load factors α_1 and α_2, and that therefore three MEAs are required in principle in each iteration step.

Because several regions with greatly differing stiffness exist now in the structure, there is a risk that ill-conditioned systems of equations arise in an FEM environment, the solution of which is associated with significant errors. The tangent modulus E_t should therefore not be chosen too small.

The main problem now is how a suitable partitioning of V in V_e and V_p can be made. There are several possibilities, but none of them, it is thought, generally gives good results. So Zarka has submitted a proposal in [3, Sect. 6.3.2], but pointed out at the same time that, at least for trusses and beams, Borhani proposed a better method in [5], but only for linearly distributed bending moments and therefore countable options for the position of plastic hinges.

Further proposals are made below that may provide some improvement in the author's opinion, but still do not fully satisfy.

Although for a first estimation of the limit load according to Sect. 8.4 the elastic zone exactly amounts to 0 % of the total volume, this proportion is now useful to increase.

One way of doing this is to assign those locations of the structure, either nodes or Gauss points, to the subvolume V_e, where the equivalent stresses in the state of the first estimate of the limit load do not exceed a certain fraction of the yield stress, for example 40 %:

$$\underline{x} \in V \quad ; \quad V_p \cup V_e = V \quad ; \quad V_p = \left\{ \underline{x} | \sigma_{\mathrm{v}(\underline{x})}^{\text{first estimate}} \geq 0.4 f_y \right\}. \tag{8.94}$$

The stress components required to form the equivalent stress can be derived from Eq. (8.27), where the minimum limit load factor, determined according to Sect. 8.4, i.e., with the assumption of $V_p = V$, is to be inserted as the load factor α, and the associated residual stresses are to be used for ρ_i. Because the residual stresses depend linearly on the load factor, they can be obtained simply by inter- or extrapolation of the residual stresses available from the MEAs already performed for the two arbitrary load factors required for the first estimate of the limit load factor according to Sect. 8.4, i.e., without performing an additional MEA.

On the basis of Eq. (8.94), the calculation procedure described in Sects. 8.2 and 8.4 is repeated, so a new asymptotic solution is obtained and two MEAs are performed for two arbitrary load factors, used for a new estimate of the limit load according to Eq. (8.33). This procedure could be repeated and so be applied iteratively.

Alternatively, it would also be worth considering performing an elastic-plastic analysis of monotonic loading either by a conventional incremental calculation, or with the STPZ according to Chap. 3, for the load level of the first estimate of the limit load gained in Sect. 8.4 (cf. Borhani [5] for trusses and beams). To this end, several MEAs are usually necessary. It is possible that the results of these MEAs are numerically erroneous because of the low hardening in the plastic zones, actually rather atypical for the use of the STPZ, and oscillate rather than converge. If, however, an acceptable solution is achieved, the associated stress distribution can be used instead of the incremental stresses from an asymptotic solution for an infinitely high load level, so providing a new basis for an analysis of the limit load with two arbitrary load factors which require two other MEAs.

The advantage of such monotonous analysis is that, if the load level is not too far away from the limit load, a good approximation for the geometry of the plastic zone and the stress state in the limit load state can be obtained. The load level applied should be chosen definitely above the still unknown limit load, which is possible by virtue of hardening, because in the subsequent limit load analysis an underestimation of V_p has more severe consequences than an overestimation.

8.7 Examples of Plastic Limit Load for Structures not Completely Plasticized

The two ways mentioned in the preceding section to improve the first estimation of the limit load obtained according to Sect. 8.4, are applied exemplarily to two examples already considered in Sect. 8.5 which have not yet led to the exact limit load.

8.7.1 Statically Indeterminate Beam

The beam clamped at one end and simply supported at the other treated in Sect. 8.5.2.2 is considered, limited to the line model.

The criterion Eq. (8.94) has the effect that the signs of the bending moments of the asymptotic solution $M^{\infty}_{(x)}$ built upon are no longer identical to the signs of the fictitious elastic solution $M^{\text{fel}}_{(x)}$, but need to be found by means of an MEA for the asymptotic state. Further, $\bar{\sigma}^{\text{fel}}_{(x,y)}$ and $\bar{M}^{\text{fel}}_{(x)}$ are to be replaced by $\mathrm{d}\sigma^{\infty}_{(x,y)}$ and $\mathrm{d}M^{\infty}_{(x)}$ in the sgn-terms of Eqs. (8.47), (8.48), (8.50), and (8.51). In addition, the residual moments depend now, unlike the calculations in Sect. 8.5.2.2, on the chosen load factors α_1 and α_2.

The initial moments $\mathrm{d}M^{\infty}_{0(x)}$, gained for the $0.4\,f_y$-criterion of Eq. (8.94) on the basis of the moments obtained for the first estimate of the minimum limit load factor in Fig. 8.20, and normalized by the fully plastic moment M_{pl}, to be used for the calculation of the asymptotic state, are shown in Fig. 8.27, along with the normalized incremental moments $\mathrm{d}M^{\infty}_{(x)}$ thus resulting.

For the arbitrarily chosen load factors $\alpha_1 = 1$ and $\alpha_2 = 2$, and utilizing the signs of $\mathrm{d}M^{\infty}_{(x)}$ instead of $\bar{M}^{\text{fel}}_{(x)}$, the normalized initial moments are obtained from Eq. (8.48) and are shown in Fig. 8.28.

Instead of Eq. (8.51), the new distribution of the limit load factor is now obtained based on Eq. (8.33), where the deviatoric residual stresses are replaced by the residual moments and f_y by M_{pl}. The result is shown in Fig. 8.29.

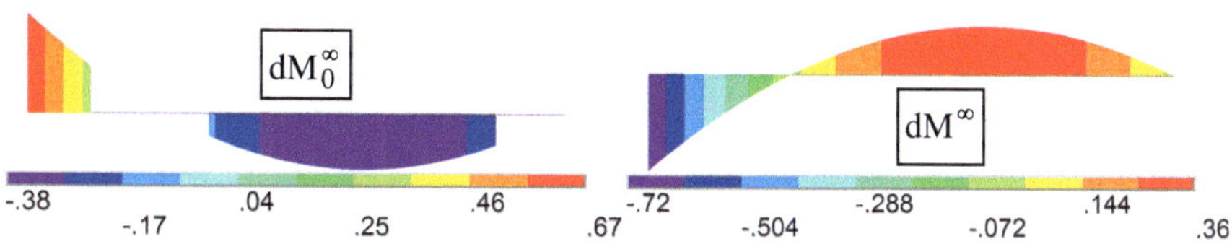

Fig. 8.27 Distribution of the initial moments for the asymptotic state (*left*); incremental moments from the $0.4\,f_y$-criterion of Eq. (8.94) (*right*)

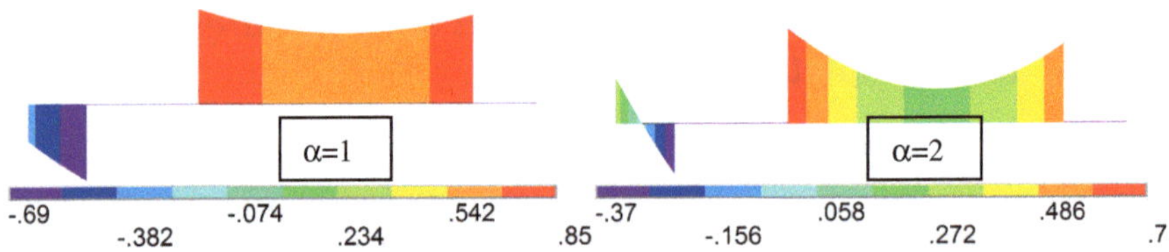

Fig. 8.28 Distribution of normalized initial moments for two chosen load factors from the 0.4 f_y-criterion of Eq. (8.94)

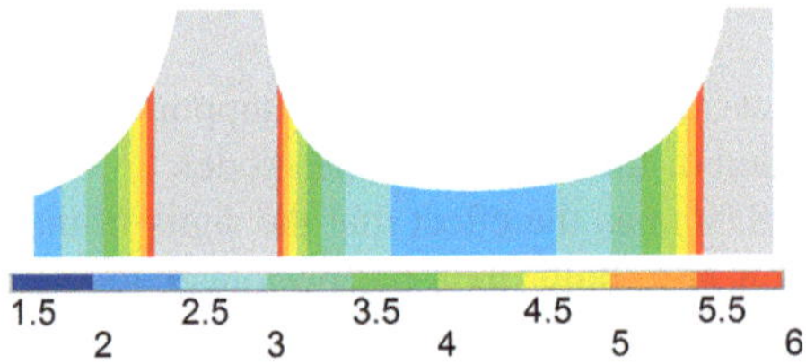

Fig. 8.29 Distribution of the improved estimation of the limit load factor for the line model of the statically indeterminate beam from the 0.4 f_y-criterion of Eq. (8.94)

The minimum value can be found at the clamped end and is

$$\min \alpha_L = 2.103. \tag{8.95}$$

The second minimum in the span is found at

$$\frac{x_m}{L} = 0.593 \tag{8.96}$$

and amounts to

$$\min \alpha_{L,\text{span}} = 2.259. \tag{8.97}$$

Thus a further approximation to the exact result (Eqs. (8.72) and (8.73)) took place, which also manifests itself in the elastic-plastic moment line of Fig. 8.30. In the span, the bending moment is now about 95 % of M_{pl}.

The arithmetic mean of the limit load factors for the clamped end and in the span (Eqs. (8.95) and (8.97)) is, at 2.180, already very close to the exact value of 2.184 (see Eq. (8.72)).

In Sect. 8.6, the possibility was mentioned of determining the finite stresses for a finite load level, chosen somewhat above the presumed limit load, instead of incremental stresses for an infinite load level. This procedure also leads here, by means of a calculation for monotonic load increase with the STPZ, to good results. For a monotonic analysis of the load level of Eq. (8.97), for example, about five MEAs are needed if a hardening modulus $E_t/E = 10^{-5}$ is adopted. As limit load

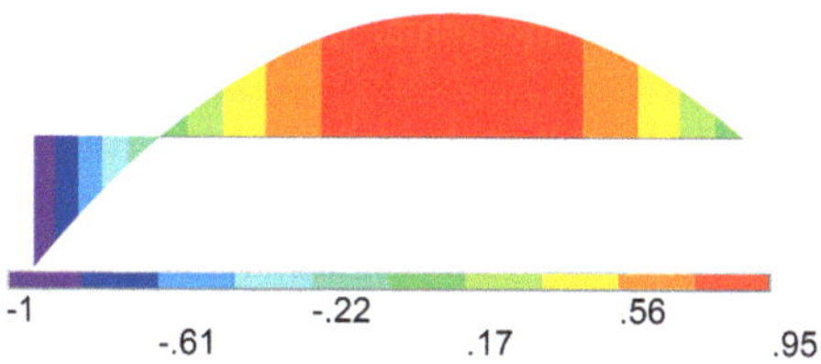

Fig. 8.30 Normalized elastic-plastic moment line for the line model of the statically indeterminate beam for the second estimate of the limit load factor, $\alpha = \min \alpha_L = 2.103$

factor at the clamped end, the value 2.177 is obtained and in the span 2.199, and thus an even better approximation of the actual limit load.

8.7.2 Disc with a Hole and Free Longitudinal Edges

To improve the first estimation of the limit load in Eq. (8.86), part of the disc is considered to remain elastic (V_e). The criterion of Eq. (8.94) is applied to the distribution of equivalent stresses for the load factor obtained as a first estimate of the limit load factor, $\alpha = 0.4807$ (see Eq. (8.86)), as depicted in Fig. 8.23. A new asymptotic analysis is required and two additional MEAs for two arbitrary load factors.

The distribution of the limit load factors thus obtained is shown in Fig. 8.31 on the right side, where V_e is cut out. Values above 0.685 are capped, so that the failure mode induced by a yield line extending from the edge of the hole inclined upward, as already addressed in Sect. 8.5.4.1, is suggested. A comparison with the distribution according to the first estimate, shown on the left side of Fig. 8.31 (identical to Fig. 8.22, but scaled differently) reveals a certain leveling of the limit load factor because of the improved estimate. The minimum value is now

$$\min \alpha_L = 0.5552 \tag{8.98}$$

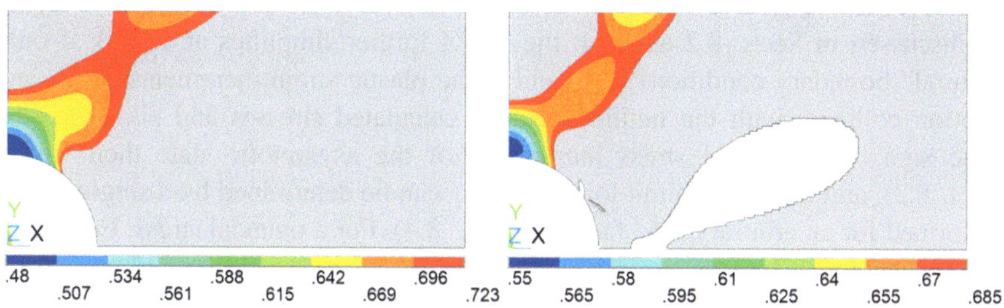

Fig. 8.31 Comparison of the distribution of limit load factors according to the first and second estimate (from the criterion Eq. (8.94))

and thus has further approached the actual limit load factor (0.6478 according to Eq. (8.87)).

In addition, the bandwidth of limit load factors has roughly halved in the relevant region, which is considered as an improvement, because a kinematic mechanism requires a constant load factor in its region. With the limit load factor after the first estimate between 0.4807 and 0.7146 (Fig. 8.31 on the left), it is, after the second estimate, between 0.5552 and 0.6707 (Fig. 8.31 on the right). The higher value is imaginable as a saddle point in the figure.

If, for example, 0.55 f_y instead of 0.40 f_y is chosen as a criterion for estimating the plastic zone in Eq. (8.94), the bandwidth of the limit load can be further reduced. In the entire region from the edge of the hole to the upper right, the limit load factor is then between 0.6133 and 0.6558.

Thus, the limit load can in principle be narrowed quite well by the STPZ. The only problem is to find an appropriate criterion for the iterative improvement of the first estimate, which can generally be used.

The method mentioned in Sect. 8.6, to replace the asymptotic solution for an infinitely high load level by a load monotonically increasing above the limit load, turns out to be unsuitable in this example, because a good result is not achievable within a few MEAs at low hardening.

8.8 Plastic Load-Bearing Capacity of a Section

So far, only the load-bearing capacities of entire structural systems were considered in this chapter. Now it is examined to what extent the STPZ can also be used to determine the load-bearing capacity of a cross-section in a bar structure.

The load-bearing capacity of a cross-section in a bar is exhausted if the cross-section is completely plasticized as a result of a given combination of section forces. Thus, the assumption $V_p = V$ (see Eq. (8.1)) is a priori exactly met, which has the consequence that the location-dependent limit load factor α_L is distributed constantly over the cross-section.

8.8.1 Fixed Axis of Rotation

As discussed in Sects. 8.2 and 8.4, the STPZ further simplifies at $V_p = V$ if only "natural" boundary conditions exist and if the plastic strain increments are at any location collinear with the fictitious elastic calculated stresses and also have the same sign. The residual stress increments of the asymptotic state then vanish (Sect. 8.2), and the plastic limit load factor α_L can be determined by a single MEA, performed for an arbitrary load factor α (Sect. 8.4). For a uniaxial stress, Eq. (8.35) is sufficient for this.

These three conditions, i.e., $V_p = V$, only "natural" boundary conditions, and collinearity with the same sign, are all satisfied simultaneously if the neutral fiber can neither move nor twist in a section, i.e., at uniaxial stress states in the absence of local and directional stress redistribution, so if the same axis of rotation is present at fictitious elastic and at elastic-plastic behavior. This is the case in the following situations:

- If the same axis of rotation can develop freely without constraints. The limit load factor calculated with the STPZ then specifies the multiple of the load level applied in the fictitious elastic calculation.
- If the axis of rotation in the single MEA required is forced to be the same as in the fictitious elastic state. Because of the constraints involved, the ratio of the various section forces applied simultaneously then differs in the limit load state from the ratio specified in the fictitious elastic calculation. The limit load factor thus determined can no longer be interpreted to provide a multiple of the originally applied combination of section forces, but represents only an auxiliary quantity for calculating the stress distribution in the plastic limit state. Therefore, the combination of section forces applied in the underlying fictitious elastic analysis is irrelevant. Actually, each of the section forces simultaneously present would get its own plastic limit load factor.

Figure 8.33 shows exemplarily the distribution of the limit load factor (see Eq. (8.35), where $E_t/E = 10^{-5}$ is used to determine ρ) for a rectangular cross-section (width B, height H), as well as the stress distribution attained in the state of the plastic limit load (see Eq. (8.27) with α_L used for α) for various configurations (Fig. 8.33a–d) of the section forces indicated in Fig. 8.32:

$$m_y = \frac{M_y}{M_{y,\text{el\_limit}}} \quad ; \quad m_z = \frac{M_z}{M_{z,\text{el\_limit}}} \tag{8.99}$$

$$M_{y,\text{el\_limit}} = f_y \frac{BH^2}{6} \quad ; \quad M_{z,\text{el\_limit}} = f_y \frac{B^2 H}{6}. \tag{8.100}$$

For uniaxial bending (Fig. 8.33a) with the elastic limit load $m_y = 1$, $\alpha_L = 1.5$ is obtained and for biaxial bending (Fig. 8.33b) with elastic limit load $m_y = m_z = 0.5$,

Fig. 8.32 Rectangular cross-section under biaxial bending with axial force

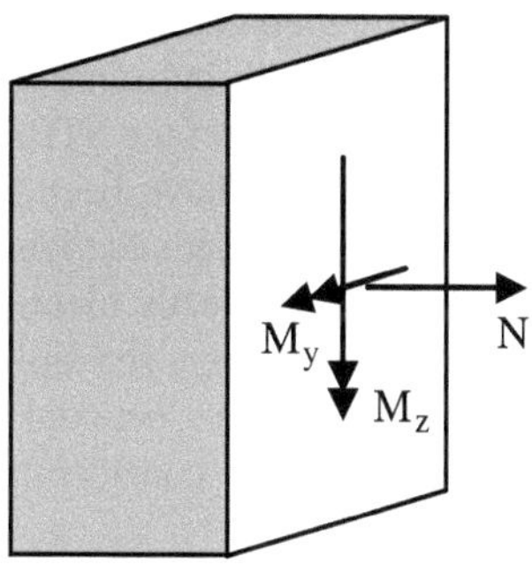

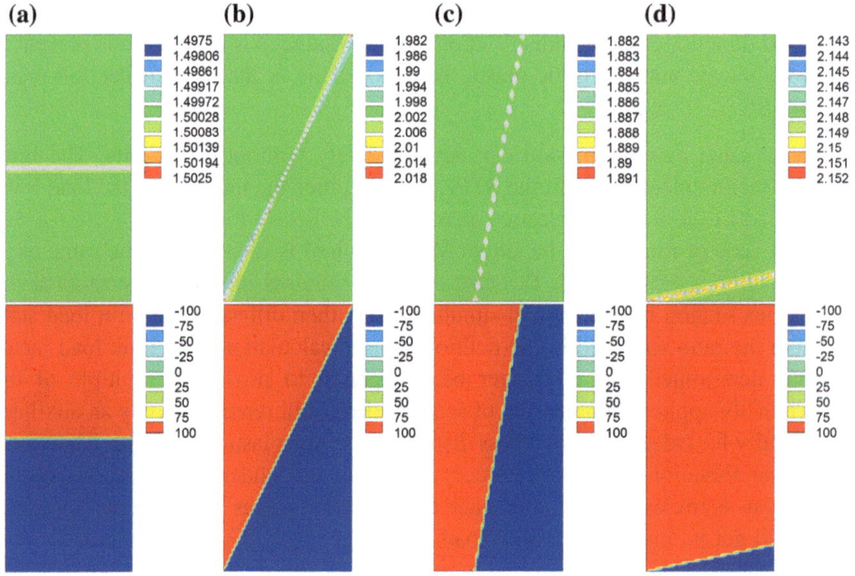

Fig. 8.33 Distribution of the plastic limit load factor (*top row*) and the stress distribution in the state of the plastic limit load (*bottom row*, $f_y = 100$): **a** uniaxial bending; **b** biaxial bending $m_y = m_z$; **c** biaxial bending with the axis of rotation inclined with 1:5.5; **d** axis of rotation is no bisectrix

$\alpha_L = 2.0$. In either case, the axis of rotation adjusts freely in the elastic-plastic state to be the same as in the elastic state. The plastic limit load factor here is the multiple of the elastic limit load.

In cases Fig. 8.33c, d, however, the prescribed axes of rotation are associated with constraints in the plastic limit state. In Fig. 8.33c, the axis of rotation (slope 1:5.5) is preserved in the elastic-plastic state to be the same as at the elastic limit behavior with $m_y = 4/15$, $m_z = 11/15$. In the plastic limit state, the corresponding section forces are $m_y = 0.3635$, $m_z = 1.4340$, so that their relation to each other has changed. In Fig. 8.33d, the fixed axis of rotation is no longer a bisectrix, which, except for the two bending moments, also induces an axial force. The three section forces are then at the plastic limit load in a different ratio to that in a fictitious elastic calculation with the same axis of rotation.

These results of the STPZ were achieved using the FEM, which, however, would not necessarily have been required. Apart from numerical errors, which thereby occur in the immediate vicinity of the zero crossing of the stresses, the exact plastic limit load is correctly identified by the STPZ with only one MEA at a fixed axis of rotation. However, the application of the STPZ is unnecessary in these cases, because the cross-sectional bearing capacity can even more easily be determined from elementary considerations.

8.8.2 Axis of Rotation not Fixed

The situation is different at an arbitrarily shaped cross-section, which is exposed to any combination of section forces, so that the position of the axis of rotation is not fixed and therefore may not coincide at the plastic limit load with the fictitious elastic calculated one.

The rectangular cross-section under oblique bending without or with axial force is again regarded as an example for the sake of clarity. In the absence of shear stresses, the Bernoulli hypothesis of plane sections remaining plane after bending holds, so that only uniaxial stresses exist that allow local, but no directional stress redistribution.

Because no residual stresses occur in the asymptotic condition because of the uniaxial stress state, just one MEA with any load factor is enough to determine the plastic limit load factor. However, in contrast to Sect. 8.8.1, although the axis of rotation is not fixed now, the so-determined plastic limit load factor is not the correct solution but only a first estimate.

Thus, for example, for the load constellation of Fig. 8.33c ($m_y = 4/15$, $m_z = 11/15$), the distribution of the plastic limit load factor represented in Fig. 8.34 is obtained, if the corresponding inclination of the axis of rotation is not fixed at 1:5.5 and may therefore differ in elastic-plastic behavior.

Consequently, the plastic limit load factor is not distributed nearly constant—an indication that the conditions mentioned in Sect. 8.8.1, which lead to the correct result with a single MEA, are not met here. In consideration of the uniaxial state of stress, this can only mean that the plastic strain increments do not at any location have the same sign as the fictitious elastic calculated stresses, so the line of zero-crossing stress in the plastic limit state is not the same as in the fictitious elastic state.

Because the objective is a uniform distribution of the plastic limit load factor, the area fraction occupied by the respective gradations of the plastic limit load factor is

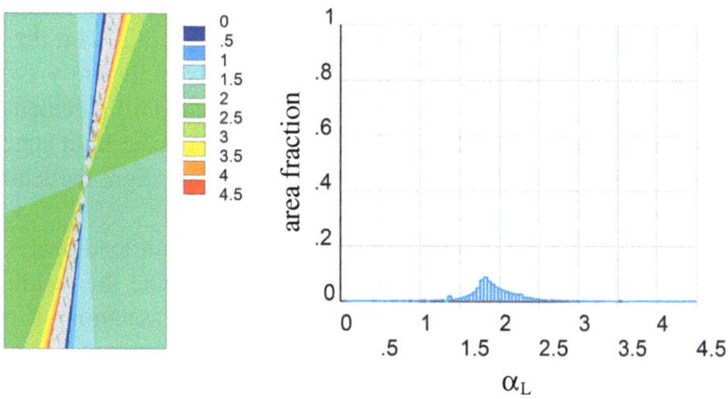

Fig. 8.34 Rectangular cross-section under biaxial bending ($m_y = 4/15$, $m_z = 11/15$); *left*: distribution of the first estimate of the plastic limit load factor; *right*: space shares of the gradations of the plastic limit load factor

of some importance. The corresponding representation on the right side of Fig. 8.34 indicates that the correct value of the plastic limit load factor is certainly somewhere between 1.5 and 2.5, and that a certain focus is close to $\alpha_L = 1.8$.

With only one MEA a satisfactory result could still not be achieved. An improvement by identifying elastic subvolumes, as mentioned in Sect. 8.6 and applied in Sect. 8.7 to a statically indeterminate beam and a disc with a hole, doesn't make sense here, because there is no elastic subvolume.

Instead, the other possibility, already mentioned in Sect. 8.6 for improving results, is used, by performing an analysis of monotonic loading with bilinear material law with low kinematic hardening to get a better idea of how the signs of the stresses at elastic-plastic behavior are distributed. These signs are then used in place of the term $\mathrm{sgn}(\mathrm{d}\sigma^{\infty})$ in Eq. (8.35).

For the monotonic loading, a load factor out of the range of load factors depicted in Fig. 8.34 is selected as the load level. Which one is specifically chosen is not so crucial, because for the further calculation process, it is not the magnitudes of the stresses in each point of the cross-section which are important, but only their signs, which should be determined as correctly as possible. However, a massive underestimation of the actual plastic limit load should be avoided, because then the distribution of stress signs would differ too much from the fully plastic state. Similarly, a load level far above the plastic limit load should also be avoided, because then the stress signs may deviate from the fully plastic state in too many points of the section, because the stress distribution approaches the distribution of the fictitious elastic state qualitatively with increasing load level in the presence of hardening.

The load level

$$\alpha = 2.0 \tag{8.101}$$

is chosen. For this, an analysis of monotonic loading is performed with the STPZ at low hardening ($E_t/E = 10^{-5}$) according to Chap. 3. After two MEAs, the iterative improvement of the stresses is not yet finished, because the plastic zone has not yet been identified correctly (as can be seen on the left of Fig. 8.35; on the right after iteration is completed), but the neutral fiber is already sufficiently twisted with respect to the fictitious elastic calculation (cf. Figure 8.33c). Incidentally, after completion of the iteration, the stress distribution in the plastic limit state is almost found by the monotonic analysis, which is produced by the slight hardening and the proximity to the plastic limit load.

After the two MEAs for monotonic loading, a plastic limit load analysis is again performed, which requires one additional MEA to determine the residual stresses for an arbitrary load factor, for example, $\alpha = 0$, again assuming a completely plasticized cross-section ($V_p = V$). However, the signs of the stresses shown in Fig. 8.35 on the left are inserted for $\mathrm{d}\sigma^{\infty}$ in Eq. (8.35) in place of the signs from an asymptotic analysis according to Sect. 8.2, and thus those of the fictitious elastic calculation.

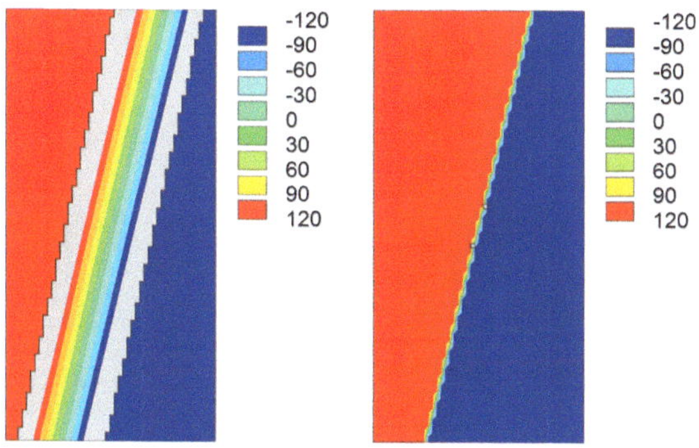

Fig. 8.35 Rectangular cross-section under biaxial bending (load levels $m_y = 4/15$, $m_z = 11/15$): stress distribution at monotonic calculation; *left*: after two MEAs; *right*: after six MEAs

The initial stresses obtained from the calculation for finite monotonic loading to be used in the subsequent MEA to calculate the plastic limit load, and their alteration relative to the asymptotic behavior, can be seen in Fig. 8.36.

The plastic limit load factors thus finally determined are shown in Fig. 8.37.

A comparison with Fig. 8.34 reveals that the quality of the distribution of plastic limit load factors depicted in Fig. 8.37 has become much better. It can be estimated now that the load factor is about

$$\alpha_L \approx 1.86 \cdots 1.90. \tag{8.102}$$

If a new calculation for monotonic load increase is carried out now, for example with $\alpha = 1.88$, serving as a basis for a third plastic limit load analysis, we can see in

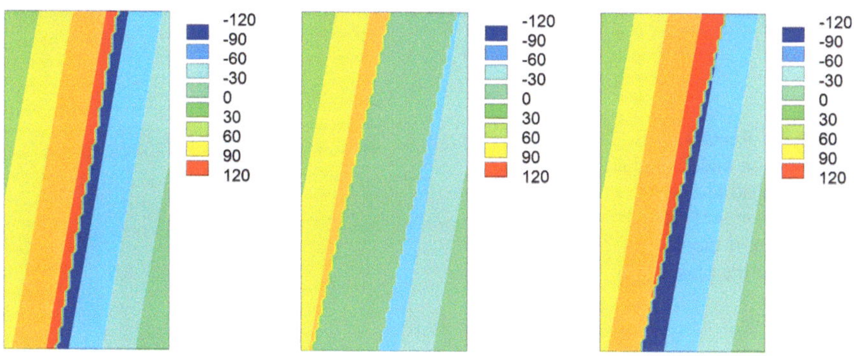

Fig. 8.36 Rectangular cross-section under biaxial bending (load levels $m_y = 4/15$, $m_z = 11/15$): initial stresses; *left*: produced by asymptotic behavior; *center*: from calculations for finite monotonic load increase with two MEAs; *right*: from calculations for finite monotonic load increase with six MEAs

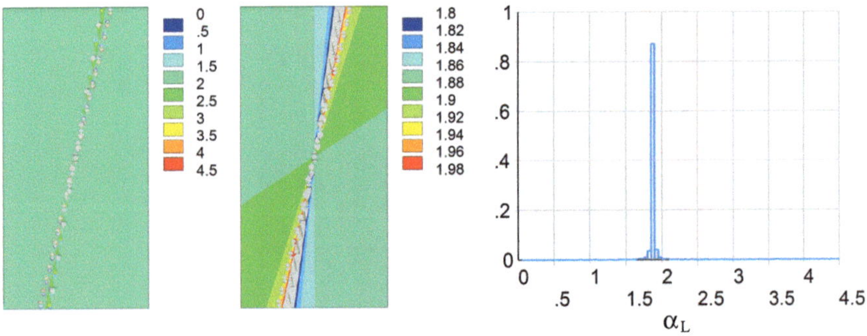

Fig. 8.37 Rectangular cross-section under biaxial bending (load levels $m_y = 4/15$, $m_z = 11/15$); *left and center*: distribution of the improved (second) estimate of the plastic limit load factor at different scales; *right*: space shares of the gradations of the plastic limit load factor

Fig. 8.38 that the estimate of the plastic limit load factor can be further improved. This amounts to

$$\alpha_L \approx 1.874 \cdots 1.876. \tag{8.103}$$

For comparison, the exact plastic limit load is given. If the plastic limit load factor is required to be the same for both moments, the plastic limit load may be obtained from the condition

$$a_L m_y = \sqrt{3 - 2a_L m_z} \quad \text{for} \quad m_y \leq m_z, \tag{8.104}$$

so that the actual plastic limit load factor is

$$\alpha_{L,\text{actual}} = 1.875. \tag{8.105}$$

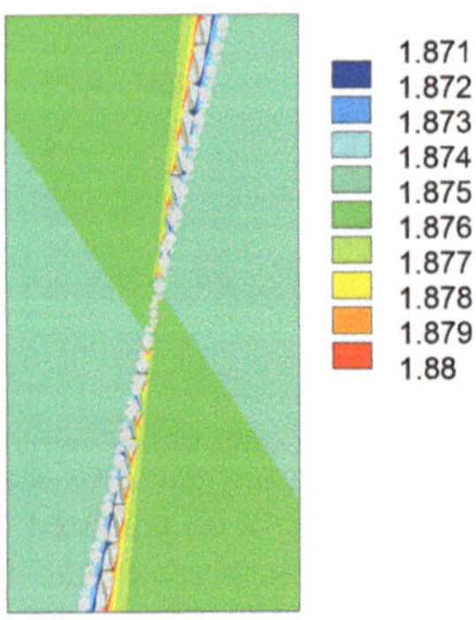

Fig. 8.38 Rectangular cross-section under biaxial bending (load levels $m_y = 4/15$, $m_z = 11/15$): distribution of the third estimate of the plastic limit load factor

Although the axis of rotation is inclined with the slope 1:5.5 in the fictitious elastic state, it is twisted to 1:4 in the state of the plastic limit load.

The STPZ has come very close to the actual plastic limit load, but a series of linear analyses was necessary. Apart from the obligatory fictitious elastic analysis, one MEA was required as a basis for the first estimate of the plastic limit load, some MEAs (two–six) for the monotonic analysis with the load factor 2.0, one MEA for the second estimate of the plastic limit load, some MEAs (two–six) for the monotonic analysis with the load factor 1.88, and an additional MEA for the third estimate of the plastic limit load.

The overall conclusion is that the plastic limit load of structures that are completely plasticized in the state of the plastic limit load, e.g., a thick-walled cylinder under internal pressure, can be found correctly with the STPZ with little effort, namely a maximum of three linear analyses, provided that the stresses in the plastic limit state have the same signs as in the fictitious elastic state at every location in the structure. Otherwise only a first estimate of the plastic limit load is initially obtained which can then be improved iteratively, whereby the computational burden increases rapidly. Possibly, the STPZ is not suited for this to the same extent as it is to determine the strain range and the accumulated strains in the state of elastic or plastic shakedown (Chaps. 4–7).

References

1. Seshadri, R., Fernando, C.P.D.: Limit loads of mechanical components and structures using the GLOSS R-node method. Trans. ASME J. Press. Vessel Technol. **114**, 201–208 (1992)
2. Jospin, R.J.: Displacement estimates of pipe elbows prior to plastic collapse loads. Nucl. Eng. Des. **178**, 165–178 (1997)
3. Zarka, J., Frelat, J., Inglebert, G., Kasmai Navidi, P.: A new approach to inelastic analyses of structures. Martinus Nijhoff Publishers, Dordrecht/Boston/Lancaster (1988) and (greatly expanded) 1990
4. Maier, G., Comi, C., Corigliani, A., Perego, U., Hübel, H.: Bounds and estimates on inelastic deformations, Commission of the European Communities, Contract RA1-0162-I and RA1-0168-D, Report EUR 16555 EN. European Commission, Brussels (1992)
5. Borhani Alamdari, B.: Nouvelles methodes de calcul de la charge de ruine et des deformations associees. Thèse de doctorat en Physique, Universite de Technologie de Compiegne (1990)
6. ANSYS Release 14.5, ANSYS Inc. Canonsburg, USA (2012)
7. Schimmöller, H.: Analytische Bestimmung von Eigenspannungszuständen auf der Grundlage der Elastizitätstheorie. Schriftenreihe Schiffbau der TU Hamburg-Harburg, Nr. 524. http://doku.b.tu-harburg.de/volltexte/2011/1048/pdf/Bericht_Nr.524_H.Schimmoeller_Analytische_Behandlung_von_Eigenspannungszustnden_auf_der_Grundlage_der_Elastizittstheorie.pdf
8. Rahimi, G.H., Alashti, R.A.: Limit load analyses of plates with a hole under in-plane loads. Sci. Iran. **12**(4), 442–454 (2005)

Index

A
Anisothermal, 159, 172
  Asymptotic state, 272

B
  Backstress, 10, 58
  Besseling model, 13, 212
  Bree tube, 32, 103, 130, 141, 175, 193, 205, 256

C
  Chaboche model, 13, 60, 161
  Continuous beam, 36
  Cross-sectional bearing capacity, 308

D
  Deviator, 4
  Directional stress redistribution, 9, 17, 70, 80, 81, 85, 147, 272
  Direct method, 51
  Disc with a hole, 87, 278, 298

E
  Elastic compensation method (ECM), 53, 271
  EMAP, 53
  Endurance limit, 45
  Equivalent stress, 5, 69
  Equivalent stress range, 98, 107, 169, 180, 262

F
  Fatigue, 2, 44
  Fictitious elastic analysis, 33, 42, 48, 58, 69, 96, 274, 308
  Flow law, 7, 58

G
  GLOSS method, 53, 271

H
  Hardening, 9, 11, 21, 160, 214
  HCF, 45

I
  Incremental analysis, 49
  Initial bending moment, 286
  Initial strain, 60, 97, 168, 235, 273
  Initial stress, 63, 97, 237, 273, 286
  Internal variable, 10
  Isothermal, 10, 166, 172, 228

K
  Kinematic shakedown theorem, 52

L
  LATIN, 53
  LCF, 2, 41, 45
  Linear matching method (LMM), 53, 271
  Local stress redistribution, 17, 91, 274, 277, 309, 311

M
  Masing, 96
  Mathematical programming, 52
  Mises equivalent stress, 5, 27, 70, 109, 215, 280
  Modified elastic analysis (MeA), 63
  Mróz model, 211
  Multilinear hardening, 13, 211, 233
  Multi-parameter loading, 150

H. Hübel, *Simplified Theory of Plastic Zones*,
DOI 10.1007/978-3-319-29875-7

N

Newton-Raphson, 50, 94
Nonlinear hardening, 13, 52, 211
Non-radial loading, 79
Normality rule, 8, 27
Nozzle, 259

O

Ohno–Wang model, 15
One-parameter loading, 95, 150
Overlay model, 13, 233

P

Perfectly plastic, 11, 19, 24, 36, 50, 63, 212, 223, 271
Plastic hinge, 17, 21, 36, 49, 75, 290
Plastic limit load, 19, 24, 46, 52

Q

Quadrilinear material law, 223, 226

R

Radial loading, 70
Radial return, 50, 79
Ratcheting, 2, 13, 22, 24, 25, 41, 45
Residual stresses, 42
RID, 24
RSDM, 53

S

Service life, 2, 44, 95
Shakedown, 22, 24, 50, 51, 95, 169
Shakedown theorems, 52
Softening, 10, 161
Static shakedown theorem, 52
Strain accumulation, 21, 45
Strain enhancement factor, 44, 48, 76, 101, 142
Strain limits, 46
Strain range, 44, 48, 95, 167, 228
Stress redistribution, 17, 131, 302
Successive projection, 151

T

Temperature-dependent hardening, 160
Temperature-dependent yield stress, 163
Temperature-dependent Young's modulus, 167
Temperature gradient, 33, 105, 260
Temperature transient, 105, 260
Theory of plastic zones, 49, 64
Three-bar model, 25, 42, 152
Transformed internal variable (TIV), 57
Tresca equivalent stress, 5, 9, 215
Trilinearization, 228, 262
Trilinear material law, 213, 214, 222, 228
Twice-Yield method, 50
Two-bar model, 99, 115, 138, 173, 184, 197, 244

Y

Yield line, 49, 299, 307
Yield surface, 5, 162

Z

Zarka's method, 53, 57, 95, 246

MIX
Papier aus verantwortungsvollen Quellen
Paper from responsible sources
FSC® C105338

If you have any concerns about our products,
you can contact us on
ProductSafety@springernature.com

In case Publisher is established outside the EU,
the EU authorized representative is:
Springer Nature Customer Service Center GmbH
Europaplatz 3, 69115 Heidelberg, Germany

Printed by Libri Plureos GmbH
in Hamburg, Germany